Carnivores of British Columbia

ROYAL BC MUSEUM HANDBOOK

Carnivores
of British Columbia

David F. Hatler, David W. Nagorsen
and Alison M. Beal

Volume 5:
The Mammals of British Columbia

ROYAL **BC** MUSEUM
Victoria, Canada

Edited and designed by Gerry Truscott, RBCM.
Layout and typesetting by Peggy Cady Graphics.
Cover design by Chris Tyrrell and Michael Carter, RBCM.
Printed in Canada by Hignell Printing.

Front-cover photograph of a Cougar by Victoria Hurst.
Back-cover photograph of an Ermine by David F. Hatler.
See page 395 for photograph and illustration copyrights and credits.

National Library of Canada Cataloguing in Publication Data
Hatler, David F. (David Francis)
    Carnivores of British Columbia

    (The mammals of British Columbia ; v. 5)

    At head of title: Royal BC Museum handbook.
    Includes bibliographical references: p. 371.
    ISBN 978-0-7726-5869-2

    1. Carnivora - British Columbia. I. Nagorsen, David W.
II. Beal, Alison M., 1953- III. Royal BC Museum. IV. Title.
V. Series: Royal BC Museum handbook.

QL737.C3H37 2008   333.95'9709711   C2008-960117-3

To
Ian McTaggart Cowan,
an extraordinary naturalist, remarkable biologist,
wonderful human being, and as good a mentor
as anyone could hope for.

The authors thank the following individuals and organizations for their generous financial support in the preparation of the manuscript for this book:

BC Ministry of Water, Land and Air Protection

British Columbia Trappers Association

Canadian Wildlife Federation

Forest Investment Account, BC Ministry of Sustainable Resource
    Management

Forest Research British Columbia (FRBC) Biodiversity Fund

Fur Council of Canada

Fur Harvesters Auction

Fur Institute of Canada

Guide-Outfitters Association of British Columbia

Habitat Conservation Trust Fund

Steve and Jean Johnson

Mountain Equipment Co-op

NGR Resort Consultants

North American Fur Auctions

Royal BC Museum

Bill and Dee Russell

Wild Fur Shippers Council

CONTENTS

PREFACE

This is the fifth in a series of six handbooks revising the Royal BC Museum's Handbook 11, *The Mammals of British Columbia* by Ian McTaggart Cowan and Charles J. Guiguet (1965), which is out of print. As with the first four volumes (*Bats of British Columbia; Opossums, Shrews and Moles of British Columbia; Hoofed Mammals of British Columbia; Rodents and Lagomorphs of British Columbia*), this handbook emphasizes identification, distribution and natural history. It is based on the large volume of published information and accumulated data that have become available over the 40 years that have elapsed since Cowan and Guiguet's pioneering work, with the specific goal of being as complete and up-to-date as possible.

There have been changes in taxonomy and nomenclature since the original publication, and one of those changes requires emphasis here. Two families of mammals that occur in British Columbia, the Phocidae (haired seals) and the Otariidae (eared seals), were formerly in a separate order, the Pinnipedia, but are now included in the Carnivora. Consistent with the original plan for this series, the five species in those two groups are not included in this book and will be covered in the final volume in this series, *The Marine Mammals of British Columbia*. It is because of that distinction that the reader will encounter phrases such as "the carnivores covered in this book", and "terrestrial carnivores" in various sections of this volume. For those who might be confused, the Sea Otter is included in this book because, even though it is a marine mammal in habit, it is a member of one of the six primarily terrestrial families of carnivores in BC.

1

When one is summarizing from a large body of literature, it is difficult to attribute sources without sacrificing readability. Nevertheless, where specific information has been provided, I consider it essential to identify the researcher(s) and, in many cases, to indicate where the research was done or the observations made. The people named in the body of the accounts are professional biologists unless stated otherwise.

The three authors have worked independently both in their careers and in preparation of this book, and the word "we" simply does not apply in those cases where personal observations and experiences are referenced. Accordingly, all uses of "I" refer to the first author unless noted otherwise. Reverting to that now, I reflect back over the long, arduous, often stressful, but always interesting process of putting together the material for this book. There were the lapses that afflict all writers, and I nod knowingly at the following lament by the 18th-century lexicographer Samuel Johnson: "the writer shall often trace his memory, at the moment of need, for that which yesterday he knew with intuitive readiness and which will come uncalled into his thoughts tomorrow." But my biggest problem, after sifting through the imposing piles of documents, revisiting old field notes, and recalling applicable memories, was in deciding what not to say and when enough was enough. The reader whose appetite for more is whetted by the accounts in this volume can be assured that, given the current interest in carnivores, the appearance of new sources and new information will be ongoing for the foreseeable future.

DFH, January 2008

GENERAL BIOLOGY

Introduction

The word "carnivore" comes from the Latin *carnis* (flesh) and *voro* (to devour), and literally means "eater of flesh". As a strictly descriptive term applied to an animal's basic food habits, it is parallel to "herbivore" (eater of plant materials), "insectivore" (eater of insects) and "omnivore" (eater of both plants and animals, including insects). A common cause for confusion is the fact that members of the mammal group "Carnivora" demonstrate a full range of food habits, from the completely carnivorous diets of cats, seals and some weasels to the basically omnivorous habits of the bears to the almost completely herbivorous nature of the Giant Panda. This book is about the six terrestrial families in the Order Carnivora that are native to British Columbia: Canidae (dog family or canids), Ursidae (bear family or ursids), Procyonidae (raccoon family or procyonids), Mephitidae (skunk family or mephitids), Mustelidae (weasel family or mustelids) and Felidae (cat family or felids).

The evolutionary history of the Carnivora is not completely clear because the fossil record is incomplete, but scientists believe that the ancestors of present forms arose from small probably insectivorous mammals in the late Eocene epoch, about 35 million years ago. Consistent with that view, those ancestral carnivores (the Miacids) were small and relatively unspecialized. The events and circumstances leading from that simple beginning to the subsequent rise and fall of large, imposing species such as Giant Short-faced Bears and Sabretooth cats and to the array of lifestyles among the carnivores still extant is fascinating, but beyond the scope of this volume. Among available sources, E.C. Pielou's book, *After the Ice Age*, is a highly readable account of climatic, geological and biological developments in North America over the last 20,000 years.

All carnivores are predators in the sense that they earn at least a part of their living by killing and eating other animals. But predation is not exclusive to the Carnivora. Hawks and owls are obvious examples among birds, but the American Robin pulling a worm from the lawn, the Barn Swallow scooping a moth out of the air and the Hairy Woodpecker extracting pine beetle larvae from a tree trunk are also practising predation. So is the Common Garter Snake swallowing the Western Toad, the toad snapping up a spider and the spider eating the fly. Further, as outlined by Robert Ardrey, Paul Shepard and others, the adaptability, cooperation and strategizing involved in predation by early Humans were probably major factors moving our species forward to the position we occupy in the world today; as suggested on a bumper sticker I once saw, "We are all descendants of successful hunters." This book, then, is not specifically about predators and predation (a very large and complicated subject), but about a group of mammals for which predation is a central feature of life.

This preliminary section describes biological and ecological features that carnivores have in common, and the variations, subtle to striking, that are characteristic of the particular groups and species covered in greater detail later in the book.

Selected References: Ardrey 1961, Eisenberg 1981, Errington 1967, Ewer 1973, Pielou 1991, Shepard 1973.

Form and Structure

> *The carnivores are animals many of whose adaptations are so obvious that, even to a superficial view, there is something admirable about them. It is no mere chance that the two animals man has taken for choice as his house guests (domestic dogs and cats) are both carnivores, despite the fact that they are more expensive to feed than vegetable eaters.* – R.F. Ewer, 1973

Of the 5080 species of placental mammals listed in the book *Mammal Species of the World* by D.E. Wilson and D.M. Reeder, relatively few have departed from either vegetation or invertebrates as their primary source of sustenance. Most of those that regularly kill and consume other vertebrates are in the Order Carnivora, which comprises 286 species, about six per cent of the total number. Although flesh is relatively easy to digest, it can be difficult to obtain. The many species that serve as prey for carnivores continue to exist in part because of structural and behavioural adaptations for detection and escape, enabling a sufficient number of individuals to reproduce before they become prey. Further, based on the flow of energy through ecosystems, one of the principles of ecology dictates that the number (or biomass) of predators that the environment can support is limited, and will be considerably smaller than the number (or biomass) of the local prey. Thus, as stated by R.F. Ewer: "the business of the killer is a highly competitive as well as a highly skilled occupation." As will be evident in the species accounts that follow, not every carnivore born will be sufficiently competitive, skilled and lucky to survive and breed successfully. Like the prey, a number sufficient to maintain the species has consistently done so, to date, for all of the carnivore species in British Columbia. The nature and range of the features contributing to their success are described below.

Size and Shape

British Columbia's terrestrial carnivores range in size from the Least Weasel, rarely more than 125 grams (the smallest carnivore in the world), to the Grizzly Bear, which may grow to 500 kg or more. Most of the species covered in this book are characterized by a distinct sexual dimorphism in size, with adult males up to twice as large as adult females. The size of a predator generally dictates the size of the prey it can regularly exploit, thus the range of sizes among the carnivores of BC ensures essentially full utilization of the province's potential mammal prey species and at least occasional use of many other animals, both vertebrates and invertebrates. Carnivores also come in several general body shapes, most of which are familiar to most people: the lithe, rather long-legged form of the cats (adapted for stealth and control when stalking), the more athletic and equally long-legged shape of the dogs (adapted for speed and endurance in pursuit), the more compact, powerful builds of bears and mustelids such as the Wolverine and American Badger (adapted to digging and/or moving heavy objects), and the short-legged, elongated shape typical of the weasels (enabling them to follow prey into confined spaces including their burrows).

Tooth and Claw: Tools of the Trade

Regardless of the size, shape and food predilections of individual species, all 21 of the carnivores covered in this book and the 265 other carnivore species world-wide share some common characteristics. The most consistent and recognizable is the presence in each of the relatively large, pointed canine teeth often referred to as fangs (figure 1). The canines are used primarily for catching and holding prey, and as daggers for killing. Most of the carnivores also have a pair of specialized teeth, called carnassials, on each side of the jaw (usually the last upper premolar and lower first molar on each side, figure 1), used for slicing flesh apart and shearing it from bone. The molars of most species are somewhat flat and used for crushing food items that require it (bones, insects and other invertebrates, plant material). That simple overview aside, there are numerous variations in tooth number, structure and arrangement among the species, all reflecting specializations for particular hunting and feeding patterns. In this book, the key to skull identification, and the skull drawing and dental formula in each species account show these variations.

In support of their dentition, carnivores have relatively heavy skulls with strongly developed facial muscles, and jaw articulations that allow rapid, powerful open and shut action but little side-to-side or rotary movement. The combined actions of molar architecture and massive jaw muscles enable a scavenging Wolverine to not only break large bones, but to actually chew them up into small bits and consume them.

The carnivores described in this book have four or five clawed digits on each foot. As with their teeth, claws vary in size, shape and function among species. Many have short, sharp claws that appear to be adapted primarily for climbing trees, but are probably also used for holding prey; included in that group are several mustelids and the American Black Bear. Others, such as the canids and the Northern River Otter, have short, stout, relatively

Figure 1. Carnivore skull showing canine (A) and carnassial (B) teeth.

blunt claws suitable for traction on smooth or slippery ground. The long, heavy front claws of American Badgers and Grizzly Bears are specifically adapted for digging, although I have also seen Grizzly Bears use theirs deftly, like fingers, to pick up small items. The small, retractile claws of the Sea Otter enable it to hold and manipulate slippery food items, but may be used primarily for combing and cleaning the fur. Claw specialization and function reaches its peak among the cats. Their sharply pointed, hooked talons are tools for capturing prey and climbing trees, and are formidable weapons for fighting – they are important enough to be kept sheathed and protected while not in use. Anyone who has teased a domestic cat with a string knows how quickly and accurately it can deploy its claws; and, while I did not see it, I shared the amazement of a friend whose cat had leaped more than a metre into the air and, with one paw, snagged a hummingbird out of the air.

Locomotion

Because they often travel widely in search of prey and then must pursue it under a variety of circumstances, carnivores are particularly mobile. As with their dentition, most have limb and other skeletal modifications to support their particular hunting and habitat specialties. Many are plantigrade or semiplantigrade, placing all or most of the soles of their feet on the ground when they walk, while others are digitigrade, walking on their toes with the bones analogous to our heels and wrists elevated and nearly perpendicular to the ground. Among those that are strictly plantigrade are the bears and raccoons, while many of the mustelids are semiplantigrade, able to rise up on their toes when required but also able to lower the heel and increase the foot's surface area when on soft surfaces such as snow. The digitigrade carnivores are those that regularly employ long-distance running or short-term speed, with canids and felids being the main examples.

With regard to specialized features associated with locomotion, the sharp, curved claws of several climbing species have already been mentioned. Other modifications for climbing include an extra muscle attachment surface on the scapula (shoulder blade) in bears, the Northern Raccoon, the American Marten and the Fisher, and the flexible ankles of the regularly arboreal mustelids such as the American Marten and the Fisher enabling them to rotate their hind paws almost 180° and descend trees headfirst – a definite advantage when in pursuit of prey. Their hind paws are also moderately prehensile, able to grasp branches as they go.

Several species have particular adaptations for travelling on snow. The American Marten and the Wolverine have disproportionately large feet and a bounding gait in which the hind feet land in the tracks made by the front feet. This gait propels the animal forward by taking advantage of the newly compacted, firmer platform the front feet have provided (figure 2). Other species, the Canada Lynx in particular and the Grey Wolf to a lesser extent, contend with deep snow by combining long legs with the lower foot-loading (weight per surface area) provided by large feet.

Three BC carnivores, (American Mink, Northern River Otter, and Sea Otter) are semiaquatic and adapted for locomotion in water. Of these, the Mink is the least well-equipped, with a stream-lined body but only partially webbed feet; the trade-off is greater quickness and agility on land, and it even climbs when circumstances require. Both otter species have conspicuously webbed feet, to a greater degree on the hind limbs of the Sea Otter (figure 3) reflecting its almost completely aquatic existence. Otters also have special adaptations for hunting underwater, including a large lung capacity and related physiological features to increase dive time.

Although the American Badger can travel great distances over land, it does so without noteworthy speed or agility. Its low-slung, squat body, massive shoulders and forelimbs, and large claws, are primarily adaptations for digging – and it digs with impressive efficiency.

The long, conspicuous tails of many carnivores also function in locomotion, providing balance for tight turns and sudden stops on land and in the trees, and both propulsion and steerage for those

Figure 2. Tracks of an American Marten: each pair of tracks results from the hind feet landing in and pushing off from the tracks left by the front feet.

species (mainly otters) that do most of their hunting in the water.

Senses

Carnivores have keen senses and large brains, both attributes essential for the complex business of finding and securing food, avoiding potential danger, and detecting and interacting with each other in diverse landscapes. With some senses that are particularly acute, their reception and perception of the world around them is different from ours. This is evident to anyone who has followed a dog through the woods. A sniff here elicits a tail wag, while a sniff there may result in intensified search activity, an

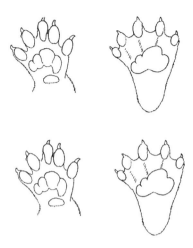

Figure 3. The feet of the Northern River Otter (above) and Sea Otter (below), showing webbed hind feet (right).

aggressive or submissive posture and response, or a clear expression of concern or fear. In a matter of minutes – and all with its nose – the dog may receive and interpret a hundred messages that are beyond the limits of human detection. The use of dogs to find or track game, to locate criminals and lost children, and to sniff out dangerous or illegal goods in luggage at airports all take advantage of that capability. A keen sense of smell is evident in the performance of a Wolverine making a beeline from some distance to a spot where it digs down through several metres of snow to the remains of an ungulate killed in an avalanche. Odours also play a large role in carnivore social behaviour, enabling or contributing to individual recognition, assessment of sexual status, mother/young interactions and territorial communication. In short, though not equally acute in all species (least in the felids), the sense of smell is highly developed and very important in the lives of carnivores.

Among the other four senses with which humans are familiar (there may be others, but how would we know?), taste is probably not particularly important to carnivores and the role of touch is poorly known. Most carnivores have stiff whiskers (vibrissae) on their muzzles and those are known to have a tactile function, perhaps helping the animals to avoid obstacles in the dark or to communicate with each other. The forepaws of the Northern Raccoon and the Sea Otter are also known to be particularly sensitive, the Raccoon's rivalling the sensitivity of the human hand. Canids hear sounds at higher frequencies than are detectable by humans. Most of the carnivores that have been tested, especially those that regularly hunt "squeaky" prey such as small mammals, are similarly equipped, detecting tones four to six times higher than the highest audible to humans. Carnivores do not appear to differ substantially from humans in detecting sounds at low frequencies.

Behavioural evidence and the physical complexity of the large eyes and associated receptors suggest that felids are more reliant upon sight than on the other senses. Sight is also important to the canids, most of which hunt in open country with wide vistas. The situation for other carnivores is less clear, but in my own experience some of the mustelids seem almost myopic when they are following a scent trail. In separate incidents over many years, I have had an Ermine, an American Marten and two American Minks stop, step back and engage their sense of sight only after encountering my shoe. Regardless of its relative importance in their daily lives, vision appears to be in black and white for most, if not all, BC carnivores. Most are active mainly at night, when colour means little and the primary need is for increased light gathering ability (large pupils, reflective layers) and retinal sensitivity.

Although it is safe to assume that all carnivore species benefit to some extent from their visual abilities, there are records of individuals continuing to function after losing their eyesight. In one case, a radio-collared blind Northern Raccoon made daily trips between foraging areas and two dens in essentially normal fashion, although more slowly, and was in good condition. Its activity patterns were, no doubt, facilitated by intimate familiarity with its home range.

Pelage

Most carnivores have two distinct layers of fur: the underfur, next to the skin, composed of fine, densely packed fibres; and the outer layer consisting of longer, heavier guard hairs. The underfur provides the primary insulation, in part by trapping pockets of air in the tiny spaces between the fibres, while the guard hairs serve as a shell to protect the underfur from abrasion and provide some water repellency to keep it dry. The guard hairs give the animal its colours and sheen.

The primary function of the fur is as insulation to keep the animal warm and reduce its energy needs. Although varying by species in efficiency, fur serves those roles remarkably well. In experiments in Alaska, researchers attempting to compare the insulation value of fur for some arctic carnivores had to scuttle their first efforts because their testing equipment failed in the cold before the animals showed any effects. They later found that Arctic Foxes were able to sleep at temperatures as low as minus 60 degrees C without having to increase metabolic heat production to stay warm; husky dogs did so at minus 45 degrees C.

Fur also serves a variety of secondary functions. Probably the most obvious is camouflage: species living in forests, where alternating bands and patches of light and shadow are common, often have pelage featuring blotches and stripes or they are dark above and pale below, while those in more open areas usually have inconspicuous solid colours (greys, browns). The most dramatic example of camouflage is that displayed by the three species of North American weasels, which turn white in snowy winter environments.

Most of the distinctive markings on a carnivore (often on the face and head), probably function in species or individual identification and facilitate social interactions. For example, it has been speculated that the white circles on the backs of a Bobcat's ears may help following kittens keep track of their mother in the thick shrub habitats where they often hunt. The conspicuous pelage of the skunks is commonly considered to be *aposematic*, meaning that it has a distinct warning function. Probably enforced by experience, other animals learn to respect and stay away from black-and-white animals.

The smooth, dense guard hairs of semiaquatic species such as the otters minimize friction and drag in the water and provide a waterproof shell for the underfur. Meanwhile, air entrapped in the underfur contributes to buoyancy.

Although the insulation function of fur is a primary factor enabling carnivores to occupy northern environments in winter, it imposes a potential overheating problem in summer. That limitation was resolved by the evolution of moulting. There are two moulting patterns among BC carnivores. The canids, bears, skunks, American Badger, Northern River Otter and Wolverine moult once each year; they shed the old coat and grow a new complement of guard hairs in the spring, gradually adding to the underfur layer until it reaches maximum thickness and density at the beginning of winter. Other species, notably the smaller mustelids (weasels, minks, martens), have two complete moults each year, one in the spring to replace the winter coat with a less dense summer model, and one in

Figure 4. A moulting American Mink, showing unusually light-coloured pelage being replaced by a new, darker summer coat.

the fall to produce the new winter pelage (figure 4). Weasels change to winter white in their fall moult; that is they produce new white hairs rather than change the colour of their summer hairs.

Selected References: Coues 1877, Eisenberg 1981, Ewer 1973, Ling 1970, Obbard 1987, Powell 1982, Weisel 2005.

Feeding Ecology

A central theme in the biology of the carnivore group as a whole is obtaining food by predation. But there are several variations in both the nature and extent of predatory activity among the various species. Most carnivores are hunters, but some, such as the bears, obtain much of their annual sustenance by grazing (green vegetation) or a form of gathering (nuts, berries). All of the carnivores at least occasionally resort to scavenging or thievery (opportunistically obtaining sustenance from animals they have not killed) and one, the Wolverine, specializes in such activity. Variations in carnivore hunting methods include stalking (felids, canids), ambush (felids, canids), long-distance pursuit (canids), teamwork (canids), pursuit in trees (some mustelids), pursuit in water (semiaquatic mustelids) and pursuit underground (American Badgers by digging, and the long, thin weasels by simply entering the burrows of small prey).

The availability of potential prey is determined largely by one or more of three factors: size, abundance and vulnerability. Generally, the larger the carnivore the larger the prey it is able to handle, although solitary species usually take smaller prey than those of the same size that hunt in groups. Of the appropriate-sized prey species in a particular area, a carnivore usually takes some more regularly than others, often the most abundant. When two prey species are equal in abundance, it may take more of the one that is easier to catch or handle than the other. Various attributes of prey species such as speed, agility, pugnacity, hiding behaviour, daily activity pattern and habitat preference may render it more or less vulnerable to predation by a particular carnivore, regardless of numbers.

In regard to available prey, the phrase, "it's a dog-eat-dog world" rings true more than most people realize. Many carnivores prey upon smaller carnivores or those with some disadvantage, and that includes cannibalism (eating one's own species). This probably occurs more often as a result of opportunism rather than systematic hunting, and when either the victim, the victor, or both are having difficulty finding other food. Because the environment cannot support an unlimited number of carnivores, starvation is a real and regular feature of their life history. A recurring theme in the species accounts is the role of nutrition in productivity and survival. A carnivore weakened by lack of food is ineffective, like a bow without an arrow. The hungrier it gets, the more frantic its search and less efficient its pursuits until, unless it is lucky and finds something to scavenge, it uses up its energy reserves and expires.

As a hedge against starvation, most carnivores store food for later use. The stored prey of Grizzly Bears and Cougars is usually an ungulate too large to be eaten in a single meal. The animal wholly or partially covers the carcass with debris and soil from the immediate area, apparently to discourage scavengers, and usually remains nearby for up to several days to defend and finish eating it. Canids may cache small prey or scraps from larger kills in den tunnels when pups are present, or under soil or snow at other times of year. Among the smaller carnivores, food storage also often follows opportunities to obtain more food than can be eaten immediately, but usually involves a number of small items rather than one large one. In many cases these caches are composed wholly or largely of a single prey species that became temporarily vulnerable. Contrary to popular notion, when the opportunity arises, most carnivores will indulge in surplus killing (slaughtering more prey than they immediately need) – that is why a raiding American Mink or Red Fox will often kill all of the chickens in a henhouse rather than just one or two. Based on long-term studies in the marshes of Iowa, Paul Errington wrote: "When they discover a wintering pool of frogs or a lakeside spring full of desperate fishes ... Minks may pack up to hundreds of pounds of frog or fish victims in the tunnels of a single snowdrift."

Surplus killing and food storing are particularly common among the mustelids, and are probably necessary for their survival. Situations such as weather conditions severe enough to prevent hunting or temporary prey scarcity could place individuals with no stored food at a severe disadvantage. With their high rates of metabolism, mustelids need to eat frequently. Fred Glover observed that, when deprived of food, Ermines initially in good condition starved to death in about 48 hours.

Some carnivores focus most of their predation on one or two prey species during most of the year (Canada Lynxes on Snowshoe Hares, Cougars on deer, Least Weasels on voles) and are considered to be specialists. Others (Coyote, Grey Wolf, Wolverine, American Marten, Fisher) are far less specific, taking whatever they encounter and can handle, and are labelled generalists. But all carnivores are intelligent and adaptable, and individuals may subsist on unusual prey or on an opportunistic basis if they need to. The following observation by Paul Errington was directed to his study species, the Muskrat, but is no less true for carnivores: "[They...] have nothing to prove, disprove, rationalize, or explain and, being to some extent free agents, they do not necessarily have to do things one way if another way will suffice. Their job is living, and they work at it full time."

Selected References: Errington 1967; Matter and Mannan 2005; O'Donoghue et al. 1995, 1998; Palomares and Caro 1999; Rosenzweig 1968; Schaller 1972.

Home Range and Social Behaviour

Most of the carnivores in British Columbia live and travel alone, except during the mating season (usually short) and when dependent young accompany their mother. Notable exceptions are Grey Wolves and Coyotes, which function in family or larger groups (packs) over much of the year.

Local populations of carnivores consist of two general categories: "residents" and "transients". The residents, mostly adults, have established secure, fairly well-defined home ranges. The benefits of familiarity gained from consistent, intensive use of an area include detailed knowledge of local prey distribution, particularly productive hunting locations, drinking-water sites, and cover and shelter structures. In contrast, transients have "no fixed address". Most are newly independent juveniles, but some are former residents displaced from their home ranges by catastrophic habitat change, failure of food supply, or (rarely) conflict with other animals. Transients constantly have to "make do" in unfamiliar terrain, finding food and safety largely by luck rather than experience. Always in search of a secure, stable home range, they most likely succeed when they are in the right place at the right time, and able to replace a resident that dies.

Theoretical analyses, such as those by Alton Harestad and Fred Bunnell, have demonstrated that, on average, home-range size is proportional to body size. But individual home ranges vary widely within species in different areas and years, and that picture is further complicated by the several different methods biologists use to measure, calculate, and express home-range size. It is beyond the scope of this book to try to sort out those difficulties and standardize results. Accordingly, the home-range sizes described in the species accounts are intended as order-of-magnitude comparisons among species, with particular reference to studies in or near BC. Humans tend to think of extensive real-estate holdings as an indication of affluence, but that is usually not the case for carnivores and other wildlife. A large home range indicates either general sparsity or a wide separation of resources, requiring the animal to expend more energy and increase its exposure to dangers while travelling between them.

Characteristically, most resident carnivores attempt to maintain all or portions of their home ranges as areas of exclusive use. They combine aggression with warning displays and signals to keep other members of their species away and minimize their competition. Resident packs of canids direct territorial behaviour primarily at neighbouring packs, but also to transients (both individuals and small groups). Because carnivores are well equipped to seriously injure each other, and because an injured carnivore has a poor chance for survival, much of their territory defense is based on mutual avoidance. Most are equipped with a variety of scent glands, and it is believed that avoidance is facilitated mostly by scent marking. Although the exact nature of the messages received by scent is difficult to assess, it is likely that an intruder knows when it is trespassing on a regularly or recently used part of a resident's territory. Even so, when the warnings transmitted in scent marks go unheeded, residents aggressively defend their territory – trespassers may be killed and those that survive learn the importance of avoidance.

Scent marking also has social functions, including information about the identity, sex and reproductive status of individuals. Especially in the canids, scents may also indicate relative group size of a local pack and the dominance status of individuals. Decades ago, Ernest Thompson Seton recognized the communications role of the special urinating spots used repeatedly by canids, referring to them as "scent telephones".

Selected References: Bekoff et al. 1984, Ewer 1973, Gese and Ruff 1997, Gittleman and Harvey 1982, Harestad and Bunnell 1979, Kitchen et al. 2000, MacDonald 1983, Muller-Schwarze 1983.

Activity and Movements

Most of the BC carnivores are mainly active during the twilight hours around dusk and dawn (crepuscular) or throughout the night (nocturnal). Individuals of some species (bears, canids, Wolverine) extend activity into the daylight hours when food is scarce or when special feeding opportunities arise. Movements by individuals can be divided into three categories: daily, mating and dispersal. Daily movements are primarily for foraging, and their extent varies by a number of factors including species, sex, age and home-range attributes. Each day a Grey Wolf pack may move dozens of kilometres to cover its large territory, while some urban Northern Raccoons travel less than a hundred metres. On average, the daily movements of transients are larger than are those of residents. Mating movements, undertaken almost exclusively by males, greatly exceed those for daily foraging and often take the animals beyond the boundaries of their home ranges. The longest movements by carnivores, hundreds of kilometres in some cases, are those undertaken by transients during dispersal, usually young-of-the-year that have left their mother's home range to search for a secure, stable area of their own.

Selected References: Chapman and Feldhamer 1982, McLellan and Hovey 2001, Murray 1987.

Resting Sites and Dens

During most of the year, individual carnivores rarely engage in foraging and territorial maintenance activities for more than 12 hours a day, and some accomplish all they need to in just 3 to 4 hours. For the rest of the day they lie resting and sleeping at sites selected for security and shelter from the weather. The large species (canids, felids, bears, large mustelids), may be in completely exposed locations in open areas when conditions are favourable, and under structures such as thickly branched trees, logs or rock ledges when they need protection from the elements. Different sites may be used on successive days, depending upon where the animal's activity ends.

Smaller species may rest in the open during mild weather conditions, but they usually go to relatively secure locations such as high in the branches of trees for climbing species or near a den entrance for others. Carnivores locate their dens in natural cavities in hollow logs and trees, under logs, stumps and windfall debris, and among rock rubble; some also use burrows in soil (often dug by prey species) and man-made structures such as attics, crawl spaces, outbuildings, haystacks, abandoned vehicles, idle farm machinery, boats and culverts. During winter in northern areas, small carnivores usually rest in structures buried under snow to take advantage of its insulating effect. Depending upon home-range size and habitat attributes, individuals may have only one resting site or den that they use continuously, or they may have several, using any one as a base of operations for days or weeks before moving on to another.

Carnivores use two other kinds of dens: for producing and housing dependent young (all species), and for winter hibernation or torpor (bears, Northern Raccoon, Striped Skunk, American Badger). Natal dens (where young are born) and maternal dens (to which young may be moved for at least part of the rearing period) are often in the same kinds of structures used for daily rest sites by most species; but canids usually occupy burrows in soil for those purposes, Wolverines often use deep, extensive systems of snow tunnels and chambers, and bears produce and rear young in their hibernation chambers. See the individual species accounts for details.

Selected References: Bull and Heater 2000, Criddle 1947, Davis 1996, Gunson and Bjorge 1979, Lariviere et al. 1999, Slough 1999, Svendsen 1982, Weir 2003, Weir et al. 2004, Zeveloff 2002.

Reproduction

In northern areas with distinct seasons, the young of most carnivores are born in the spring and therefore have most of the year's snow-free period to grow and develop before facing their first winter. For the canids, felids and the Northern Raccoon, spring births follow late winter matings and straightforward gestation periods of 50 to 70 days. For all but one of the other 14 species in BC (Least Weasel), the time between mating and birth is extended by a phenomenon known as "delayed implantation", in which the embryo floats freely in the uterus in a state of arrested development for a time before attaching to the uterine wall (implanting) and resuming normal growth. That early condition is a clear case of being "a little bit pregnant", a condition not available to humans and most other animals.

Although the reason for delayed implantation has long been a subject for debate among biologists, it is generally assumed that it evolved to retain the optimum birthing time (spring) while also providing for a mating season at the most suitable timing relative to social, logistical or environmental conditions. For species that mate in the summer, the implantation delay ranges from 5 or 6 months (bears) to 11 months (Fisher). Other species (American Mink, Striped Skunk) mate in winter and the period of delay is relatively shorter and more variable, from about 10 days to 3 months.

Another feature of reproduction among many carnivores is induced ovulation, in which the release of eggs from the ovary is stimulated by mating activity rather than occurring spontaneously on a regular cycle. One explanation for this feature is that these animals have large home ranges and do not often encounter the opposite sex; induced ovulation works better than the chance of a male arriving at the right time or waiting for the female to ovulate. A more esoteric explanation, proposed by Serge Lariviere and Stephen Ferguson, is that induced ovulation ensures that the most environmentally or genetically fit males do most of the breeding. In any case, the rather rough-and-tumble (sometimes violent) and prolonged matings described in the accounts for several species are likely manifestations of induced ovulation.

The 21 BC carnivores differ considerably in their reproductive potential, the number of young that individual females can produce over time. A species' reproductive potential depends on: age at first breeding (as young as a few months in some of the weasels and as old as six to eight years in bears), length of pregnancy (up to a year in species with delayed implantation, and 4 to 12 weeks in those without), litter size (almost always one in the Sea Otter, ranging up to an average of six or more in weasels, skunks and some canids) and litter frequency (from once annually in most species to once every two or three years in bears).

Females do not always meet their full biological potential in the production of young, and the level of nutrition is often the primary factor involved. An animal may fail to obtain sufficient food for a variety of reasons, including depletion of supply because of overpopulation or competition with other species, severe weather effects on access or supply, major habitat changes (human or natural disturbances), inadequate habitat selection by inexperienced or insecure individuals, or any number of similar factors that may depress prey populations. The resulting inadequate nutrition may result in delayed maturity (first reproduction at an older age), reduced litter size, reproductive failure (inability to carry a pregnancy to term) or reduced juvenile survival.

All of the BC carnivores, except the Sea Otter, are "altricial" (born in an undeveloped and helpless state), and require considerable parental care between birth and independence. In canids both parents and sometimes other members of the group care for their young; in all other species only the mother cares for her young. mustelid mothers look after their offspring for only a few months (until fall or early winter of the first year); felids, skunks and the Northern Raccoon extend their care through the first winter; and bears remain with their cubs for two or three years.

Selected References: Deems and Pursley 1983, Ewer 1973, Hunter and Lemieux 1996, Lariviere and Ferguson 2003, Mead 1981, Mead and Wright 1983, Sadleir 1969.

Mortality, Health and Longevity

Most carnivores do not live particularly long. The oldest known individuals for each species are captive animals that benefit from regular feeding and veterinary care. Small mustelids and skunks rarely live beyond five years, although American Martens as old as 15 years have been documented both in captivity and in the wild. The larger mustelids and the Raccoon live about 12 to 15 years in the wild; maximum ages in captivity range from 15 years in the

American Badger to 25 years in the River Otter. Captive canids may live 16 to 18 years, and wild specimens (both Grey Wolf and Coyote) have been documented as old as 14 years of age, though most rarely attain even 10 years. The longest-lived of the BC carnivores are the felids and bears, all species in those two groups attaining ages of 20 to over 30 years old in captivity; in the wild, felids can live 12 to 16 years (about half the age they reach in captivity), while wild bears can live as long as those observed in captivity (Black Bear - 24 years, Grizzly Bear - 29 years).

The main causes of death for wild carnivores are predation by other animals, natural accidents, starvation, disease, and various human causes such as hunting, trapping, animal control, vehicular accidents and poisoning from pollution. For most species, animals in their first year experience the highest rates of mortality. Juveniles often find themselves in marginal or unfamiliar habitats where they are exposed to predation or human influences, or have difficulty obtaining food. The poor health of an animal is commonly associated with inadequate nutrition, which may be due to competition with other carnivores, prey declines (including those caused by predation), the effects of weather (access to prey, increased energy requirements), habitat changes (including those caused by humans) or, in the case of young animals, inexperience.

Carnivores host many parasites and are subject to a large variety of diseases; but, as in humans, most affect individuals and do not threaten populations. The solitary lifestyles and relatively low population densities of most species are not conducive to the spread of diseases or parasites. Secure, well-fed animals may carry parasites or disease organisms without apparent ill effect. Outbreaks occur primarily when transmission is enhanced (high general population level or local density), or the general condition of animals is compromised by poor food supply or bad weather. Such effects are particularly applicable to species that reach artificially high densities in urban and rural environments (Northern Raccoon, Striped Skunk, Red Fox, Coyote).

Human Health Considerations

Some individual carnivores carry diseases or parasites that can be transmitted to humans or domestic pets, with potentially serious consequences. Those conditions are described in the species accounts, but the most important to human health are highlighted here as a preliminary caution to readers who may handle or come into contact with live carnivores, their body wastes and fluids, or unprocessed dead specimens in whole or in part (viscera, raw pelts, road kill). The BC species of greatest concern are the Coyote, Grey Wolf, Red Fox, the Northern Raccoon and the Striped Skunk.

Some of the more serious conditions that may affect humans result from accidental ingestion of the microscopic eggs of intestinal parasites. Passed in carnivore droppings and transported in dust, soil or water, the eggs are durable and can be viable for months. In the case of Hydatid Disease, eggs of the canid tapeworm *Echinococcus granulosus* inhaled or ingested develop into larvae that may cause large, hollow cysts in the internal organs, especially the liver and lungs. Transmitted similarly, the larvae of the Raccoon Roundworm may migrate to the central nervous system causing serious illness and occasionally blindness or even death. A person handling a canid or Raccoon specimen should wear protective gloves, wash hands carefully afterward and avoid breathing dust that may be raised from the animal's feet or pelt.

Rabies, a disease commonly associated with carnivores, is a major human health concern in some parts of North America, but has not yet been an issue in British Columbia. The only recorded incidence among carnivores in this province was in May and June 2004, when four young Striped Skunks from Stanley Park (Vancouver) were diagnosed with a strain of rabies from bats. There were no known human infections, and subsequent monitoring indicated that other animals in the area were not infected. Rabies is usually contracted by humans when they are bitten by infected animals, of which the primary wild carriers among carnivores on this continent are Northern Raccoons, Striped Skunks and Red Foxes. All three species often attain high densities in urban and rural areas, resulting in both increased potential for the spread of the disease and increased exposure to humans. Despite the generally positive record to date, complacency about rabies in BC is not advisable. In the early 1980s there was a major outbreak of the disease among Skunks in Alberta, and the potential for cross-country movement of infected Raccoons either accidentally as stowaways in transport trucks or rail cars carrying produce, or intentionally as pets, is a continuing cause for concern.

Other zoonoses (diseases that may be transmitted from carnivores to humans) include tularemia, brucellosis and leptospirosis, which may be contracted by ingestion or inhalation of their bacterial agents, or by entry of the bacteria through open sores. Keep these and other diseases in mind when handling dead animals or live ones that are obviously ailing. Finally, people who occasionally eat carnivores (Black Bear, Cougar, Raccoon) may, if the meat is not adequately cooked, contract trichinosis. Digesting infected meat releases cysts of the Trichina Worm, which develop into (microscopic) adults in the host's digestive tract. The adults produce larvae (up to 1500 per adult female), which migrate through the host's blood and lymphatic systems and encyst in muscle tissue. Infected humans can be develop long-term muscular pain and stiffness.

Selected References: Addison et al. 1987, Gardiner 2000, Krebs et al. 1995, Lindstrom et al. 1994, Rosatte et al. 1997.

Carnivores in British Columbia

History and Biogeography

The geological and climatic events that shaped the rugged topography of British Columbia were generally not conducive to good fossil preservation, so we cannot know as much about the history of wildlife in the province as we might like to. There is enough of a record, however, from bones recovered during excavations (gravel pits, wells, mining operations) and from natural erosion of river banks and cliffs, to confirm that BC experienced some of the same patterns observed elsewhere in North America. There is evidence that various mammoths, mastodons, bisons and wild horses were present in several areas of the province prior to the last (Wisconsin) Pleistocene glaciation, and there are also records of exotic forms such as bear-sized ground sloths, wild asses and muskoxen. Presumably there were carnivores present to prey on or scavenge those animals, but the fossil record is deficient on that point. Remains of carnivores present in BC prior to the peak of the Wisconsin glaciation have been found at only one site, a sea cave recently discovered at Port Eliza on the west coast of Vancouver Island. Among approximately 4000 vertebrate bone and tooth fragments recovered from the cave was a leg bone from either the extinct Noble Marten that lived during the ice age or a large form of the American Marten. Carnivore tooth marks, possibly from a canid, were also found on one of the bones in the cave. The Port Eliza remains are dated at 16,000 to 18,000 years ago, a time when ice was advancing over most of Vancouver Island and the environment around the cave was evidently open tundra with patches of trees. The presence of vole remains in the cave, as well as bones of familiar large mammals such as the Mountain Goat, suggests that the comparatively harsh landscape of that time had sufficient prey to support carnivore populations.

Whatever the province-wide fauna list might have been, it was essentially wiped clean about 15,000 years ago when most of the province was covered by a layer of ice that was up to two kilometres thick. The carnivores that existed in BC prior to that glaciation either died out completely or survived in ice-free areas (refugia) elsewhere. Some areas of the province were deglaciated as early as 13,000 years BP (before present) and all of the province (except for mountain glaciers) was essentially ice free by 9000 years BP. Therefore, the carnivore species now inhabiting BC have become established (or re-established) here during the past 9–13 millennia, dispersing from Beringia, the major northern refugium that extended from present Alaska across the Yukon Territory, from the refugium south of the icecap in what is now the continental United States, or possibly in some cases from smaller coastal refugia. There is growing evidence that parts of the continental shelf on the BC coast escaped the glacial ice and that new land was exposed with a lower sea level. Any fossils from that period would now be submerged in the ocean. Some biologists have speculated that a few mammals survived on nunataks, mountain tops protruding above the thick ice, but nunataks would have been sparsely vegetated, isolated islands surrounded by ice. It seems unlikely that carnivores and populations of their associated prey, particularly in the case of the larger species, would have persisted for several thousand years on tiny islands in a harsh glacial landscape.

The oldest carnivore fossil remains from post-glacial BC are of the American Black Bear, recently discovered in a cave on the Queen Charlotte Islands and dated at 10,450 to 14,390 years BP. American Black Bear skulls dating from about 10,000 years ago (figure 5) were also found in a cave on northern Vancouver Island. The skulls and teeth of those fossil Black Bears are much larger than those of mod-

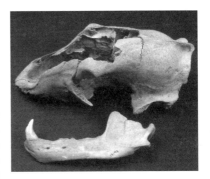

Figure 5. Skull of fossil Black Bear, carbon dated at 9800 years before present, from Windy Link Cave, Vancouver Island, BC.

ern island forms, a common feature of mammals from the late Pleistocene. More complete surveys of those coastal caves may reveal the remains of other mammalian species, including other carnivores.

The richest post-glacial fossil record obtained to date is from the Charlie Lake cave, near Fort St John in northeastern BC, which has provided an extraordinary collection of vertebrate remains dating from as long ago as 10,780 years BP. Although most of the mammalian bones and teeth examined were from rodents and lagomorphs, 12 species of carnivores have also been identified. The most commonly encountered were Red Fox, Striped Skunk and unidentified small mustelids, but the cave remains also included Grey Wolf, American Black Bear, American Marten, Fisher, Ermine, Least Weasel, Mink, Wolverine, Cougar and Canada Lynx. The abundant remains of the Striped Skunk and the small mustelids suggests that those animals may have used the Charlie Lake cave for denning. The bulk of the carnivores were in recent deposits that date from the past 1000 to 2000 years, but the oldest was the Least Weasel, found in deposits dating from about 7000 to 8000 years ago. The presence of many prey species in the Charlie Lake cave, including several small mammals and large ungulates such as Elk and Bison, suggests that the surrounding landscape had long supported a diverse community of carnivores.

Recolonization of British Columbia after the Wisconsin ice receded is a fascinating subject, and one that will likely be better understood in the next few decades based on DNA studies such as those recently pioneered by Joseph Cook, Melissa Fleming, Karen Stone and others. Some species, such as Grizzly Bear, Grey Wolf and Wolverine, likely came from refugia to the north, while the Cougar, Bobcat and Coyote almost certainly colonized from the south. Others, such as the Ermine and American Marten, may be descendants of animals from both directions or from coastal refugia. Meanwhile, some questions remain: Why do the Ermine and American Marten, two "upland" species, occur on the Queen Charlotte Islands, while the semi-aquatic American Mink does not? What is the relationship between the American Badger, currently at the extreme northern edge of its range in southern BC, and a fossil specimen of a badger (on display at the Canadian Museum of Nature in Ottawa) found near Old Crow in the Yukon Territory?

Although some (possibly most) of the 21 carnivore species now inhabiting BC may have been here before, their post-glacial appearance was essentially a new start in a new environment. The language describing the colonization of new areas tends to obscure how it occurs. Terms such as "invasion" and "range expansion" conjure images of groups of animals moving intentionally and relentlessly toward a goal destination ("Kootenays or bust"?). In reality, the process is much more subtle and gradual, involving a dynamic interplay of population processes and environment; as described by the zoogeographer P.J. Darlington:

> The area occupied by a population (range) is usually not uniform but is more favorable in some parts than in others, and the ratio of reproduction to death must often vary accordingly. Conditions are likely to be most favorable near the center of the area, and there is likely to be an unfavorable marginal zone where death exceeds reproduction but where the population is maintained by excess of reproduction near the center of the area and gradual shifting of individuals toward the margins. The actual limits of range will then be determined ... by a constantly fluctuating equilibrium between tendency to spread at the center of the range and tendency to recede at the margins. Any change, in either the population itself or the environment, which increases the ratio of reproduction to death will cause spreading and a change which decreases the reproduction/death ratio will cause receding – dying back from marginal areas.

It is important to understand that such processes are constantly underway, and that the post-Wisconsin distribution of carnivores and other wildlife in BC will always be in a state of flux. As will be evident in later sections of this book, a number of species, including Northern Raccoon, Long-tailed Weasel, Cougar and Bobcat have increased their respective ranges in the province within the last few decades.

The Landscape and Current Carnivore Distribution
With a total area of about 950,000 km^2, British Columbia is a large and environmentally diverse province. It spans 11 degrees of latitude and 25 degrees of longitude, from the wet coastal islands to the rugged and relatively dry uplands of the Rocky Mountains, and from the comparatively temperate environments along the United States border to areas regularly exposed to subarctic influences in the north. The province's diversity is further enhanced by

its physical features, dominated by a series of north-south oriented mountain ranges (figure 6). Those ranges play a major role in determining local climate by intercepting Pacific weather systems as they move eastward, creating alternating wet and dry belts on their west and east sides, respectively. The wettest regions are along the coast, especially the western slopes of the Coast Mountains and the outer shores of Vancouver Island and the Queen Charlotte Islands. East of the Coast Mountains, the rain shadow effect has resulted in a large arid region known as the Interior Plateau. The most arid conditions are found in the southern interior, in the Thompson River and Okanagan valleys. Other wet-dry belts are associated with the Cassiar Mountains, Rocky Mountains and Columbia Ranges (Cariboo, Monashee, Selkirk and Purcell mountains). The effects of latitude and altitude at a particular location combine to determine the proportion of a year's precipitation that falls as snow, a factor of considerable importance in the lives of BC mammals, including carnivores and their prey. Approximately 75 per cent of the province's land area lies above an elevation of 1000 metres.

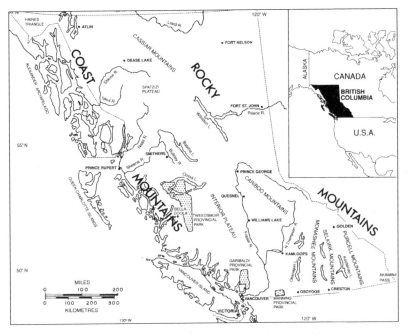

Figure 6. General geographic features of British Columbia.

Reflecting the effects of climate and growing conditions, the vegetation of the province is predominantly coniferous forest, but with patches of deciduous tree cover (sometimes extensive) in the fire-driven ecosystems of the central interior and the north, and in riparian habitats along streams and lakes. Grassland and shrub-steppe habitats occur primarily in the more arid areas of the province, especially in the southern interior. Open alpine tundra and scrubby willow-birch habitats are common in northern BC and at high elevations in the south.

To facilitate regional and local descriptions of British Columbia's biodiversity, provincial wildlife and habitat managers use two ecological classification systems. The first, used extensively by the Ministry of Forests, comprises 14 biogeoclimatic zones delineated by their relatively similar climatic and vegetation features (you can view or download the latest update of the BC biogeoclimatic zone map from the BC Ministry of Forests by going to http://www.for.gov.bc.ca/hre/becweb/resources/maps/map_download.html. In mountainous regions of the province, those zones and their various sub-zones are strongly related to elevation. In general, the carnivores are less tied to particular vegetation communities than are most of the species they prey upon. Many are opportunistic and mobile, and individuals or local populations may occupy a number of different habitats and exploit different prey in those areas over time.

The second system is the Ministry of Environment's Ecoregion Classification, which charts separate geographic areas with relatively consistent climate and topography. The ten ecoprovinces that have been established for the province (figure 7) are further subdivided into smaller units (30 ecoregions and 78 ecosections), which are used in finer scale regional and local management deliberations. The 21 carnivore species covered in this book occur in nine of the ten ecoprovinces (all except the completely oceanic Northeast Pacific Ecoprovince). Based on location records compiled for this book, the terrestrial ecoprovince supporting the lowest known number of the BC carnivore species (14 of 21) is the Taiga Plains in the extreme northeast corner (table 1). In each of the other eight, at least minor occurrences have been documented for 15 to 20 of the provincial carnivores. Further, 12 of the 21 species regularly occur or have at least been detected in all nine terrestrial ecoprovinces and only 4 (Northern Raccoon, Sea Otter, American Badger and Western Spotted Skunk) have been detected in fewer than six. These figures

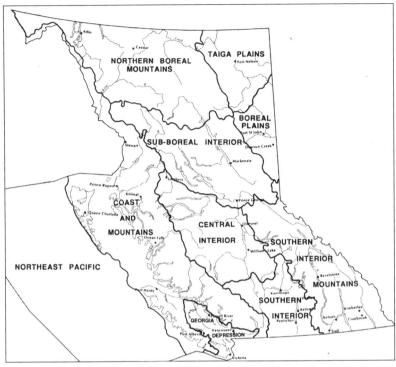

Figure 7. The 10 ecoprovinces of British Columbia.

underscore the adaptability, ubiquity and overall importance of this group of mammals in British Columbia.

Selected References: Clague 1981, Cowan 1941, Darlington 1957, Demarchi et al. 1990, Driver 1988, Fleming and Cook 2002, Harington 1977, Meidinger and Pojar 1991, Nagorsen et al. 1995, Pielou 1991, Ramsay et al. 2004, Stone and Cook 2000, Ward et al. 2003.

Conservation and Management

As valued resources, feared predators, conspicuous competitors, and common subjects of myth, legend and ceremony, carnivores have figured prominently in human history. The utilitarian and reverential aspects of our relationships with the various species are discussed in the individual species accounts under the heading

Table 1.

Distribution of carnivores in the nine terrestrial ecoprovinces of British Columbia.

Legend:
  P = Regularly present
  (p) = Minor or occasional presence.
  ? = Possible (unconfirmed) occurrence.

ECOPROVINCES	Coast and Mountains	Georgia Depression	Southern Interior	Central Interior	Southern Interior Mountains	Sub-boreal Interior	Northern Boreal Mountains	Boreal Plains	Taiga Plain
Coyote	P	P	P	P	P	P	P	P	P
Grey Wolf	P	(p)	P	P	P	P	P	P	P
Red Fox	P	P	P	P	P	P	P	P	P
Black Bear	P	P	P	P	P	P	P	P	P
Grizzly Bear	P	(p)	P	P	P	P	P	P	P
Northern Raccoon	P	P	P		P				
Sea Otter	P	(p)							
Wolverine	P	(p)	P	P	P	P	P	P	P
Northern River Otter	P	P	P	P	P	P	P	P	P
American Marten	P	P	P	P	P	P	P	P	P
Fisher	(p)	(p)	P	P	P	P	P	P	P
Ermine	P	P	P	P	P	P	P	P	P
Long-tailed Weasel	(p)	?	P	P	P	(p)			
Least Weasel	P		?	P	?	P	P	P	P
American Mink	P	P	P	P	P	P	P	P	P
American Badger			P	P	P				
Striped Skunk	P	P	P	P	P	P	(p)	(p)	
Western Spotted Skunk	(p)	P							
Cougar	P	P	P	P	P	P	(p)	(p)	(p)
Canada Lynx	(p)		P	P	P	P	P	P	P
Bobcat	(p)	P	P	P	P	(p)			

"Human Uses". The value of many has long centred on their use in the fur trade, BC's first industry.

As in other areas of North America, conservation and management of carnivores in British Columbia were not on the agenda for more than a century following European contact. Among the first resources exported from the province, and fortunately the only carnivore casualty of exploitation prior to management, was the Sea Otter. Pelts obtained at Nootka Sound, Vancouver Island, by Captain Cook's expedition in 1778 were sold for high prices in China, providing the basis for a lucrative market that resulted in another "gold rush" of sorts. Although the Beaver was the staple item in the fur trade over most of Canada at the time, the Sea Otter became the most prized and valuable of all furs. British and American interests competed fiercely, both hunting directly and trading with coastal inhabitants, securing up to 15,000 pelts annually from BC and Alaskan shores between 1785 and 1814. By 1880, the annual harvest had dwindled to less than 100 and, by 1911, when the International Fur Seal Convention gave the species complete protection, the Sea Otter was already essentially extirpated from BC.

Meanwhile, provincial legislators had long since identified the importance of wildlife in the lives of British Columbians, passing an "Act Providing for the Preservation of Game" in 1859. Directed primarily to species commonly used as food, especially ungulates and certain birds, this legislation particularly recognized the need for hunting closures during the period when animals bear and rear their young. But there is no evidence that conservation or management measures were directed to any of the carnivore species until years later. Seasonal closures were instituted for American Martens and Northern River Otters in 1896, for American Black Bears in 1909, and for Grizzly Bears and several furbearer species between 1914 and 1924. Furbearer conservation received an added boost in 1926 with establishment of the Registered Trapline System (RTS), which provided individual trappers with exclusive-use areas so that they could manage and husband furbearers for long-term sustainable use. Eighty years later, the RTS is still in place, and the continued widespread distribution and abundance of furbearer species is evidence of its success.

Conspicuously absent from any conservation consideration through the first two-thirds of the 20th century were the Cougar, Grey Wolf and Coyote – all three regarded as detrimental to human interests. Referred to in agency annual reports under categories such as "pests", "vermin" and "noxious animals", these three had bounties on their heads for half a century, until the mid 1950s, and were intensively hunted, trapped and poisoned with extermination as the ultimate goal. Reflecting increased scientific information and changing public attitudes, the Wildlife Act of 1966 officially categorized the Grey Wolf and Cougar as "wildlife" and established a protective management regime in the form of specified open seasons, bag limit, and restrictions on harvest methods. Similar status was afforded to the Coyote in 1971.

All 21 carnivores covered in this book are categorized and managed as wildlife in BC, with conservation of species as the highest management priority. Conforming to ecological principles, predators are less abundant than their prey. That leads to some species, such as Wolverine and Fisher, being naturally rare (occurring at low density over the landscape). "Rare" in that sense describes a biological fact, but is sometimes interpreted by the uninformed as "endangered". Benefitting from the attention afforded to harvested wildlife by both managers and users, most of the province's carnivores are not considered at risk. The exceptions, listed in table 2, include four species and two subspecies on the provincial Red List and three species and one subspecies on the Blue List. Two of those (American Badger and Queen Charlottes Ermine) are listed as endangered or threatened at the national level by COSEWIC, and are subject to management authority under the federal Species at Risk Act (SARA).

The species accounts give details on the status of the species at risk, but the news for BC carnivores is generally good. The Sea Otter is back and thriving following a reintroduction program in the early 1970s, and a combination of recovery efforts and new information has resulted in a more optimistic picture for the American Badger. The Queen Charlottes Ermine is listed on a precautionary basis because of its apparent rarity and genetic uniqueness, but is not known to be any less widely distributed or abundant now than at any other time since glaciation. The Vancouver Island Wolverine appears on the Red List because it has not been detected in the province in decades, its subspecies designation is equivocal and evidence for previous existence of a viable population is lacking. The carnivores on the Blue List are those that, because of relative rarity and certain biological characteristics, are considered more likely to become endangered and, therefore, require closer monitoring. In short, they are potentially but not actually at risk.

Table 2. Carnivores considered to be at risk in British Columbia as of February 2008. The provincial Red List includes species that are extirpated, endangered or threatened in BC, while the Blue List identifies those considered vulnerable because of their biological characteristics or the existence of known threats to which they are sensitive. In the federal list by COSEWIC (Committee on the Status of Endangered Wildlife in Canada), Endangered refers to "a wildlife population facing imminent extirpation or extinction", Threatened indicates "a wildlife species likely to become endangered if limiting factors are not reversed" and Special Concern is equivalent to the provincial Blue List.

TAXON	RANKING	
	BC	COSEWIC
SPECIES		
Grizzly Bear (*Ursus arctos*)	Blue List	Special Concern
Sea Otter (*Enhydra lutris*)	Red List	Special Concern
Wolverine (*Gulo gulo*)	Blue List	Special Concern
Fisher (*Martes pennanti*)	Blue List	None
Least Weasel (Mustela nivelis)	Blue List	None
American Badger (*Taxidea taxus*) *	Red List	Endangered
SUBSPECIES		
Vancouver Isl. Wolverine (*Gulo gulo vancouverensis*)	Red List	Special Concern
Vancouver Isl. Ermine (*Mustela erminea anguinae*)	Blue List	None
Queen Charlottes Ermine (*Mustela erminea haidarum*)	Red List	Threatened

* The indicated rankings are for *T.t. jeffersoni*, believed to be the only subspecies of American Badger in BC.

Urban mythology holds that carnivores thrive only in wilderness, and that human presence is always detrimental. In fact, some species have expanded their ranges or increased in numbers either because of human developments (Coyote, Northern Raccoon, Striped Skunk) or despite them (Black Bear, Bobcat, Long-tailed Weasel). Carnivores species are opportunistic and adaptable, and individuals learn to either ignore or avoid humans if doing so meets their needs. Dispersing juveniles and displaced adults find living situations that work for them, whether people are there or not, and so we get Northern Raccoons in attics, Striped Skunks under porches, American Minks and Northern River Otters in boats and boathouses, Coyotes and Red Foxes on golf courses and in municipal parks, and Black Bears, Cougars and Bobcats in back yards. Consequently, although "the only good predator is a dead predator" will never again be a guiding principle, selective control to protect human life and property (including pets and livestock) from marauding individuals will always be an important element in carnivore management. Predator control to protect or enhance prey populations (especially ungulates) has been undertaken in the past, but is highly controversial and is now largely reserved for use only in aiding recovery of species at risk.

In British Columbia, conservation and management of all but one carnivore covered in this book are the responsibility of the provincial government, under the auspices of the Ministry of Environment. The exception is the Sea Otter, for which, as a marine mammal, management responsibility is shared with federal authority under the Department of Fisheries and Oceans. Management and protection of wildlife habitat is primarily under the authority of the Ministry of Forests, with advice from the Ministry of the Environment. Maintaining biodiversity, an objective of habitat protection, is assumed to confer benefits to carnivores, but the needs of individual species are rarely addressed directly, except for those of Grizzly Bears. It is now known that retention of logging debris (logs, stumps, branches) in cut-over areas, though generally offensive to the untrained eye, is an important habitat protection measure for forest-dwelling species such as American Martens and Fishers.

Selected References: Conservation Data Centre 2007, Darling 2000, Hancock 1987, Hatter 1987, Lopez 1978.

Studying Carnivores

Early carnivore research was largely limited to laboratory analyses of dead specimens harvested by hunters and trappers or collected by researchers, and to observations of live individuals kept in captivity. Those studies provided foundational information on topics such as taxonomy, morphology, reproduction, food habits, behaviour and physiology, and were augmented by anecdotal, mostly opportunistic observations of free-ranging animals. Actual field research on the carnivores has been, in many ways, the "last frontier" of mammal study because the subjects are intelligent and evasive, mostly nocturnal, and often relatively sparsely distributed, making data collection difficult. Of the approximately 370 publications listed in the references section at the back of this book, only about 30 were available in the early 1960s when Cowan and Guiguet wrote *The Mammals of British Columbia.*

A research technique commonly used in early field studies of carnivores was snow-tracking. In the areas and seasons it could be used, snow tracking provided new information on movements, home range, aspects of predation, habitat use and, sometimes, social interactions. The title of Raymond Schofield's 1960 report, "A thousand miles of fox trails in Michigan's Ruffed Grouse range", gives an idea of the effort expended in some of those studies. Another common technique, adapted from bird-banding, involved live-trapping and marking individuals with ear-tags to obtain, by subsequent recaptures, more sex- and age-specific information on movements, ranges and habitat use, as well as data on population size and composition. Tagging studies were best suited to the smaller carnivores, for which adequate sample sizes could be maintained. American Marten studies undertaken by Vern Hawley, Fletcher Newby and Richard Weckwerth in Montana in the 1950s were notable successes among the tagging-based projects, and served as both inspiration and model for my own studies on American Minks on Vancouver Island in the 1960s and early 1970s.

For the larger species, particularly the Grey Wolf, the most successful of the early ecological studies relied upon observations from small aircraft, mostly in winter. The aerial method provided for a more mechanized version of snow-tracking, enabling efficient coverage of the large distances often travelled by wolf packs, and for direct observations relating to social and predatory behaviour and population dynamics. At Isle Royale in Lake Superior, local wolves became habituated to the research aircraft overhead and went about their business normally, providing unprecedented information in a series of classic studies by researchers David Mech, Rolf Peterson and others.

The technological breakthrough in carnivore research was radio-tracking (telemetry) which came into common use in the early 1970s. Radio-tracking enabled researchers to contact study animals regularly and on demand rather than occasionally and by chance. Combined with the use of aircraft, telemetry provided the first reliable means to keep track of and obtain consistent data from the larger, wide-ranging species (Grey Wolf, Cougar, Grizzly Bear) in all seasons. Radio-tracking produced more detailed information on movements, ranges, activity patterns and habitat use than had been available previously, and also provided the first significant data on the causes and timing of natural mortality among carnivores. On the negative side, telemetry-based observations were biased to daytime and relatively fair weather, since it was usually not practical for researchers to be afield after dark or in storms. In a proliferation of ground-based telemetry studies in the 1980s and 1990s, biologists learned a lot about where species such as American Martens and Fishers rested during the day, but relatively little about what the animals did during their most active periods at night. Further, the frequent flying required for aerial telemetry was expensive and often dangerous.

Some of those problems are being overcome by the next technological advance in telemetry, the use of Global Positioning System (GPS) transmitters. These devices can be programmed to record and store location coordinates at specified intervals (as often as every few minutes). The coordinates are downloaded to a computer, via satellite, at intervals and times convenient to the user. The time is near at hand when the snowshoes will be left hanging in the shed and the biologist will follow the fox a thousand kilometres in just a

Figure 8. Immobilized wolf showing radio-collar.

few minutes, dry shod and with coffee cup in hand. Unfortunately, that will tell us only where the fox lived, not how.

Researchers usually attach standard VHF (Very High Frequency) and GPS transmitters to animals on durable collars (figure 8) designed for that purpose. Some species, such as Northern River Otter, Fisher and Wolverine, do not retain collars well and for those, a special transmitter package designed to be implanted in the animals' body cavity has been developed and is now in common use.

Tagging and telemetry studies require capturing an animal. The methods used depend on the species and local conditions to maximize efficiency and minimize animal damage, as required by the Canadian Council on Animal Care. For carnivores up to the size of a Fisher, the most common capture device is a commercially available cage trap (figure 9). For the wary and powerful Wolverine, biologists use a variation of the cage trap built on site from logs (figure 10). Some of the larger species, particularly Grey Wolves and Grizzly Bears, may be captured in open terrain by shooting them with drug-loaded darts from a helicopter, and all three of the wild cats in British Columbia have been subdued by darts after they have been treed by hounds. But the most common capture method for several species, involves the use of devices that catch and restrain the animal by the foot.

Readers aware of "ban the leghold trap" campaigns will be interested to know that the trap described and portrayed in associated fund-raising literature has, in fact, been banned for use in BC since 1983. Since the early 1980s, more than $13 million has been invested in trap research and development in this country, mostly

Figure 9. Northern Raccoon in a cage trap.

Figure 10. Cage trap made of logs to capture Wolverines.

in programs administered by the Fur Institute of Canada, and in 1997, Canada was a signatory of the Agreement on Interna-tional Humane Trapping Standards (AIHTS) with Russia and the European Union. The AIHTS specifies that trapping for any purpose, whether commercially for fur, for control of problem wildlife or for securing animals for research, must use approved humane devices. Such approved foothold traps and foothold snares have now been used successfully in research and translocation projects on American Badgers,

Figure 11. A spring-activated cable snare used to capture bears alive.

Northern River Otters, and all of the canids and felids. For research on bears, the most common capture device is a spring-activated leg snare using heavy cable.

Detailed assessment of population size has always been difficult for the carnivores, even within the boundaries of relatively small study areas. Measures of relative abundance and population trend have been obtained by various methods ranging from track counts along transects or comparisons of animal visitation rates at artificial scent stations to the use of questionnaires directed to people with regular or consistent opportunities to observe carnivores and their sign. An example of the last method is the case in North Dakota, in which Red Fox populations were monitored, in part, by question-naire responses from rural mail carriers. For determining the actual population densities listed in the species accounts in this book, the most common method for most species has been extrapolation from the ratio of marked to unmarked individuals among animals recaptured or observed subsequent to initial tagging. A relatively new technique, pioneered in BC by John Woods and his co-workers for use on Grizzly Bears, is the use of DNA analysis to identify the minimum number of individuals in an area. For that method, it is necessary only to collect some hair, which is usually done by arranging light entanglement in barbed wire when the bear approaches to investigate a scent package. The method has been adapted to smaller carnivores (American Marten, Fisher, Canada Lynx) using either barbed wire or glue strips. In a recent study in the Cariboo region of BC, Corinna Hoodicoff used Velcro pads placed at burrow entrances to capture hair from American Badgers.

Selected References: Cowan 1987, Hatler and Beal 2005, Stebler 1939, Woods et al. 1999, Zielinski et al. 2006.

Taxonomy and Nomenclature

The modern system of biological classification developed in 1758 by the Swedish botanist Karl von Linné (better known as Linnaeus) uses a hierarchy of taxonomic categories (e.g., class, order, family, genus, species), with the species as the basic unit. Each species has a unique, two-part scientific name (binomial) consisting of the genus name followed by the species modifier. By convention, the binomial is written in italics with the genus name capitalized and the species modifier in lower case. For example, the scientific name for the Ermine is *Mustela erminea*. Species that share a number of similar traits are usually grouped in the same genus. Fifteen species are recognized in the genus *Mustela*, of which three occur in BC.

Many species have distinct geographic races or subspecies that are formally recognized by taxonomists with a three-part name (trinomial). For example, *Mustela erminea haidarum*, a form of Ermine restricted to the Queen Charlotte Islands, is one of 20 North American subspecies. Taxonomists group species and genera into higher categories based on their presumed relationships. The 286 known species of carnivores (order Carnivora) worldwide are grouped by taxonomists into 12 families; 7 of those occur in BC.

The scientific names for mammal species and higher taxonomic categories used in this book (see Checklist) are as listed in *Mammal Species of the World* (Wilson and Reeder 2005) and the *Revised Checklist of North American Mammals North of Mexico, 2003* (Baker et al. 2003). Subspecies names are mostly taken from *The Mammals of British Columbia: A Taxonomic Catalogue* (Nagorsen 1990). Those publications provide detailed information on the taxonomy of BC species, including citations to original subspecies descriptions. There is no universally accepted list of common names for mammals equivalent to the American Ornithologists Union's checklist of bird names. All but one of the primary common names listed in this book were taken from the *Revised Checklist of North American Mammals North of Mexico, 2003*. The exception is use of the name Cougar rather than Mountain Lion or Puma for *Puma concolor*. As listed in the species accounts, many carnivores have several common names.

Taxonomy is a dynamic science and there have been changes in the taxonomy and nomenclature of carnivores since the original *Mammals of British Columbia* was published in 1965. Powerful new taxonomic tools, particularly the use of DNA analysis, are providing new insights into mammalian relationships. The skunks, which were traditionally grouped with the weasels in the family Mustelidae, show significant genetic and morphological differences from that group and are now treated by most mammalogists as a separate family, the Mephitidae. Morphological and genetic differences now support the placement of the Cougar in its own genus, *Puma*, rather than the genus *Felis*, and the New World otters in the genus *Lontra* rather than *Lutra* (the latter name now restricted to Old World otters), and the American Mink in its own genus, *Neovison*, rather than *Mustela*.

In contrast to those changes in higher taxonomic categories, species-level taxonomy has been stable with no changes in the number of recognized carnivore species in the province since publication of *The Mammals of British Columbia*. But the species names of the Wolverine, Red Fox and Least Weasel have all been changed, reflecting the view that North American and Old World forms of those species are no longer considered to be separate species. Future genetic studies may result in taxonomic changes. For example, there are two highly divergent DNA groups in the American Marten, both occuring in BC, that may prove to be distinct species. Interestingly, some early taxonomists using skull structure had come to a similar conclusion. More genetic research is needed to resolve the species status of these two forms, especially in areas where they come into contact.

Although subspecies are recognized for most of the BC carnivores, the attendant taxonomy is controversial. Many of the subspecies listed in the Taxonomy section of the species accounts were described 50 to 100 years ago, often on the basis of minor differences in pelage colour or skull features. Some authorities have suggested that it may be inappropriate to recognize subspecies in the carnivores and other mammals that range over large distances. Nevertheless, a surprising result of DNA studies is that even mobile species, such as the Grizzly Bear, have distinct genetic lineages or

DNA groups that occupy different parts of their range. The geographic distribution of those DNA groups, however, is often inconsistent with traditional subspecies boundaries. The genetic data also suggest that the number of subspecies recognized for many carnivores is probably excessive. Historically, North American populations of the Cougar had been grouped into 15 subspecies, but a recent DNA study found only one genetic group in North America. Similarly, American Black Bears from western Canada fall into two DNA groups, raising doubts about the validity of the five subspecies currently recognized in BC. Clarifying subspecies taxonomy is particularly important for species such as the American Badger, Ermine and Wolverine. One or more of their subspecies are either listed under Canada's Species at Risk Act (SARA) or appear on the province's Red or Blue lists of mammals considered to be at risk. Modern taxonomic studies are needed to verify that subspecies named represent genetically unique populations that actually warrant special conservation measures.

Selected References: Baker et al. 2003, Nagorsen 1990, Wilson and Reeder 2005.

CHECKLIST OF SPECIES

This is a list of the scientific and common names for the 21 native wild carnivores of British Columbia covered in this book.* The six carnivore families are arranged according to their generally accepted phylogenetic order. Genera within a family and species within genera are listed in alphabetical order by their scientific names.

Order Carnivora: Carnivores

Family Canidae: Canids (Dogs)
Canis latrans Say   Coyote
Canis lupus Linnaeus   Grey Wolf
Vulpes vulpes (Linnaeus)   Red Fox

Family Ursidae: Ursids (Bears)
Ursus americanus Pallas   American Black Bear
Ursus arctos Linnaeus   Grizzly Bear

Family Procyonidae: Procyonids (Raccoons)
Procyon lotor (Linnaeus)   Northern Raccoon

---

* Three species of Eared Seals (Family Otariidae) and two species of Haired Seals (Family Phocidae) found in BC are not listed. They will be covered in the handbook *Marine Mammals of British Columbia*.

Family Mustelidae: mustelids (Weasels and Allies)

Enhydra lutris (Linnaeus)	Sea Otter
Gulo gulo (Linnaeus)	Wolverine
Lontra canadensis (Schreber)	Northern River Otter
Martes americana (Turton)	American Marten
Martes pennanti (Erxleben)	Fisher
Mustela erminea Linnaeus	Ermine
Mustela frenata Lichenstein	Long-tailed Weasel
Mustela nivalis Linnaeus	Least Weasel
Neovison vison Schreber	American Mink
Taxidea taxus (Schreber)	American Badger

Family Mephitidae: Mephitids (Skunks)

Mephitis mephitis (Schreber)	Striped Skunk
Spilogale gracilis Merriam	Western Spotted Skunk

Family Felidae: Felids (Cats)

Puma concolor (Linnaeus)	Cougar
Lynx canadensis Kerr	Canada Lynx
Lynx rufus (Schreber)	Bobcat

IDENTIFICATION KEYS

This section provides two independent keys, one for whole animals and the other for skulls. Although the whole-animal key may be helpful for identifying some species from sightings, it is intended primarily for use with restrained or anesthetized live animals, dead animals or museum study skins. The key to skulls is intended for use with cleaned skulls, either prepared as museum specimens or found in the field.

Both keys are dichotomous, showing diagnostic characteristics arranged into pairs of statements (couplets). Each couplet consists of two alternative choices labelled "a" or "b". The user begins with the first couplet and selects the statement (either 1a or 1b) that best fits the specimen at hand. The selected statement provides direction to the next applicable couplet. For example, in the following whole-animal key, if statement 1b is true (selected), the user is directed to couplet 3; if statement 3b is true, the user is directed to couplet 6; and so on. By working through the couplets, step-by-step, one eventually arrives at a species identification. The user should then consult the account for that species (page number provided) to verify that the identification is consistent with its description and range map.

An introductory section at the beginning of each key summarizes the diagnostic traits that are the most important for identifying the 21 carnivore species covered in this book. The intent was to avoid subjective characteristics (e.g., slightly smaller than or slightly darker than), instead emphasizing discrete features such as presence or absence, absolute size differences, and absolute colour or pelage pattern differences.

Key to Whole Animals

The most diagnostic external traits for identifying species are pelage colour and markings, ear size, relative length of the tail, hind-foot length, and head-and-body length. Figure 12 shows the standard body measurements that are routinely taken for carnivore specimens. Note that head-and-body length is not an independent measurement, but obtained by subtracting tail length from total length. All weights and measurements used in this handbook are metric. The symbol > indicates "greater than" and < indicates "less than". For example, "hind-foot length < 200 mm" means that the hind foot is less than 200 millimetres long.

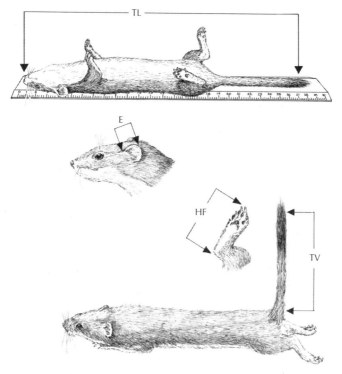

Figure 12. Body measurements used in identification keys and species accounts measurements: TL= total length (front of jaw to end of last tail vertebra, not including hairs); TV=tail vertebrae length (base of tail to end of last tail vertebra, not including hairs); HF= hind foot length (heel to end of longest claw); E=ear length (notch inside of ear to outermost edge, not including hairs).

NOTE: Some carnivores may carry diseases or parasites that are transmissible to humans. Take appropriate precautions while handling live animals or fresh specimens (see Human Health Considerations, page 20, in the General Biology section).

1a. Head-and-body length > 130 cm; tail shorter than hind-foot length 2

1b. Head-and-body length < 130 cm; tail longer than hind-foot length 3

2a. Face not dish-shaped in side view; no prominent hump on the shoulders (figure 13); middle claws on front feet < 55 mm long ....American Black Bear, page 109

2b. Face dish-shaped in side view; prominent hump on the shoulders (figure 14); middle claws on front feet > 55 mm long ....Grizzly Bear, page 132

3a. Cat-like in appearance 4

3b. Not cat-like in appearance 6

4a. Total length < 100 cm; tail-vertebrae length < 50% of head-and-body length; ears have distinct tufts 5

4b. Total length > 100 cm; tail-vertebrae length > 50% head-and-body length; ears lack distinct tufts ....Cougar, page 324

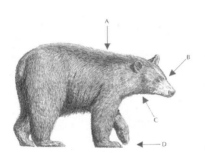 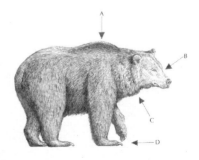

Figure 13. Black Bear features: A – no hump; B – face not dished; C – no neck ruff; D – short claws.

Figure 14. Grizzly Bear features: A – prominent hump; B – dished face; C – usually has neck ruff; D – long claws.

Figure 15. Tails of Bobcat (left) and Canada Lynx.

5a.   Tip of the tail has a dark distal band and several proximal bands on the dorsal surface (figure 15, left), and whitish on the underside; ear tufts < 30 mm long; hind-foot length < 200 mm ....Bobcat, page 350

5b.   Tip of the tail completely black on the dorsal and ventral surfaces (figure 15, right); ear tufts > 30 mm long; hind-foot length > 200 mm ....Canada Lynx, page 338

6a.   Dog-like in appearance 7
6b.   Not dog-like in appearance 9

7a.   Total length > 100 cm; hind-foot length > 150 mm; tail usually lacks a prominent white tip 8
7b.   Total length < 100 cm; hind-foot length < 150 mm; tail usually has a prominent white tip ....Red Fox, page 96

8a.   Total length > 150 cm; hind-foot length > 200 mm; width of nose-pad > 30 mm ....Grey Wolf, page 77
8b.   Total length < 150 cm; hind-foot length < 200 mm; width of nose-pad < 30 mm ....Coyote, page 61

9a.   Pelage black with contrasting multiple white body stripes .10
9b.   Pelage not black and lacks white body stripes 11

Figure 16. Striped Skunk back pattern.

Figure 17. Western Spotted Skunk back pattern.

10a. Two broad, continuous white stripes on the dorsum extending from behind the head to the rump (figure 16) .... .Striped Skunk, page 303

10b. Four white stripes or bands on the dorsum extending from behind the head to the mid back; three discontinuous bands or series of spots on the lower back and rump (figure 17) .... .Western Spotted Skunk, page 316

11a. Toes on the hind foot prominently webbed .... .12

11b. Toes on the hind foot not prominently webbed .... .13

12a. Hind foot flat, modified into a flipper; tail uniform in width (figure 18, left); tail-vertebrae length < 30% of body length .... .Sea Otter, page 171

12b. Hind foot not modified into a flipper; tail tapers wider at the base (figure 18, right); tail-vertebrae length > 50% of body length .... .Northern River Otter, page 196

13a. Hind-foot length > 150 mm; dorsal pelage marked with a dark brown-blackish saddle bordered by two lighter bands extending from the head to the base of tail . . . .
. . . . .Wolverine, page 182
13b. Hind-foot length < 150 mm; dorsal pelage lacks a dark saddle bordered by lighter bands 14

14a. Face has a black mask; tail has seven or eight black rings
. . . . .Northern Raccoon, page 154
14b. Face lacks a black mask; tail has no rings 15

15a. Face has distinctive white markings and a narrow white stripe extends from the nose to the shoulders; longest claws on the front feet > 30 mm . . . . . . .American Badger, page 291
15b. Face has no distinctive white markings and no white stripe from the nose to the shoulders; longest claws on the front feet < 30 mm 16

Figure 18. At right, tails of Sea Otter (left) and Northern River Otter.

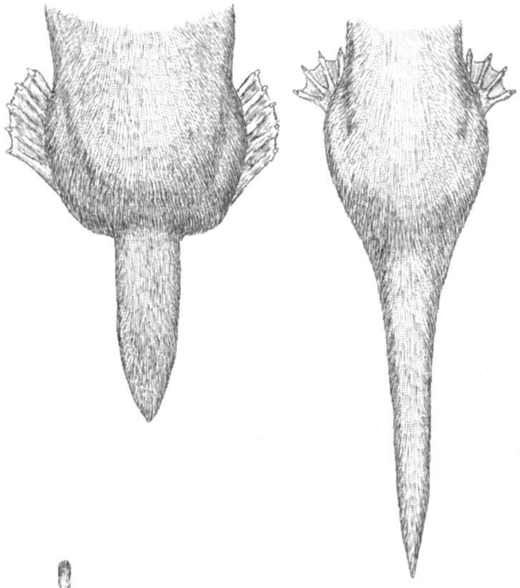

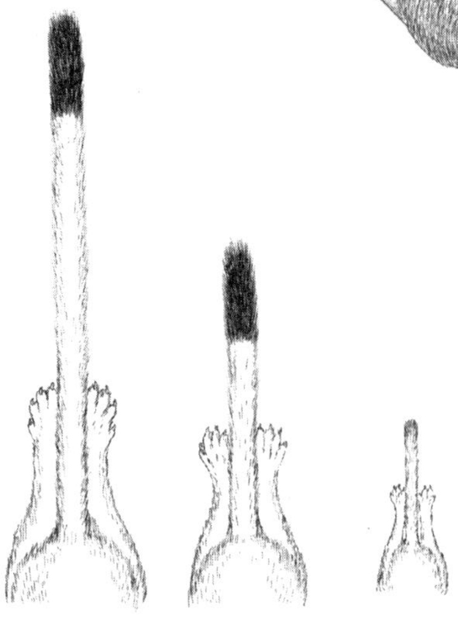

Figure 19. Tails of Long-tailed Weasel (left), Ermine (middle) and Least Weasel.

16a. Ear prominent, > 35 mm in length. 17
16b. Ear not prominent, < 35 mm in length 18

17a. Total length > 700 mm; tail-vertebrae length > 300 mm
 . . . . .Fisher, page 227
17b. Total length < 700 mm; tail-vertebrae length < 300 mm . . . . .
 . . . . .American Marten, page 210

18a. Ventral pelage similar in colour to dorsal pelage and
 uniformly brown except for a few scattered white markings;
 winter pelage brown . . . . .American Mink, page 276
18b. Ventral pelage white to pale brown and contrasts strikingly
 with dorsal pelage; winter pelage usually white (except for
 some animals in normally snow-free areas in southwestern
 BC) 19

19a. Tail-vertebrae length > 20% of the total length and > 25% of
 the head-and-body length; tail has a prominent black tip
 (figure 19, left and middle) 20
19b. Tail-vertebrae length < 20% of the total length and < 25% of
 the head-and-body length; tail lacks a prominent black tip,
 but may have a few black hairs (figure 19, right)
 . . . . .Least Weasel, page 268

20a. Tail-vertebrae length > 33% of the total length and > 50% of
 the head-and-body length (figure 19, left)
 . . . . .Long-tailed Weasel, page 257
20b. Tail-vertebrae length < 33% of the total length and < 50% of
 the head-and-body length (figure 19, middle)
 . . . . .Ermine, page 243

Key to Skulls

This key, which relies on cranial and dental traits, is designed to be used on cleaned skulls (i.e., with flesh and other tissues removed) and museum specimens. You will require calipers for taking cranial and dental measurements. Diagnostic traits for identifying species include dental formulae (number of incisors or cheek teeth* in the upper and lower jaws), cranial measurements (see figure 20), dental measurements and skull shape. The features described pertain to adult specimens with fully developed skulls and permanent dentition.

1a. Margin of bony palate extends only slightly beyond the last tooth of the upper toothrow (figure 21) 2

1b. Margin of bony palate extends well beyond the last tooth of the upper toothrow (figure 22) 6

2a. Condylobasal length > 100 mm; each lower jaw has seven cheek teeth; each upper jaw has six cheek teeth 3

2b. Condylobasal length < 100 mm; each lower jaw has five cheek teeth; each upper jaw has four cheek teeth 5

Figure 20. Skull measurements used in identification keys: A–C = basal length; B–C = condylobasal length; C–E = skull length; D–D1 = zygomatic breadth.

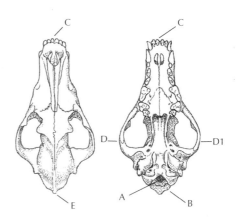

---

* Because distinguishing between premolars and molars can be difficult, this key lumps them together under the general term "cheek teeth", referring to all the teeth behind the canines.

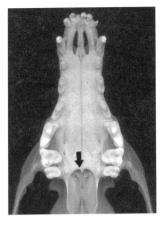

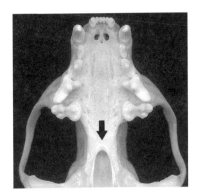

Figure 21.   Figure 22.

3a. Condylobasal length > 160 mm; postorbital processes thick and lack shallow depressions on their dorsal surface (figure 23) 4

3b. Condylobasal length < 160 mm; postorbital processes thin, with shallow depressions on their dorsal surface (figure 24) ....Red Fox, page 96

4a. Skull length > 215 mm; zygomatic breadth > 115 mm ....Grey Wolf, page 77

4b. Skull length < 215 mm; zygomatic breadth < 115 mm ....Coyote, page 61

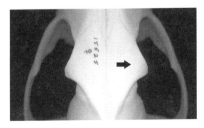

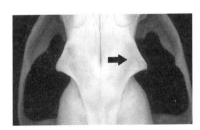

Figure 23.   Figure 24.

5a. Condylobasal length > 60 mm; dorsal part of the skull convex (figure 25); mastoid region of the skull not flared in dorsal view (figure 27) . . . . .Striped Skunk, page 303

5b. Condylobasal length < 60 mm; dorsal part of the skull flat (figure 26); mastoid region of the skull flared in dorsal view (figure 28) . . . . .Western Spotted Skunk, page 316

6a. Basal length > 200 mm; last upper cheek tooth rectangular and much longer than its width 7

6b. Basal length < 200 mm; last upper cheek tooth either triangular or rectangular and wider than its length 8

7a. Last upper cheek tooth (M2) > 30 mm long (C–D) and > 16 mm wide (A–B); last three upper cheek teeth (P4–M2) > 60 mm long (figure 29) . . . . .Grizzly Bear, page 132

7b. Last upper cheek tooth (M2) < 30 mm long (C–D) and < 16 mm wide (A–B); last three upper cheek teeth (P4–M2) < 60 mm long . . . . .American Black Bear, page 109

8a. Six cheek teeth in each upper jaw (figure 30) . . . . .Northern Raccoon, page 154

8b. Fewer than six cheek teeth in each upper jaw 9

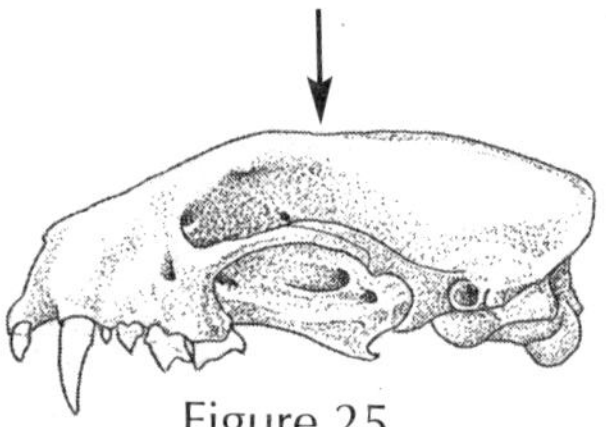

Figure 25.

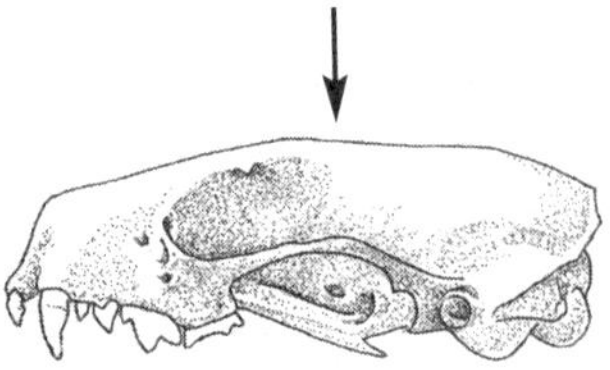

Figure 26.

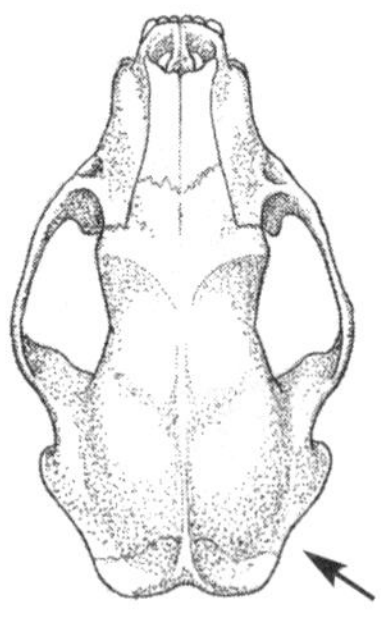

Figure 27.

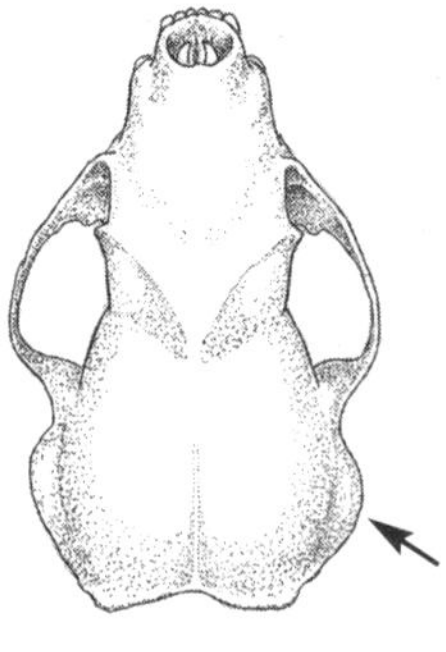

Figure 28.

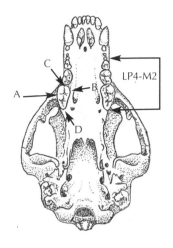

Figure 29. Skull measurements to distinguish bear species: A–B = width of last upper molar (M2); C–D = length of last upper molar (M2); LP4–M2 = length of last three upper cheek teeth (P4 to M2).

9a. Five or six cheek teeth in each lower jaw (figure 31) 10

9a. Three cheek teeth in each lower jaw (figure 32) 19

10a. Five cheek teeth in each upper jaw 11

10b. Four cheek teeth in each upper jaw .14

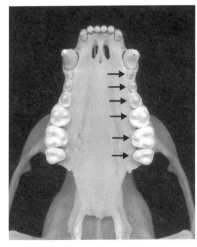

Figure 30.

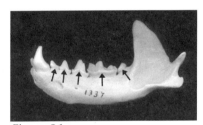

Figure 31.

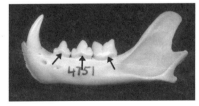

Figure 32.

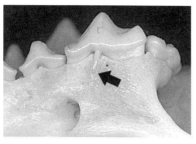 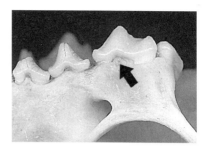

Figure 33.   Figure 34.

11a. Postorbital processes well developed and pointed; infraorbital foramen oblong and > 8 mm; auditory bullae flat
........................Northern River Otter, page 196

11b. Postorbital processes blunt and not well developed; infraorbital foramen ovoid and < 8 mm; auditory bullae not flat 12

12a. Condylobasal length > 125 mm; zygomatic breadth > 90 mm
...............................Wolverine, page 182

12b. Condylobasal length < 125 mm; zygomatic breadth < 90 mm
....13

13a. Condylobasal length > 90 mm; fourth upper cheek tooth (P4) has a median rootlet visible in lateral view (figure 33)
....Fisher, page 227

13b. Condylobasal length < 90 mm; fourth upper cheek tooth (P4) has no visible median rootlet (figure 34)
...........................American Marten, page 210

14a. Two incisors in each lower jaw; last two upper cheek teeth extremely flat and modified for crushing (figure 35)
...............................Sea Otter, page 171

14b. Three incisors in each lower jaw; last two upper cheek teeth not flat or modified for crushing (figure 36) 15

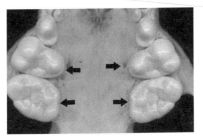

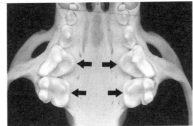

Figure 35.   Figure 36.

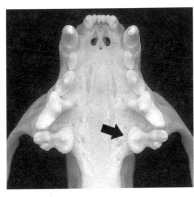

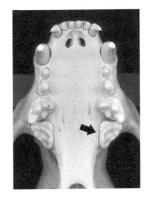

Figure 37   Figure 38.

15a. Condylobasal length < 100. mm; last upper cheek tooth (M1) dumb-bell shaped (figure 37) 16

15b. Condylobasal length > 100 mm; last upper cheek tooth (M1) triangular (figure 38) . . . . .American Badger, page 291

16a. Condylobasal length > 55 mm; third upper cheek tooth (P4) > 6 mm long . . . . .American Mink, page 276

16b. Condylobasal length < 55 mm; third upper cheek tooth (P4) < 6 mm long 17

17a. Condylobasal length > 32 mm 18

17b. Condylobasal length < 32 mm . . . . . . .Least Weasel, page 268

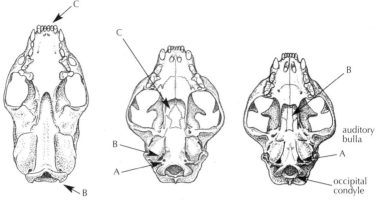

Figure 39.   Figure 40.   Figure 41.

18a. Postglenoid length (A–C) < 48% of condylobasal length (B–C)
(figure 39) . . . . .Long-tailed Weasel, page 257
18b. Postglenoid length > 48% of condylobasal length
. . . . .Ermine, page 243

19a. Four cheek teeth in each upper jaw; basal length > 150 mm
. . . . .Cougar, page 324
19b. Three cheek teeth in each upper jaw; basal length < 150 mm
. 20

20a. Foramina between the auditory bulla and occipital condyle
form two separate openings (A,B); presphenoid bone (C)
flares widely in posterior portion (figure 40)
. . . . .Canada Lynx, page 338
20b. Foramina between the auditory bulla and occipital condyle is
fused into a single opening (A); presphenoid bone (B) does
not flare (figure 41) . . . . .Bobcat, page 350

Selected References: Glass 1973, Hall 1951, Hoffman and Pattie
1968, Ingles 1965, Jones and Manning 1992.

SPECIES ACCOUNTS

The species accounts for British Columbia's 21 terrestrial carnivore species follow the same order as in the checklist. Each account includes drawings of the whole animal and its skull to depict general features; diagnostic features important for identification are illustrated in the identification keys. The text for each account is based on extensive review of scientific literature available through 2006. I attempted to include observations and research results from BC wherever possible, but many species have been little studied in this province and it was therefore necessary to fill gaps with findings from other areas. The information in the species accounts is divided into nine sections, described here.

Other Common Names lists alternative common English names, regularly used French names and the Provincial Species Code which is now routinely used as a convenient shorthand by wildlife managers and researchers in BC.

Description gives the relative size and shape of the species, followed by the actual measurements and weights, dental formula, pelage length and colour patterns, and other distinctive traits, including the relative length and shape of body parts such as the tail, legs and ears. A concluding sub-section labelled *Identification* helps distinguish the subject species from similar animals.

Each species description begins with a statement to convey a general impression of size, usually in comparison to a common domestic animal such as a house cat or a particular breed of dog. Unless stated otherwise, the four specific measurements listed are those normally taken for museum specimens: total length, tail length, hind-foot length and ear from notch (see figure 12). Because differences in size (dimorphism) between the sexes and between subspecies are common among the carnivores, the measurements and weights for those two categories are listed separately. All linear measurements are in millimetres (mm); weights are in grams (g) for species less than 10,000 g, and in kilograms (kg) for species 10 kg (10,000 g) or heavier. The values given are the mean (average), range (in parentheses), and sample size (number of individuals measured). For example, the tail length for male Coyotes is given as "342 (300-387) n=18" to show that the average length of the tail is 342 mm, based on measurements of 18 individuals whose tails ranged from 300 to 387 mm in length.

The intent was to use measurements and weights only from adult (fully grown) specimens. But the young of many of the species (especially the mustelids), grow quickly and are near adult size (length, but usually not weight) by late fall of their first year, at age six to eight months. Thus, it is likely that the many mammalogists who have contributed to the database over the years have not all interpreted "adult" the same, and that some measurements at the lower end of the range were from juveniles. The most variable measurement for most species is tail length, which contributes significantly to the range of values for total length. Weights are even more variable than body measurements for most species. The lower values in the range may have been either from juveniles, very young adults or animals in poor physical condition, and the upper values from pregnant females or animals with abundant body fat and full digestive tracts. For example, an adult Grey Wolf may have 10 kg or more of meat in its stomach after gorging on a fresh kill, temporarily increasing its weight by 20 per cent or more.

The dental formula denotes the number of teeth in one side of the head. The first value is the number of teeth in the upper jaw and the second is the number in the lower jaw. For example, for the Coyote, "incisors 3/3, canines 1/1, premolars 4/4, molars 2/3" indicates three upper and three lower incisors, one upper and one lower canine, four upper and four lower premolars, and two upper and three lower molars.

Distribution and Habitat summarizes the World, North American and Canadian distribution of the species and gives details on its range in BC. It also provides an overview of the habitats most commonly used by each species, with particular reference to foraging, daily resting and maternal activity.

A provincial range map incorporates information from museum

records and sightings. The black dots are locations for BC specimens in natural history museums across North America and documented locations (sightings and unambiguous "sign") from
literature, biological surveys, interviews with and reports from
knowledgeable people, and personal observations.

The range maps for 14 species also show grey dots signifying
harvest data. Since 1985, the provincial wildlife management
agency (Ministry of Environment) has maintained a Fur Harvest
Database, which tracks pelt sales from each of approximately 2900
registered traplines on Crown land in BC. The grey dots represent
the centres of individual trapline areas in which a harvest of the
subject species has been recorded in at least 3 of the 18 years for
which data were available, unless indicated otherwise. For Cougar
and Grizzly Bear, grey dots also indicate supplementary location
points for a variety of government kill records (hunter harvest, animal control, road kills) compiled since 1975.

All specimen, observation and harvest locations represent point-
in-time encounters between the subject species and humans willing
and able to document them. Thus, an absence of location points on
a map area may simply represent the rarity of humans there, which
is particularly likely in some of the more remote and rugged
regions of BC. In the same vein, clusters of location points on the
maps generally reflect areas of long-term, regular human activity,
and should not be interpreted as species hot-spots or areas of concentration. Finally, the ranges of some species are changing as a
result of habitat and climate changes, and an area of absence today
may not be tomorrow. For all of these reasons, readers of this book
are encouraged to report occurrences outside of the distributions
shown in the range maps.

Natural History covers aspects of the species' biology and ecology
under six sub-headings: *Feeding Ecology* – diet, nutritional requirements, prey hunting methods; *Home Range and Social Behaviour* –
size, use and defense of individual activity areas, security status of
individuals, extent of grouping tendencies; *Activity and Movements*
– timing and extent of travel for daily foraging and dispersal;
Reproduction – mating season, duration of pregnancy and birth
dates, litter size, factors affecting productivity, parental care, development of young; *Health and Mortality* – parasites, diseases, causes
and extent of mortalities, longevity; and *Abundance* – population
densities, current trends in BC.

Human Uses describes how humans have interacted with BC's
carnivores. First Peoples have traditionally hunted several species

for subsistence and used various body parts for ceremonies and other purposes. Other British Columbians have found commercial and industrial uses for carnivores.

Taxonomy contains an updated list of subspecies in BC, based on a review of applicable studies since the publication of Cowan and Guiguet's *The Mammals of British Columbia*. It also has a discussion of any taxonomic problems and questions that remain.

Conservation Status and Management identifies species and subspecies considered at risk, outlines the management framework for those not-at-risk, and discusses the various issues associated with these animals (e.g., predation on pets, livestock, species-at-risk, and other wildlife; damage to property and injury of humans; transmission of diseases and parasites).

Remarks includes interesting anecdotes and facts that do not fit readily into the other sections.

Selected References lists publications considered seminal for the species in western North America, works specifically relating to studies in BC, major review documents, and other significant or interesting papers. The list is from literature searches undertaken through December 2006 and is not exhaustive, but it should provide a good starting point for readers desiring more information.

Coyote   *Canis latrans*

Other Common Names: American Jackal, Brush Wolf, God's Dog, Prairie Wolf, Yodeler. Provincial Species Code: M-CALA.

Description

The Coyote, a member of the dog family, has the general appearance of a German Shepherd but is about half the size and much more lightly built. It has a narrow, pointed muzzle, prominent pointed ears, long slender legs, relatively small feet, and a long, bushy, usually black-tipped tail. Coyote pelage is usually heavy and thick in winter, but may be thin and flat-looking in summer. The basic body colour is a grizzled grey with buffy undertones, and often with a line of darker hairs down the back. The throat and belly are pale to white, and the forelegs and feet are a mixture of cinnamon and white. Most individuals have a thicker mane of silvery, black-tipped guard hairs on the back above the shoulders. Coyotes in the northern half of British Columbia tend to be paler than those in the south. Males are somewhat larger than females, averaging about 15 per cent heavier.

Measurements:

total length (mm):
    male: 1179 (1070-1360) n=18
    female: 1111 (1020-1200) n=20

tail vertebrae (mm):
    male: 342 (300-387) n=18
    female: 326 (283-392) n=19

hind foot (mm):
    male: 195 (182-214) n=18
    female: 184 (161-200) n=18

ear (mm):
    male: 119 (110-127) n=13
    female: 113 (105-122) n=12

weight (kg):
    male: 12.0 (9.3-18.1) n=15
    female: 10.6 (8.6-12.9) n=13

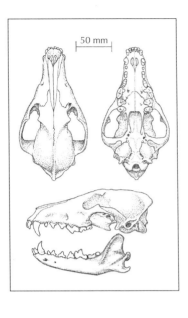

Dental Formula:
    incisors: 3/3
    canines: 1/1
    premolars: 4/4
    molars: 2/3

Identification:
Coyotes are distinguishable from Grey Wolves primarily by their smaller size, more slender build, proportionately smaller feet, longer and more pointed ears, and narrower (pointed) muzzle. Coyotes also lack the prominent facial ruff found on most wolves. Difficulty distinguishing between Coyotes and Red Foxes is likely only for foxes of the cross phase (see page 96), as some of those have an overall buffy or grey tone. Foxes are more slender and dainty than Coyotes, with a prominently white-tipped rather than black-tipped tail, and usually with black rather than buff or cinnamon on the backs of the ears and on the lower legs and feet. Hybrids between Coyotes and domestic dogs may be difficult to distinguish from pure Coyotes unless they have features of recognizable breeds, such as the floppy ears of a Springer Spaniel.

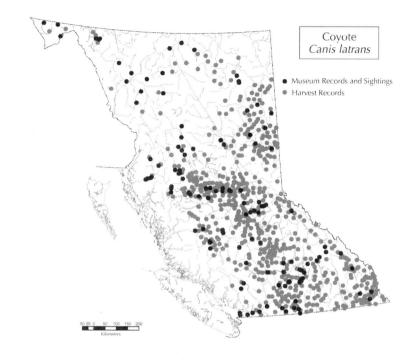

Coyote
Canis latrans

● Museum Records and Sightings
● Harvest Records

Distribution and Habitat

Originally a species of open plains, prairies and deserts in central and western areas of North America, the Coyote has greatly expanded its range over the past 150 years and now occurs almost continent-wide, from Costa Rica to interior Alaska. It is not uniformly distributed over the entire area, but is generally present wherever it is not impeded by snow depth or Grey Wolf occurrence. In British Columbia, Coyotes can be found throughout much of the interior, particularly in and around open grasslands, parklands and agricultural areas, but are generally absent from the wet coastal forests west of the Coast Mountains and from all of the coastal islands. Although it first appeared in the lower Fraser River valley in the 1930s, its expansion into all available habitats including core urban centres in that area has occurred primarily in the last two decades.

The Coyote is a highly adaptable species, regularly occurring in a variety of habitats from valley bottoms to the timberline, but usually in open or sparsely treed areas. Its continental range expansion over the past century has been facilitated at least in part by human activities that have cleared or opened forests, and probably also by changing (warming) climate. Coyotes are able to exist in close proximity to humans. They thrive in rural and agricultural areas, and have become increasingly common in urban areas where they occupy habitat patches such as riparian areas, parks, vacant lots and strips along roadways. In BC, their occurrence appears to be limited primarily by high precipitation, particularly in winter. In some areas, Coyotes may avoid areas of deep, soft snow with seasonal movements to lower elevations, by travelling in thick canopy forests that intercept much of the snowfall, or by restricting activities to "rain shadow" areas or slopes that readily shed snow because of their exposure to sun and wind.

During periods of inactivity, Coyotes occupy a variety of bed sites in open areas or under tree canopies or other structures when they need protection from sun, precipitation or wind. For birthing and rearing of young, Coyotes often make dens in underground burrows in the sides of knolls, banks or steep slopes; they also use other protected sites, such as rock caves, dense thickets, blowdown tangles and the hollows under the roots of large trees. They may fully excavate a burrow, or enlarge and modify one originally made by other animals, such as American Badgers, Red Foxes and marmots. Although a pair may use the same den year after year, it is also common for adults to move their pups to a number of different dens in any one year. Most such moves are short, measurable in hundreds of metres, but one male reportedly moved four pups, one at a time, a distance of 8 km.

Natural History

Feeding Ecology
Across the continent, Coyotes subsist mostly on small to medium-sized mammals, particularly voles, mice, ground squirrels, marmots, rabbits and hares. But the Coyote is the ultimate generalist and opportunist, and the list of all the species it has been known to eat is extensive. Animals it cannot kill, especially those larger and stronger, become available to it at times in the form of carrion. Included in the broad list of regularly eaten items are fruits and seeds (including cereal grains), grasshoppers and beetles, various amphibians and reptiles, large numbers of birds (including eggs and young) and, where Coyotes live near humans, a variety of

pets, poultry, livestock and garbage. Simply put, if a Coyote cannot easily find its preferred prey (such as voles or hares) it will find something else. Coyotes in the Cascade Range in Oregon ate fruits more than any other food in summer and fall; Dale Toweill and Robert Anthony found fruits in 291 (63%) of 462 scats they examined, including blackberries, various raspberries, huckleberries, elderberries, Oregon grapes, Choke Cherries, and Hairy Manzanitas. Coyotes have also been seen wading into water to hunt for aquatic animals such as tadpoles, crayfish and fish. In one case, Joseph Springer observed three Coyotes fishing in a pool left by receding water of the Columbia River in Washington. Although not efficient in their efforts, two eventually caught Carps about 50 cm long.

In the only British Columbia study, Knut Atkinson examined 862 Coyote scats from the lower Fraser River valley near Chilliwack. He found that Townsend's Voles, known to occur in that area at densities of 200 to 1000 animals per hectare, were the primary prey at over 65 per cent of the total. Also in the diet were other small rodents (about 4%), other wild mammals (Northern Raccoon, Muskrat, North American Opossum, Black-tailed Deer, about 5%), small birds (2%), plant material (especially fruits such as apples and plums, but also grasses, 10%), domestic livestock (mostly sheep, 4%) and miscellaneous items (insects, paper, cloth, plastic, rubber, less than 5%).

With acute vision, hearing and sense of smell, and great agility and speed (up to 70 km per hour), the Coyote is well-equipped for locating and obtaining food. Its physical attributes are enhanced by a level of intelligence that enables the animal to be flexible and sometimes even innovative in its hunting tactics. Coyotes can often be seen hunting voles and mice in pastures and meadows, by responding to sounds under snow or grass and pouncing forward to pin the prey to the ground with their forepaws. Coyotes also hunt by lying in wait along known prey travelways, by associating with other predators (such as American Badgers and Golden Eagles) to catch ground squirrels or hares that the other predators have flushed, and by closely following large ungulates, such as Elk and Bison, to catch small mammals disturbed and exposed by the herbivores' foraging activities. Coyotes seen among herds of ranging cattle usually have voles rather than beef in mind. Some have learned to exploit the short-term hunting opportunities that occur during farm mowing or swathing operations, when rodents and small birds are exposed by passing machinery. Near Smithers, a

Coyote followed close behind my tractor and mower for most of a day, gathering five or six prey items at a time, taking them away, and then reappearing to repeat the process. Although wary of my presence, it appeared to understand the dynamics of the situation, approaching to within three metres of the moving machine, but quickly dashing away when I slowed or stopped.

There are many examples of such individual ingenuity, but literature accounts on the species also contain numerous reports of pairs or groups of Coyotes capturing prey by teamwork: driving prey animals to others in ambush; distracting prey while others sneak up for attack; engaging a female while others catch its young; distracting another predator while others steal its prey; pursuing prey alternately, one (rested) Coyote at a time, until the prey is exhausted; and ganging up on a species too large or formidable for one Coyote to handle. One spring day several years ago, in the Bulkley Valley, I watched two adult Coyotes attempt to distract a female Black Bear from her two small cubs. When the Coyotes saw the bears they separated, one moving directly toward them and the other circling around behind. The apparent plan went awry when the female bear, perhaps having previous experience, sent the cubs up a nearby tree when she detected the first Coyote. Anne Rathbun and her associates watched an example of ganging up in Wyoming, where three Coyotes killed a yearling American Badger.

As with most carnivores, when prey is particularly abundant or vulnerable, Coyotes may kill more than required to meet their immediate needs. Such surplus killing is detrimental to agricultural interests when the prey is sheep or turkeys, but beneficial when it is rodents. For the Coyote, surplus kills may serve as insurance against leaner times for both the individual involved or for other members of the pack. It may leave surplus kills in place, transport them to another location for caching (usually by burying in snow or soil), or carry them back to a den. In an impressive study in the Yukon, Mark O'Donoghue and his co-workers monitored 27 caches of snowshoe hares made by Coyotes. The Coyotes returned and consumed 14 of the carcasses later in the winter, and it was evident that some of the returning animals remembered the location of a carcass rather than happening upon it by accident. Carrying extra prey to a den is probably most common when pups are present. A road-killed adult male I examined near Houston (north-central BC) during the pup-rearing season had been carrying seven voles in its mouth and had nine more in its stomach (figure 42).

Figure 42. Road-killed Coyote, showing the number of voles it had eaten or was carrying at the time of its death.

Home Range and Social Behaviour

Coyote populations generally consist of stable, cohesive groups (packs) and solitary individuals. A pack consists of a mated pair of adults and their progeny, in some cases only the current year's pups and in others one or more older animals from previous litters. Packs generally operate within a stable home range that is separate from those of neighbouring packs. Solitary Coyotes, on the other hand, usually have no secure home range and, therefore, travel widely, continually trespassing through established pack territories, competing for space and resources. Most transients are dispersing young animals (less than two years old), but some are either infirm, old (eight years or more) or both.

In a pack, the mated pair (alpha male and alpha female) is the core and dominant social unit. Their pair bond is strong, and is maintained throughout the year and, in most cases, for life. The two alpha animals and other pack members often travel and forage separately in their territory, especially in summer when dependent pups are present and most food items are small. Pack sizes usually range from three to five animals, but there are records of up to ten.

Packs stay together conspicuously in winter, especially where larger prey, such as live deer or ungulate carrion, is available. While some researchers suggest that winter packs facilitate the handling of larger prey and the defense of large carcasses, Francois Messier and Cyrille Barrette propose the reverse, based on their studies in southern Quebec, that the availability of large carcasses facilitates the existence of larger packs. The subordinate (beta) pack members generally do not contribute significantly to either handling or defending large food sources, but benefit because of the volume of food available; the alpha pair tolerates rather than recruits them.

A pack defends its territory against other Coyotes directly by aggression and indirectly by scent marking and howling. Aggression is necessary to enforce the effect of marking by scent and sign, making intruders aware of the consequences of trespass, although encounters appear to be brief and ritualized. I found no accounts of Coyotes being killed or seriously injured in territorial disputes. In most cases observed by Eric Gese in Yellowstone Park, an alpha animal from the resident pack pressed the attack and the intruder quickly submitted and withdrew. Scent marking along and near territorial boundaries is conducted primarily by alpha animals, especially males, and usually involves urination and associated behaviours such as ground scratching. The leg-lifted urination posture, which enables the marking of higher objects for better dispersion of the scent, is characteristic of alpha males. Howling is performed only by pack members, mostly alpha males, usually along territory borders.

When population levels are high, most areas capable of supporting Coyote packs are occupied. The only opportunity for a transient to attain resident or alpha status in a pack is for it to happen upon and be able to take advantage of a vacancy created by the death of one or both alpha animals. Upon the death of one alpha adult, the other takes a new mate, a beta animal from its own or another pack, or a transient. The loss of both alpha adults provides an opportunity for transients to pair and start a new pack, particularly if no beta animals are available, but may also result in an area take-over by a neighbouring pack. In a case observed by Eric Gese, an alpha male died and the alpha female left the territory temporarily. By the time she returned with a new mate, less than a month later, a neighbouring pack had expanded its territory into the area she had vacated.

Home-range size varies geographically, seasonally and with changes in primary prey abundance, but averages 10–20 km^2 in the best habitats (prairies and plains). In the lower Fraser River valley of British Columbia, researchers Knut Atkinson and David Shackleton documented relatively small home ranges, from 1.6 to 11.1 km^2 and averaging 5.6 km^2. Small home ranges generally indicate high food availability and, at that time, the extensive agricultural lands in the area were known to support high densities of Townsend's Voles. Transients range over much larger areas, up to 182 km^2 in Atkinson and Shackleton's study.

Activity and Movements

Coyotes are most active around sunrise and sunset, and during the night, but their activity patterns are often dictated by the behaviour of their major local prey. Daily movements of residents vary seasonally, depending on habitat, prey abundance and other factors, but averaged 4 km in one study in the American Midwest. The largest movements are usually those undertaken by transients, particularly juveniles during dispersal from the natal home range. Most juveniles (both sexes) disperse between late autumn and late winter when they are 7 to 11 months old, but some remain in the family territory for another year or more before dispersing. The distance of such movements varies considerably among habitats and areas, averaging just 7 km in a California study area, 21 km in central Alberta, 42 km in Jasper National Park and 59 km in Colorado. Dispersal movements often exceed 100 km; a juvenile male ear-tagged by Richard Rosatte in the city of Niagara Falls, Ontario, was recovered seven months later near Windsor, a straight-line distance of 320 km.

Adult females made the longest known individual Coyote movements: 323 km in Iowa and 544 km in the Canadian prairies. In the latter case, the animal was radio-collared by Canadian Wildlife Service biologists Lu Carbyn and Paul Paquet in southeastern Manitoba in early January 1983. The Coyote was monitored and found in the same general area 22 times over the next 11 months. It then made the long trek to its final capture location in central Saskatchewan between late November and mid February, travelling at a minimum rate of 6.4 km per day.

Reproduction

Coyotes generally breed between January and March, and bear young in April or May after a gestation period of 60–63 days. Atkinson and Shackleton, in their lower Fraser River valley study, documented 30 per cent of the breeding over a three-year period occuring in January, 50 per cent in February and the remaining 20 per cent in March to early April. Litters average 5 or 6 pups in most areas, but can be higher when food is abundant, 12 or more in some cases. Both sexes can breed in their first winter at 9–10 months of age. The proportion of females that breed and give birth varies in different areas and different years, depending upon food availability, population density and social organization. Usually, only the alpha female produces a litter, but when prey is abundant and the Coyote population is low, most adult females and up to 75 per cent of female yearlings produce litters. In that context, there is some evidence that both pregnancy rates and litter sizes may be highest in areas where Coyotes are heavily hunted or trapped or intensively controlled.

Both alpha adults – and sometimes other pack members – care for their young. Coyote pups are blind and helpless at birth, weighing 250–275 grams and sporting a dull grey, woolly pelage. They grow at a rate of about 300 grams per week and develop quickly: at about two weeks of age their eyes open; by three weeks they begin to accept solid food and venture outside the den; at five to seven weeks they are weaned; and at six months they have grown to nearly full size and become largely independent.

Health and Mortality
Coyotes are host to a wide variety of parasites, internally more than 50 species of flukes, roundworms and tapeworms, and externally various lice, fleas, ticks and mites. While parasites may be detrimental to individuals, most do not cause significant harm to a population. A possible exception is the Mange Mite, which causes hair loss, illness and death (especially in juveniles) in some winters. Mange is usually associated with a high Coyote population, probably because it is more readily spread from animal to animal under those conditions. Infectious diseases such as canine distemper and canine parvovirus, known to cause mortality in Coyotes, are also most prevalent at high population densities. Where Coyotes occur in close association with humans, there is some potential for disease transmission, especially mange and distemper, to and from domestic dogs and other pets.

The Coyote has not been a species of great concern in relation to transmission of diseases to humans, but that could change with its increasing occurrence and growing numbers in urban habitats. Among the diseases of potential interest are rabies, tularemia and bubonic plague, all of which have been reported in Coyotes, but not yet in British Columbia. There is also a minor risk of Hydatid Disease, which is caused by the larval form of a tapeworm commonly carried by canids in some areas (see Grey Wolf, page 88).

Cougars are the primary natural predators of adult-sized Coyotes (yearling and older) in western North America, killing them as prey or, occasionally, to establish ownership of an ungulate carcass. Of five cougar-killed Coyotes documented by Diane Boyd and Bart O'Gara in Montana, one was intact near a deer carcass, one had been partially consumed and three totally consumed. The extent of mortality due to predation is not well known, but is probably highest in mountain and forest habitats occupied by the larger ungulate predators. Of 18 radio-collared Coyotes monitored in the Flathead area of northwestern Montana, 6 were killed by cougars and 2 by wolves in a three-year period (1994–97). I am also aware of a Coyote being killed by a Wolverine. As related to me by Carl Gitscheff, a trapper in the Dawson Creek area, tracks indicated that the two animals met and fought, and the Wolverine prevailed. There was no evidence of a dispute over food; the Coyote carcass was intact, but the Wolverine may have left it on Gitscheff's approach.

Wolves have only recently become a factor in Coyote life history in some areas, appearing by natural recolonization in the Flathead region and by reintroduction in Yellowstone National Park. Yellowstone Coyotes had attained a fairly high ecological position and high density in the 60-year absence of wolves, but a recent study by Adam Switalski found that to be changing. From 1997 to 2000, two to five years after the wolf reintroduction, eight of nine Coyote packs lost one or more alpha adults, and the population studied fell by 25 to 33 per cent each winter. Some of those losses were known and most were believed to be wolf kills, and the evidence indicated that most resulted from disputes over ungulate carcasses rather than predation.

Coyote pups are vulnerable to predation by a number of other species, including Bobcats, American Badgers, bears, other Coyotes and Golden Eagles. Most mortality near human settlements occurs from hunting, trapping, predator control and accidents (particularly road kills). Although Coyotes as old as 14 years have been documented in wild populations, relatively few live past 8 years. The maximum recorded age in captivity was 18 years.

Abundance

The highest known densities of Coyotes, up to 20 per 10 km^2 in early summer after young are born, have been found in the open habitats of the American southwest where winters are mild and prey is diverse and abundant. Even in those areas, and in the best habitats farther north, estimates of 2 to 4 Coyotes per 10 km² are more common. There have been no detailed Coyote population studies in British Columbia, but it is a certainty that densities vary considerably across the diverse landscape of this province, and over time. Where Coyotes rely heavily upon Snowshoe Hares, their numbers may fluctuate considerably following cyclic changes in hare numbers. In studies led by Arlen Todd in central Alberta, Coyote densities declined from 4.4 per 10 km² in 1970–71, when hares were abundant, to 0.8 per 10 km² in 1974–75 after the hare population had crashed. That more-than-fivefold change within a hare cycle is probably typical for Coyotes in similar habitats and circumstances in the Peace River region of BC. Studies by Mark O'Donoghue and associates in the Kluane Lake area (Yukon Territory), also revealed a large (sixfold) change in Coyote populations between peak and low hare populations, but at much lower densities (0.15 to 0.90 per 10 km²). Those figures are likely applicable to similar areas in northwestern BC.

In other BC locations where Coyotes are common, including the southern Skeena, southern Omineca, Cariboo, Thompson-Okanagan and Kootenay regions, population changes are not as extreme or predictable as in the north, and are probably affected more by varying winter snow conditions than by prey abundance. In those areas, densities likely reach a high of about 4 Coyotes per 10 km², but not over large areas or for long periods. Due to combinations of mild winters, diverse and abundant food supplies in agricultural areas, and greater protection due to urban sensitivities, Coyotes began to increase in the lower Fraser River valley in the 1980s. Spreading into urban and suburban centres throughout that area, they appeared to fill all available local habitat by the end of the 1990s, and likely the population has peaked and then levelled off or declined somewhat since then.

Human Uses
In some aboriginal societies, particularly in Central America and the American southwest, the Coyote was considered a "medicine" animal and is featured in many legends. It was probably not present in most of British Columbia before Europeans arrived, and there is no record of any kind of traditional relationship in this province. The primary human use of the species over the past two centuries has been in the fur trade. Coyote fur is used for coats and hats, and in

trim on garments made from other materials. The continental harvest is often about 400,000 pelts per year, but in BC it rarely exceeds 5000 and has averaged under 1000 since the mid-1990s.

Taxonomy

Based on skull and pelage traits, 19 subspecies have been recognized, with 2 (*incolatus* and *lestes*) occurring in British Columbia. Phil Youngman noted that the skull characters that separate those two subspecies are subjective and recommended that the subspecies be combined. Further, although their study did not include samples from BC, Niles Lehman and Robert Wayne found that genetic variation among Coyote populations across North America showed no obvious patterns consistent with described subspecies. In short, there appears to be no basis for recognizing separate Coyote subspecies.

In eastern North America the taxonomy of the Coyote and the Grey Wolf has been complicated by recent hybridization between the two, producing animals intermediate in traits, but that appears to be rare in the west. In DNA studies, K. Pilgrim and colleagues found no evidence for Coyote-Grey Wolf hybridization among samples from Alberta, BC and Montana.

Conservation Status and Management

The Coyote is one of the few species that has expanded its North American range over the past century, in pace with the spread of human settlements and developments. It has spread despite attempts to control or to eradicate it because of its taste for livestock, pets and game animals. In generally unsuccessful efforts to reduce Coyote numbers, the species has been trapped, shot and poisoned by professional predator hunters on government payrolls, and has had bounties placed on it. Both methods were used in British Columbia until the early 1960s, but to no avail. The Coyote is presently managed in this province as both a game species (since 1966) and a furbearer (since 1976), and can therefore be legally harvested by both licensed hunters and trappers.

Agricultural depredations by Coyotes are most often directed to domestic sheep, but individuals also cause havoc on poultry farms in some areas and occasionally at beef and dairy ranches during calving time. Coyotes have become an increasing threat to domestic pets in urban areas, especially cats and small dogs. More seriously they occasionally chase, threaten or bite humans. Coyotes have attacked children in western North America, even killing one child in California. Of four other serious attacks reported by biologist Lu Carbyn in 1989, three occurred in national parks (two in Jasper and one in Yellowstone) and the other near Creston in southern BC. Since then, there have been at least two more attacks in the Lower Mainland. Researchers believe that most Coyote attacks on humans have occurred during times when prey is scarce, or by individuals that have been fed by or otherwise habituated to humans.

Coyote predation on wild ungulates is mainly directed to newborns (fawns, lambs) in spring and summer, and is a limiting factor for local populations of those animals in some areas. In a study by Susan Lingle in southern Alberta, 40 of 47 radio-collared Mule Deer fawns disappeared in their first year; Lingle attributed 34 of those losses (85%) to Coyote predation. In the Junction Wildlife Management Area along the Chilcotin River in BC, Daryll Hebert and Scott Harrison documented increased Bighorn Sheep lamb survival, from 12–18 per 100 ewes to 56 per 100 ewes, after an experimental Coyote removal program. In areas with winter snow, small mammals may be relatively unavailable and predation on ungulates, including adults, is more common in that season. Such predation usually involves packs rather than individuals, and is most successful when escape conditions for prey animals are poor (when they are constrained by deep snow or may be chased to areas of unsure footing such as on frozen lakes and ponds), or are in poor condition (as in the case of a 15-year-old cow Elk killed by two Coyotes in Yellowstone Park).

Coyote predation is also of concern where it involves species at risk. For example, Coyotes expanded their range onto the Gaspe Peninsula in Quebec in the early 1980s and, according to biologists Michel Crete and Alain Desrosiers, by the end of the decade Coyote predation on calves was threatening the survival of a remnant Caribou herd. Following a Coyote control program in the early 1990s, the number of calves surviving until fall in one subpopulation increased from an average of 3.3 per 100 Caribou cows in 1987–89 to 40 per 100 in 1992. In BC, the recent increase of Coyotes in the lower Fraser River valley is of potential concern in relation to the conservation of several species at risk: Pacific Water Shrew, Trowbridge's Shrew, Townsend's Mole, Mountain Beaver, Coastal Giant Salamander and Oregon Spotted Frog and local subspecies of Snowshoe Hare and Southern Red-backed Vole.

Remarks

The Coyote is as well known for its voice as for its activities. Indeed, the name is derived from an Aztec word *coyotl* meaning "barking dog", which is also the literal translation of the Latin name *Canis latrans*. The characteristic "yip-yipping" call was described by one observer as "a howl which the Coyote chased, caught and bit into small pieces". Prairie folklore has the Coyote "howling at the moon", but University of Regina researchers led by Darren Bender demonstrated that Coyotes actually howl more often on nights with little or no moonlight. Overall, as stated by H.E. Anthony of the American Museum of Natural History,

> *To many who have heard the ecstatic little prairie wolf greet their camp-fire from out of the dusk, or have arisen at break of dawn and heard his frenzied hymn to the sun, a West without the Coyote seems colourless and flat.*

The Coyote is the only one of BC's three wild canids that is exclusively North American, as both the Grey Wolf and the Red Fox also occur in Europe and Asia. It can successfully interbreed with wolves and domestic dogs, producing fertile offspring; but its relationship with wolves is more precarious than that would imply. Coyotes are usually less abundant in areas with high wolf numbers, and ongoing changes in Coyote numbers in the Yellowstone Park area after the reintroduction of wolves is instructive on that point. Although Coyotes may benefit from carrion left at wolf kills (especially in winter), they must compete with the wolves for smaller prey species and always risk becoming wolf prey themselves. Coyotes have an opposite relationship with Red Foxes, outcompeting, displacing and in some cases killing them (see Red Fox, page 105).

A familiar cinema cartoon depicts a Coyote employing a number of tricks to outwit and catch a desert bird (Road Runner) – and always failing miserably. In the real world, the Coyote would probably succeed quickly, and without the use of dynamite. Popular literature, has many stories of Coyote resourcefulness and durability, often told with grudging admiration by sheep ranchers and professional predator hunters. In his book on the subject, Charles Cadieux reports that government hunters in the United States killed at least 3,612,220 Coyotes between 1937 and 1981, and that in 1982 "there were more Coyotes in the United States than ever before". That large body count was only the number of verified-deaths, not including animals that walked away from a poison station to die, and Cadieux suggests that the actual total might have been twice as high. His book concludes with a tribute to the Coyote:

> *It is my personal belief that when the last human has fallen, and the last skull lies on the irradiated earth, a Coyote will come trotting out of some safe place. Don't ask me where he'll come from; but I believe that he will survive as he has always survived. The Coyote will trot in his furtive, skulking manner, to the skull. He will approach it carefully with the caution borne of millennia of avoiding steel traps and snares and pitfalls. He will cautiously sniff it. His educated nose will tell him that he no longer has anything to fear from this bleached remnant of a once great civilization. Taking a few short steps to get in the exact position, he will lift his leg.*

Selected References: Atkinson and Shackleton 1991; Bekoff and Wells 1980; Bowen 1982; Cadieux 1983; Carbyn 1989; Dobie 1961; Gese and Ruff 1997, 1998; Hatler et al 2003a; Hebert and Harrison 1988; Leydet 1977; Messier and Barrette 1982; Switalski 2003; Todd and Keith 1983; Voight and Berg 1987; Windberg 1995; Young and Jackson 1951.

Grey Wolf

Canis lupus

Other Common Names: Timber Wolf, Tundra Wolf, Wolf; *Loup, Loup Gris*. Provincial Species Code: M-CALU.

Description

The Grey Wolf is the largest wild member of the canid (dog) family, and the largest known specimens have come from northwestern North America. Its general appearance is similar to that of a German Shepherd dog, but with longer fur, a bushier tail, and proportionately longer legs and larger feet. The overall build is heavy, with the male in particular having a broad head and chest, and thick legs. The muzzle and nose-pad are wide; short, rounded ears are usually erect; and the face most often appears even wider than it is because of a prominent ruff of facial hairs on the cheeks. The eyes are yellow, intense and penetrating. Pelage colour varies considerably, from nearly pure white (rare in BC) to uniformly dark sooty grey to coal black. Despite the name, few wolves are actually grey, although that is the general colour label applied to paler animals with the more typical brindled pelage mixed with light grey, black or white. Pale (grey phase) individuals usually have lighter coloured legs and undersides. The Grey Wolf also has a patch of long hairs on the back over the shoulders that rises as a distinct "mane" when the animal takes a threatening or defensive posture. Adult males are generally larger than adult females, averaging

35–50 kg versus 30–40 kg in most areas. The largest Grey Wolf on record, a male from Alaska, weighed 79.4 kg.

Measurements:

	All Specimens	Subspecies	
		nubilus	*occidentalis*
total length (mm):			
male:	1711 (1500-1930) n=27	1669 (n=8)	1729 (n=19)
female:	1675 (1490-1870) n=20	1642 (n=2)	1678 (n=18)
tail vertebrae (mm):			
male:	468 (400-575) n=25	488 (n=8)	459 (n=17)
female:	449 (382-520) n=18	409 (n=2)	454 (n=16)
hind foot (mm):			
male:	284 (240-326) n=24	284 (n=6)	285 (n=18)
female:	272 (220-285) n=19	272 (n=2)	272 (n=17)
ear (mm):			
male:	128 (115-140) n=7	none	128 (n=7)
female:	120 (120-120) n=2	none	120 (n=2)
weight (kg):			
male:	39.3 (29.0-50.0) n=20	31.5 (n=4)	41.3 (n=16)
female:	37.8 (32.0-45.0) n=11	none	37.8 (n=11)

Dental Formula:
   incisors: 3/3
   canines: 1/1
   premolars: 4/4
   molars: 2/3

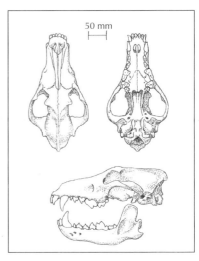

50 mm

Identification
In the field, Grey Wolves can only be confused with Coyotes and some Domestic Dogs. Confusion with Coyotes would likely arise only for grey-phase wolves, because white or black Coyotes are exceedingly rare. Grey Wolves are much larger

and more heavily built than Coyotes and have proportionately larger feet, shorter and less pointed ears, and broader muzzle and nose. The wolf also has a conspicuous facial ruff. Hybrids between Grey Wolves and Domestic Dogs may be difficult to distinguish from pure wolves unless they have features of recognizable breeds such as the pug nose of a Bulldog or the floppy ears of a spaniel.

Distribution and Habitat

The Grey Wolf is a circumpolar species, originally occurring north of 20° North Latitude over most of North America, including the Arctic islands and throughout Eurasia. The species was extirpated from most of the continental United States with the spread of European settlement and, until recently, the only viable population was in the vicinity of Lake Superior, in Isle Royale National Park and the northern forests of Minnesota, Wisconsin and Michigan. The species has now returned to the American west, as a result of a reintroduction to the Yellowstone area and natural colonization of northern Montana and Idaho by immigrants from British Columbia.

In Canada, wolves never occurred on Prince Edward Island and were reportedly extinct in Nova Scotia and New Brunswick by 1870 and on mainland Newfoundland by 1911. But the species is present over most of the rest of Canada, except for the more settled areas along the southern border and unforested portions of the prairie provinces. The Grey Wolf currently ranges throughout most of BC, with the most continuous distribution and highest densities in the central and northern portions of the province. After many years of absence or very low density, wolves reappeared on Vancouver Island during the 1970s and are now common over most of the island. They also regularly occur on many other islands along the coast, but not on the Queen Charlottes. Wolves re-occupied most of the southern Rocky Mountains and East Kootenay area during the 1980s and 1990s, and are beginning to appear in areas to the west. In February 2005, trapper Pete Wise saw a black wolf on the Coldstream Ranch near Vernon; and in August 2005, Grindrod resident Richard Kahut and his son encountered a family group (two adults and four large pups) near Mabel Lake, east of Enderby.

Grey Wolves are highly adaptable and occurred historically in most habitats in the Northern Hemisphere, from coastal rainforest to open prairie to the high Arctic, but not in the desert and chaparral habitats of the American southwest or the hot lowland forests of

Grey Wolf
Canis lupus

● Museum Records and Sightings
● Harvest Records

the southeast. The primary requisite in most areas, regardless of climate and vegetation, has always been a sufficient supply of large ungulate prey. In North America, that was satisfied primarily on the plains and prairies by Bison, in temperate and boreal forests by Mule Deer, White-tailed Deer, Elk and Moose, and in the far north by Caribou. The distribution and abundance of prey is still an important determinant of distribution, and BC wolves currently use most habitats at all elevations over all of the province except the dry southern interior and the highly developed urban and agricultural areas of the southwest.

During winter in areas with heavy snowfall, wolves frequent ungulate winter ranges, travelling and hunting along frozen lakes and rivers, but also using forests with closed canopies and open slopes that shed snow as a result of exposure to sun and wind. In deep snow areas, wolves leave distinct narrow trails caused both by packs travelling in single file, and by individuals and packs reusing the same routes. They often travel on back roads and trails made by humans, but generally avoid heavily-used roads and human settlements.

When inactive, wolves rest in open sites or take shelter under a tree canopy or other structure if necessary. When producing and rearing young they usually den in burrows in the sides of knolls, banks or steep slopes; they either fully excavate the den themselves or modify a burrow made by other animals. Don Eastman and I found an active wolf den on Level

Figure 43. Grey Wolf den on Level Mountain, northern British Columbia.

Mountain (Telegraph Creek area) in June 1979 (figure 43) at 1425 metres elevation on an open south-facing slope that provided a broad vista of surrounding terrain. In contrast, and illustrating the adaptability of the species, a den I examined in July 1975 on a ranch near Smithers (elevation 600 metres) was under a large pile of wood debris at an old saw-mill site.

A mated pair may use the same den year after year. A den in Jasper National Park, monitored for 15 years by Canadian Wildlife Service biologist Lu Carbyn, was used by wolves in 8 of those years. That den had two entrances separated by more than 4 metres of tunnel, and the nest chamber in a 3.4 metres side passage was 1.1 metres wide, 0.4 metres high, and more than 1 metre underground. Other dens receive long-term use because there are few alternatives. On Ellesmere Island, Northwest Territories, where much of the land is underlain by permafrost and not amenable to burrowing, biologist David Mech observed a rock-cave den that was used in three consecutive years and, based on carbon dating of bones found nearby, had been used by wolves off and on for more than 700 years.

Natural History

Feeding Ecology

The Grey Wolf preys primarily on ungulates, but as the top carnivore in ecosystems where it occurs, it preys opportunistically on a wide variety of other species. Wolves often travel and hunt alone during the snow-free season, focusing on small to medium-sized prey such as voles, ground squirrels, Snowshoe Hares, ptarmigans, grouse, Beavers and young ungulates. On September 4, 1998, in the Cayoosh Range near Lillooet, I watched a large male catch and gulp down two Water Voles, and then kill and eat most of a juvenile Hoary Marmot in just a few minutes hunting along a small subalpine creek. The only detailed study of Grey Wolf food habits in British Columbia was undertaken by Barbara Scott and David Shackleton on Vancouver Island, where Black-tailed Deer, Elk and Beaver were the most important foods, in that order, year-round. The list of what wolves eat pretty much reflects what is available. John Theberge and Thomas Cottrell examined summer wolf scats from two dens in the Kluane Park area, Yukon Territory. Moose was the primary food item at both, but the second and third most frequent items were Beaver and Snowshoe Hare at one, and Arctic Ground Squirrel and voles at the other.

The versatility of wolves is also shown by observations of them raiding Canada Goose nests and killing moulting adults on the Copper River Delta, Alaska, and even eating fish when circumstances permit. In one case, a wolf caught 16 whitefish in one hour at a rapids near Great Slave Lake, Northwest Territories. Closer to home, Chris Darimont and his associates documented wolf fishing behaviour at a salmon spawning creek on the BC coast near Bella Bella. They found more than 700 salmon carcasses they believed were caught by wolves, and postulated that coastal wolves have a regular, seasonal relationship with salmon. I have also seen evidence, on the Skeena and Taku river systems in northern BC, of wolves scavenging on salmon carcasses (spawned-out) in late fall and early winter.

The primary winter prey of Grey Wolf packs are Moose or Caribou in the north and Elk or deer in the south, but some packs may temporarily focus on other species such as Mountain Sheep and Mountain Goats. Wolves reintroduced to Yellowstone National Park had no experience with Bison but learned how to hunt them and included them in their rounds and on their menu more quickly than biologists had predicted. Consistent with their opportunistic nature, wolves prey upon whatever will provide them protein and that occasionally includes other carnivores. My observations of River Otter and Lynx predation by wolves are detailed in the accounts for those species. Wolves have also been known to kill American Martens, American Minks, Wolverines, Coyotes, Black Bears and Cougars, although sometimes not consuming the carcasses. All of the published records for Black Bears involved wolves killing them in their dens or after forcing them out. Biologist Brian Horejsi and his co-workers found an adult female bear carcass almost completely consumed and her two cubs in a nearby tree; they judged that the cubs had a poor chance of survival. Wolves also compete with other carnivores for food, but counter-balance that somewhat by providing scavenging opportunities for carrion at large ungulate kill sites. Ungulate carcasses can be a major focal point for carnivore activity in winter (figure 44).

Grey Wolves are highly intelligent and are well equipped to locate, catch, and subdue their prey. They have keen senses of vision, smell and hearing, high endurance for long pursuits, quickness and agility in close quarters, top running speeds of 55–70 km/hour in short bursts, and powerful jaws.

It usually takes a pack to bring down an adult of the larger prey species; but individual wolves frequently subdue medium and smaller sized ungulates (Caribou, Mule Dear, White-tailed Deer, Mountain Sheep, Mountain Goats). Packs hunt any size ungulate, sometimes one or two wolves herding prey to the waiting pack. When two or more wolves hunt together, one animal can distract while the other(s) slip in for the kill. Wolves have been reported more than once killing Moose calves despite their mother's defense, particularly when twins are involved, and in one case killing a Polar Bear cub. For large, potentially dangerous prey such as Moose, wolves often use a slash and run method: one darts in from behind to draw blood while the Moose is facing another pack member, then retreats when the Moose whirls in response. Sometimes this method takes hours or even days to kill the prey, after it has been weakened by loss of blood and exhaustion.

Figure 44. Wolf-killed Moose, a centre of Carnivore activity in the northern wilderness.

There is some truth to the suggestion that wolves tend to prey on the young, sick or old. When pursuing a herd of ungulates, they usually catch the one that falls behind, and young animals are less able than adults to resist attack. When wolves take a healthy adult it is usually because they recognize a momentary advantage, a mistake made by the prey or the effects of terrain or other environmental features. As demonstrated by Michael Nelson and David Mech in Minnesota and by David Huggard in Banff National Park, one of the most important factors in winter is snow depth. Wolf predation rates and feeding opportunities are highest when prey animals have more difficulty escaping in deep snow.

Despite the predatory capabilities of wolves, most of their attacks on large ungulates are unsuccessful. In studies at Isle Royale, less than ten per cent of their contacts with Moose ended in a kill. As reported by David Mech, the wolves usually aborted an attack when the Moose turned to defend itself. In my experience, Moose also escape predation by taking to water, where their longer legs give them an advantage. Some prey species, such as Caribou and deer, commonly escape by outrunning their pursuers, and Mountain Sheep and Mountain Goats run to cliffs (escape terrain) where they secure a mobility advantage. Researchers have calculated that individual wolves consume from 2 to 7 kg of meat per day, but as suggested above, that rarely comes at a consistent, steady pace. With kill rates estimated in one area at the equivalent of one Moose every 7 to 16 days in summer and 5 to 11 days in winter, wolves may go several days on an empty stomach before gorging (10 kg or more) when a kill is made. They are physiologically better adapted than most carnivores to such a boom-and-bust lifestyle.

Grey Wolves sometimes cache prey remains, burying them in snow or soil. And they indulge in surplus or excess killing, in most cases where several prey animals are vulnerable due to unfavourable conditions, such as deep snow. Biologist Frank Miller and his associates documented excess killing of Caribou calves. The wolves ran through the herds, killing calves at will and leaving many of them in place. On one occasion, the researchers found 34 dead calves in a 3 km^2 area – 17 were only partially consumed and the other 17 completely intact. The Caribou calves are most vulnerable to such predation in the first week or two after birth.

Home Range and Social Behaviour

The Grey Wolf is the most social of BC's carnivores, in that the year-round functional units of populations are resident packs rather than individuals. Packs are usually family-based, the core consisting of a pair of adults and their offspring of various ages although radio-tracking studies have confirmed that some packs include unrelated animals. Pack sizes rarely exceed 8 wolves by late winter in areas where the primary prey is deer, but do so fairly commonly where larger ungulates, such as Moose or Elk, predominate. Half of 24 packs closely monitored by Warren Ballard and co-workers in Alaska consisted of 10 or more wolves (to a maximum of 20), and 10 of 73 packs observed or inferred by John Elliott in the Muskwa area of northern BC had 10 or more (maximum 27). The largest pack for which I have documentation was observed twice along the upper Stikine River in Spatsizi Provincial Park. Pilot Ron Bruns spotted it first, counting at least 36 wolves on March 3, 1976. Flying with Bruns three days later, Don Eastman, Tony Sinclair and I were treated to the remarkable sight of 29 wolves strung out in single file along the river; it is likely that some had already left the river ice before we saw them. The nature and composition of such large packs has not been determined, but it seems evident that more than one successfully breeding female is present.

Packs maintain a social structure in which the breeding adults are the alpha (top-ranking) individuals. Other pack members (non-breeders and offspring) are subordinate to the alpha animals, but may have their own dominance-based rankings below alpha, especially at kills. When food is in short supply, either generally or in relation to the size of a particular kill, lower ranking members do not get fed and may eventually have to leave the pack to forage for themselves. Outside of the pack, the rest of a wolf population is made up of such displaced animals, mostly younger ones, alone, in pairs or in splinter packs of 3–5 animals or more. Those animals either wander as transients over large areas or settle (temporarily or permanently) in areas that are unoccupied or are rarely visited by resident packs.

Resident wolf packs are territorial, conducting all their activities in well-defined territories that they defend against other wolves. Aggressive encounters in which wolves have been killed have been reported from many areas, but most territorial defense is in the form of scent marking with urine and feces, particularly along the boundaries between pack territories. Scent marking and howling, both of which are done primarily by alpha animals, serve as warnings to transients and help neighbouring packs keep track of each other and avoid direct encounters. Territory size varies both regionally and seasonally, depending mostly upon the abundance and distribution of prey. In northeastern Alberta, three packs monitored by Ron Bjorge and John Gunson had annual territories of 296, 542, and 878 km², and generally ranged over larger areas in winter than in summer. Those territories are considerably larger than summer territory sizes reported by Ian Hatter for Vancouver Island (57–112 km²), but much smaller than documented for eleven packs in south-central Alaska (943–2541 km²: average 1645 km²).

Activity and Movements

Grey Wolves are highly mobile, travelling extensively over their large territories. In his Isle Royale studies, David Mech recorded a one-month distance of 443 km covered by one pack, and one individual moved 48 km in one day. Trotting at 8 to 10 km per hour, wolves commonly cover daily distances of 25 km or more; I have seen a number of examples during aerial surveys in northern BC. Such mobility is essential for maintaining territory boundaries, and to enable the animals to keep track of prey distribution and predation or scavenging opportunities. But wolves may move little for days or weeks when they have a large ungulate carcass or other concentrated food source.

As with most species, the longest documented movements have been made by transient animals during dispersal. Although transient wolves are mostly young ones that have left the pack they were born into, the transient category also includes non-pack individuals of all ages and small non-breeding groups that don't have established territories. Some of the long-distance movements that have been recorded for radio-collared wolves are: young, sexually mature male travelling 670 km from the Great Slave Lake area in Northwest Territories to a location near Cold Lake, Alberta; two-year old male and yearling male together, probably accompanied by two other untagged animals travelling 732 km from the Nelchina Basin to the Brooks Range in Alaska; a yearling or young adult male travelling 886 km from northern Minnesota to northeastern Saskatchewan. Those are all straight-line (map) distances, and the actual distances travelled on the ground were certainly much greater.

Although the above records only involved males, studies in most areas have shown that females move similar distances. One account provides reference to dispersal of 840 km by a female Grey Wolf in Montana, and describes the movements of an adult female from the Canadian Rockies near Banff through southern Alberta to southeastern BC, then eastward to Browning, Montana, then west to Kellogg, Idaho, and back north to Canal Flats, BC, where it was killed – the total distance travelled likely exceeded 1000 km. From their study of dispersal in a colonizing wolf population in the Glacier National Park (Montana) area, Diane Boyd and Daniel Pletscher suggest that while the immediate factor causing an animal (usually a subordinate) to disperse may be food shortage or social strife, the ultimate result is reproductive opportunity. A female I radio-collared in Spatsizi Park, a member of an apparently non-breeding pack of six, was located on the Spatsizi Plateau near Hyland Post on 20 of 22 tracking flights between April 1990 and March 1992. She then moved north out of the park and was relocated in August east of Dease Lake, 90 km north of her former range in Spatsizi. Remaining in the Dease Lake area at least until the study ended 10 months later, she was alone when seen in December, was with another, larger wolf (believed to be a male) in February and April, and was located (but not seen) in or near what appeared to be a den in the side of an esker in June 1993.

Reproduction

Female Grey Wolves reach sexual maturity at two years, but usually do not breed until age three. In stable resident packs, the alpha female is the primary and usually the only breeder, such that one litter per pack is the rule. In some cases however, a pack has raised more than one litter in a year, and transients also occasionally breed and produce pups. As documented by Rodney Boertje and Bob Stephenson in Alaska and David Mech in Minnesota, productivity is strongly related to nutrition. In years with low abundance or availability of prey, the proportion of reproducing females is lower and litter sizes are smaller than when food is abundant.

Wolves breed in late winter, usually February or March. There are no published records for British Columbia, and the only applicable sighting I am aware of is my own. On March 3 1980, while engaged in an aerial survey along the Nahlin River north of Telegraph Creek, I observed a pair of wolves on the ice at mid-river standing rear-to-rear in the copulatory tie typical of mating canids.

Wolf pups are born in April or May after a 63-day gestation period. Litters as large as 11 have been documented, but most have 4 to 7 pups (average around 5). Wolf pups are blind and helpless at birth, weighing about 500 grams. Their eyes open at about two weeks, they begin to explore areas outside the den at three weeks and they take solid food at about five weeks. Pups born in a resident pack, fed and protected by both parents and sometimes by other pack members, have a better chance of surviving than those born outside a pack. Adult wolves have been seen in aggressive encounters with Wolverines, Black Bears and Grizzly Bears at active dens, but whether those carnivores came for the pups or the carrion the adults had brought to them is unknown.

Wolf pups do not venture far from the den during their first two to three months, but by late summer they begin moving to a series of alternate locations that biologists refer to as "rendezvous sites", which they occupy while the adults are hunting. In September 1998, I happened upon a rendezvous site in the Cayoosh Mountains near Lillooet, that was occupied by five half-grown pups. It was a lovely little subalpine meadow with a pond at its centre, and tracks indicated that the pups were spending a lot of time near and in the pond.

By late fall, pups are nearly full size and begin travelling with the pack. Most young wolves disperse in the spring, as yearlings, but some may be forced to leave earlier if food is scarce.

Health and Mortality

Grey Wolves are host to a number of parasites, including at least 57 species of flukes, tapeworms, roundworms and spiny-headed worms and at least 11 species of lice, fleas, ticks and mites, but most are not of any regular consequence to wolf populations. An infestation of the Mange Mite causes hair loss and results in reduced physical condition and the death of some wolves (especially juveniles) in some winters. Ian McTaggart Cowan observed a mangy wolf in Banff National Park that was almost half naked, and in such poor condition that it weighed only 17 kg. A young female wolf that I found in Spatsizi Park in March 1985, though not as severely affected by mange, was nearly dead and and weighed just 19 kg. Mange is usually associated with a high population density, probably because it can more readily spread from animal to animal.

A tapeworm commonly found in wolves, *Echinococcus granulosus*, is the agent of Hydatid Disease. A number of mammals, including humans, can develop this disease by the accidental ingestion of tapeworm eggs, which are passed in canid droppings. The eggs can be viable for months in dust or soil. Once inhaled or ingested, the developing larvae cause hollow cysts in the internal organs, especially the liver and lungs. In humans, this is a potentially serious disease and is thought to be acquired most often from domestic dogs that were fed uncooked, infected ungulate organs.

Rabies can occur in wolves, and some authorities believe that the animal's reputation as a fearsome, slathering beast may have arisen primarily from historic human encounters with rabid animals. Although serious, rabies has not been demonstrated to be widespread or chronic anywhere, and has not yet been detected in a wolf from British Columbia. Other diseases in wolves include canine distemper, encephalitis, listeriosis, cancer and arthritis, but the frequency of their occurrence and effects in wild populations are not known. Rolf Peterson and co-workers in Alaska circumstantially linked the decline of one wolf pack to distemper after documenting the deaths of two of its members to the disease.

Grey Wolves are relatively short-lived for animals of their size, with few living beyond 10 years in the wild. Their lifestyle exposes them to many dangers, particularly those associated with attacks on large prey. There are several published accounts of wolves being killed by prey, usually Moose, but also Muskox, Elk and White-tailed Deer. Most deaths resulted from being kicked by the prey, but one young female wolf in Minnesota bled to death after being gored by a White-tailed buck that the pack had caught and subsequently killed. When he was a conservation officer in Cassiar in the 1970s, Brian Baldwin observed an adult female Moose killing two wolves in quick succession, pummelling them into the ground with a rapid flailing of her forefeet. Researchers have also documented many healed injuries suffered by wolves, including broken ribs, broken legs and minor skull fractures. Of over 1200 skulls examined by Bob Rausch in Alaska, 25 per cent had sustained compression fractures, and M. Phillips recorded a similar finding (22%) for an unknown number of samples from Minnesota. It is likely that injuries are a common feature of the prey-handling education process for young, inexperienced wolves, and not all are lucky or agile enough to graduate. An advantage of being a pack member is that skeletal injuries are less likely to be fatal than they would be for a lone animal, because the pack members can still eat food obtained by the pack. In the pack of 29 I observed on the Stikine River in 1976, one had a broken hind leg.

Wolves also suffer serious or fatal injuries while hunting smaller prey. In the Spatsizi Park area a wolf had a serious run-in with a Porcupine and as related by Shawn Boot, the wolf was so debilitated that it was reduced to eating horse droppings in the local guide-outfitter's corral. The wolf remained at the corral for two days, until the people there judged it unlikely to survive and shot it. In a post-mortem examination I counted 92 quills in its jaws, 39 in its muzzle and nose, 107 inside its mouth, and small numbers on its chest and front paws. In another BC incident, Wildlife Branch employees Ken Child, Ken Fujino and Milt Warren observed perhaps the most bizarre record of a fatal mistake one evening in July 1977. Camped in the Finlay Range in north central BC, they saw a wolf chasing a band of Stone's Sheep on a nearby mountainside. A ewe and lamb that were hard-pressed by the wolf stumbled and fell off a 75-metre cliff. Intent on confirming the fate of the sheep, the men hiked to the site where they found not two but three carcasses. The young female wolf, probably in her haste to claim her prey, had also fallen to her death.

Such fatal accidents are uncommon, as are deaths from predation. Grizzly Bears sometimes find and kill pups in dens and occasionally kill adult wolves defending their pups or a kill.

Among radio-collared wolves, one of the most common causes of death has been aggressive encounters between wolves, probably related to territorial defense. In three telemetry studies, the proportion of fighting deaths was 3 of 6 in an expanding population in northwestern Montana, 6 of 14 during a population decline in northeastern Minnesota, and 9 of 12 during a severe population decline on the relatively confined area of Isle Royale, Michigan. Actual observations of such mortality are rare, but Paul Marhenke and three companions in Denali National Park, Alaska, watched four wolves drive an old adult male wolf from a Caribou carcass, catching and killing it after a chase of about 100 metres.

During an aerial survey in March 1993, I observed a dead wolf on the ice in the middle of the Spatsizi River and was able to visit the kill site on foot a short time later. Fresh tracks revealed that the dead wolf, a young adult male, had been walking upstream and had met two other wolves walking downstream. Scuffle marks and blood in the area suggested that the dead wolf had broken away twice, but had been caught and pulled down each time. After killing the lone wolf with a number of bites on the body and throat, the other wolves returned upstream.

Despite their well-deserved reputation as efficient and effective predators, another common cause of death among wolves is malnutrition. When ungulates are in short supply, usually in winter, lower ranking members of the wolf pack do not get enough to eat and many (especially pups) do not survive. In David Mech's northern Minnesota studies, the number of wolf deaths attributable to malnutrition equalled that to aggression (6 of 14) and all three of the documented mortalities not due to aggression at Isle Royale were malnutrition-related. Even when prey is abundant, young dispersing wolves and older transients may not be able to subdue large prey. Relying primarily upon small prey and scavenging opportunities, transient wolves do not have as high a survival rate as resident pack members. Wolves emaciated by mange likely reach such poor condition because of malnutrition, which makes them more susceptible to the parasite infestation. Finally, in some areas, the primary mortality of wolves is human-caused, through trapping, hunting and predator-control activities, but also from road and railway accidents. Documented mortalities of that nature in the various studies referred to above were as low as zero (at Isle Royale), but constituted 10 of 24 wolf mortalities (41%) in northeastern Minnesota, and 36 of 43 (84%) in northwestern Montana. The wolves most vulnerable to human-caused deaths are transients and the younger animals in packs.

Abundance

Grey Wolf numbers vary across their range and over time, mostly in relation to prey abundance and availability, except where human exploitation is a factor. Some of the most intensive wolf-population studies have been in Alaska, where density estimates of 10 to 20 wolves per 1000 km^2 appear to be typical for large boreal areas in which Moose is the primary winter prey. Farther south, where White-tailed Deer are the main winter prey, Doug Pimlott and associates recorded densities of 30–40 wolves per 1000 km^2 in Algonquin Park (Ontario) in the early 1960s, and Todd Fuller found up to 59 per 1000 km^2 (average 44.5 over 6 years) for populations in Minnesota during a period of increase during the 1980s. The highest documented density was 91 wolves per 1000 km^2 on Isle Royale, an island national park in Lake Superior where the only prey are Moose and Beavers. That density in a total population of 50 wolves on the island was the peak attained in 1980 after several years of increase, and it could not be sustained. As reported by Rolf Peterson and Rick Page, the Isle Royale wolf population had crashed to just 14 animals (25 wolves per 1000 km^2) by 1982, but then increased again and stabilized at less than half the peak density (about 40 per 1000 km^2) through 1986.

In British Columbia, there have been no detailed wolf-population studies over large areas or long time periods, but there have been numerous short-term surveys in the north and some more intensive local studies on Vancouver Island. The northern surveys were mostly in the early to mid 1980s and related to concerns about predation effects on ungulates (especially Caribou and Moose). Based on winter aerial counts of wolves and tracks, density estimates by provincial government biologist John Elliott and co-workers ranged up to 10 wolves per 1000 km² in the Horseranch Range (near Cassiar), 19 in the Kechika area (Liard River drainage) and 39 in the Muskwa River area (northern Rockies), and I estimated up to 29 wolves per 1000 km² over a four-year period on Level Mountain. The estimate in the more game-rich central portion of the Muskwa area during the 1983–84 winter was 303 wolves, of which 209 were actually seen on the survey. The calculated density for that core area (6800 km²) was 44 wolves per 1000 km². On northern Vancouver Island, studies made from the mid 1980s to the mid 1990s were directed primarily to concerns about wolf predation on Black-tailed Deer. Estimated wolf densities there were higher than on the northern mainland, ranging from 43 per 1000 km² in the Nimpkish area (studied by Knut Atkinson and Doug Janz) to 59 per 1000 km² in the Adam River drainage (Barbara Scott and David Shackelton's study area). As reported to biologist John Theberge for a review on the status of wolves in Canada, the total BC wolf population was estimated at 8100 in the early 1990s, and was believed to be increasing.

Human Uses

The relationship of humans with Grey Wolves is long and complex. Suffice it to say that wolves are prominent in human myth and legend going far back into ancient history in the Old World, and that both "good" and "bad" images are involved. In North America, the wolf was an important animal in the ceremonies of many aboriginal societies, and recent scholars have drawn interesting parallels between wolf packs and tribal communities. Further, as is well known, the two major clan divisions of aboriginal groups along the northwest coast, including British Columbia, were the Wolf and the Raven.

Before the arrival of Europeans, aboriginal peoples used the pelts of wolves to make robes and ceremonial objects. In the last 150 years, the primary use has been in the fur trade. Winter wolf pelts are thick and heavy, and although sometimes used for trim on parkas or full length coats, their primary market in recent decades has been for taxidermy.

Taxonomy

Grey Wolf taxonomy is both complex and highly controversial, with taxonomists unable to agree on the number of valid subspecies or whether some subspecies are actually separate species.

Those issues are complicated by changes in the species' original range, coupled with local extinctions of some forms and reinvasion of others. In their classic book *Wolves of North America*, Stanley Young and Edward Goldman described 24 subspecies, with 6 occurring in British Columbia. In the original *Mammals of British Columbia*, Cowan and Guiguet reduced the number in BC to 3. And in the most recent (1995) taxomic revision, Ronald Nowak grouped North American Grey Wolf populations into 5 subspecies, with just 2 occurring in BC: *occidentalis* and *nubilus*.

Genetic studies from several areas of western North America have revealed patterns that are not entirely consistent with Nowak's groups. For example, Byron Weckworth and his colleagues found evidence for a distinct genetic group on the Alexander Archipelago and Alaska Panhandle that is consistent with a former subspecies *ligoni*, which Nowak lumped with *nubilus*.

An unresolved issue is the taxonomic status of the Vancouver Island population. Characterized by dark pelage and large robust teeth, those wolves were originally classified as a distinct subspecies *crassodon*. But Laura Friis demonstrated that this subspecies had been altered by the immigration of wolves from the mainland during the 1970s and 1980s. Although, the Grey Wolves currently occupying Vancouver Island cannot be reliably distinguished by morphology from those on the adjacent mainland coast, a DNA study by Michael Roy and colleagues revealed some minor genetic differences between those two groups. More research comparing Grey Wolves from Vancouver Island with those from other coastal areas is needed to assess the genetic distinctness of the Vancouver Island population. Until further work dictates otherwise, Grey Wolf subspecies classification follows Nowak's determinations.

Canis lupus nubilus Say – a vast range from southeast Alaska and coastal and southern BC, across much of the western and central United States, and eastern Canada. This subspecies differs from *occidentalis* mostly by its larger skull size.

Canis lupus occidentalis Richardson – western Canada and Alaska. In BC it occupies the eastern and northern regions of the province. The former subspecies *columbianus*, as listed in the original *Mammals of British Columbia,* is now included in *occidentalis*.

Conservation Status and Management

Grey Wolves are consummate predators and that, combined with their high reproductive potential, high dispersal capability, high intelligence and adaptability to a broad range of habitats, has often put them in conflict with human interests. Despite popular perception fostered by mythology and fairy tales, that conflict has generally not involved predatory attacks on humans. There are only a few reliable records of wild, non-rabid wolves injuring humans in North America, including a case on Vancouver Island in which a sleeping camper was bitten by a wolf that had been fed by and habituated to humans in that area. More recently, an incident from northern Saskatchewan has proved to be the first continental record of a human being killed by wolves; again, habituated animals were involved.

On this continent generally, and British Columbia specifically, the primary issue has been predation on livestock and wild ungulates. Of livestock, wolves prey mostly on cattle, particularly calves on open range in the first three months after birth, but they also take adult cattle, horses, and sheep in some areas and seasons. Currently in BC, wolf predation is thought to be a main factor limiting the size of certain ungulate populations, including Black-tailed Deer on Vancouver Island, some populations of Moose in the north and Caribou in several areas. Predation by wolves is also a subject of concern in relation to recovery of the endangered Vancouver Island Marmot.

Historical attempts to control wolves on a large scale in BC included bounty programs (terminated in the mid 1950s), and extensive government-sanctioned poisoning by aerial drops of large baits (discontinued in the early 1960s). Most official wolf-control programs since then have been relatively local in nature, focusing on specific problem individuals or packs in livestock areas and particular regions (sometimes fairly large) for ungulate enhancement. All have been controversial, particularly those done using aerial shooting to manage populations of wild ungulates, and managers are currently experimenting with alternative methods such as capture and sterilization of dominant animals to reduce wolf numbers. The Grey Wolf in BC is currently designated as both a game animal (since 1966) and a furbearer (since 1976), and can therefore be legally harvested by both licensed hunters and trappers, subject to restrictions by seasons, bag limits and methods.

Remarks

Much of the war against wolves in the western United States was carried out between about 1890 and 1930. During that time a number of individual wolves attained legendary status either for the extent of their depredations, their ability to evade hunters and trappers, or both. Given names such as Lobo the King of Currumpaw, the Custer Wolf and Three Toes of the Apishapa, such wolves contributed much to public perception of wolves prior to scientific study and helped justify the need for extermination. However, in an examination of historical accounts for 59 of those famous wolves, Philip Gipson and Warren Ballard concluded that most were exaggerated. The publication that arguably did more to change human attitudes about wolves than any other up to the late 1950s, Farley Mowat's *Never Cry Wolf*, faced similar criticism. Canadian Wildlife Service biologist A.W.F. Banfield, who hired (and fired) Mowat fairly early in the project that the novel describes, noted in a review that some of the author's facts were real (taken from scientific reports) and others were invented, and that Mowat's account of his role in the wolf study was consistent with his actual occupation as a fiction writer.

Selected References: Atkinson and Janz 1994; Ballard and Gipson 2000; Ballard et al. 1987; Banfield 1964; Bergerud and Elliott 1986; Boyd and Pletscher 1999; Bryan et al. 2006; Carbyn 1983; Darimont et al. 2003; Friis 1985; Gipson and Ballard 1998; Hatler et al. 2003; Hatter 1988; Hayes et al. 2003; Hoffos 1987; Janz and Hatter 1986; Jones and Mason 1983; Kunkel and Pletscher 2000; Lopez 1978; Mech 1970; Messier 1985; Murie 1944; Nowak 1995; Peterson and Page 1988; Pletscher et al. 1997; Scott and Shackleton 1980, 1982; Seip 1992; Theberge 1991; Weckworth et al. 2005.

Red Fox   *Vulpes vulpes*

Other Common Names: Black Fox, Cross Fox, Fox, Silver Fox, Sly Reynard; *Renard Roux, Renard Royal*. Provincial Species Code: M-VUVU.

Description
The Red Fox is a small member of the dog family, about the size of a Cocker Spaniel, but slimmer. It has a slender, pointed muzzle, prominent erect ears, long, slender legs, and small feet. Although coloration varies considerably, it is usually categorized in three general colour phases – red, cross and silver – all of which may appear among members of a single litter. In the red phase, which is the most common, individuals vary from a pale coppery golden colour to a rich, dark cherry red on the back and sides, usually with paler underparts (white to grey), and most have contrasting black muzzles, ears and paws. The cross phase is grey to brownish in basic body colour, with darker face and legs, and gets its name from a cross pattern formed by dark (usually black) hairs down the upper back and across the shoulders. The cross is usually set off by lighter,

yellowish to pale-red patches on the neck and shoulders. While the silver-phase fox is generally black all over, light-tipped guard hairs on the back and upper sides give it a silvery appearance. All three colour phases have bushy white-tipped tails that make up over one third of the adult animals' total length. Adult males are generally larger and average about 15 per cent heavier than females. Juveniles in their first autumn are near full-size and almost as heavy as adults.

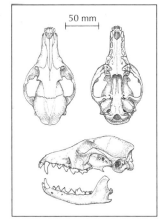

Measurements:

	All Specimens	Subspecies	
		abietorum	*cascadensis*
total length (mm):			
male:	1094 (965-1170) n=13	1127 (n=10)	987 (n=3)
female:	1062 (1023-1119) n=8	1062 (n=8)	none
tail vertebrae (mm):			
male:	416 (350-478) n=13	432 (n=10)	360 (n=3)
female:	404 (378-427) n=9	404 (n=9)	none
hind foot (mm):			
male:	179 (150-200) n=13	187 (n=10)	156 (n=3)
female:	176 (161-188) n=8	176 (n=8)	none
ear (mm):			
male:	100 (89-103) n=11	101 (n=10)	89 (n=1)
female:	98 (96-100) n=7	98 (n=7)	none
weight (g):			
male:	5407 (4529-6350) n=12	5568 (n=10)	4603 (n=2)
female:	4436 (3750-4990) n=9	4436 (n=9)	none

Dental Formula:
incisors: 3/3
canines: 1/1
premolars: 4/4
molars: 2/3

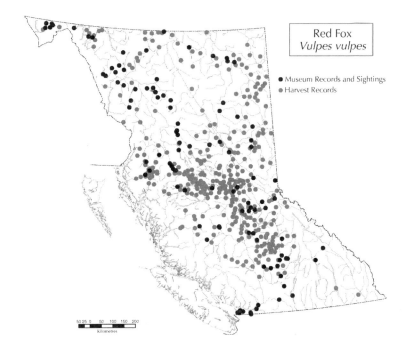

Red Fox
Vulpes vulpes

● Museum Records and Sightings
● Harvest Records

50 25 0   50   100   150   200
kilometres

Identification

In British Columbia, the Red Fox is likely to be confused only with the Coyote, and only when the fox is in a cross colour phase, which may have an overall buffy or grey tone, like a Coyote. Foxes are more slender and dainty than Coyotes, have a white-tipped rather than black-tipped tail, and usually have black rather than buff or cinnamon on the backs of the ears and on the lower legs and feet.

Distribution and Habitat

The Red Fox is the most widely distributed carnivore in the world, occurring naturally throughout most of North America and much of Europe and Asia, and as an introduced species in Australia. In a recent review paper, Jan Kamler and Warren Ballard suggest that most of the foxes in the central and eastern United States, and possibly also in much of eastern Canada are "non-native", descending from European Red Foxes that were introduced by settlers in the 1800s. In British Columbia, the species is found over much of the mainland, but is most common in the central interior and northern portions of the province. The Red Fox does not regularly occur in the wet coastal forests west of the Coast Range, or naturally on any of the coastal islands, although it does appear in agricultural and suburban habitats in the south, including the Lower Mainland. Recent observations of interest in the south include an active den in a small riparian area behind London Drugs in Vernon, and a black (silver phase) vixen with four black pups that were a local attraction in the Sun Peaks Ski Resort area near Kamloops in summer 2005. There are historical records from Vancouver Island, but those certainly involve animals that were released or escaped from fur farms.

Red Foxes usually occupy relatively open habitats, often interspersed with patches of trees or shrubs. The species is highly adaptable, subsisting in and around natural meadows and grasslands in southern BC, large expanses of alpine tundra and subalpine parkland in the north and, closer to civilization, in a variety of agricultural and urban areas. In the urban environment foxes commonly use remnant natural habitats such as parks and riparian areas, vegetated recreational facilities such as soccer fields and golf courses, and road-side features such as boulevards, ditches, and berms. Some urban foxes adopt a more plebian lifestyle, living and foraging in alleys, backyards and industrial areas.

One of the most important habitat considerations for Red Foxes is the presence or absence of Coyotes. Foxes tend to do better in areas where Coyotes do not occur or are at least temporarily rare. Coyotes out-compete, kill, or displace them. Because the Red Fox has relatively low foot-loading and is able to stay on the surface in deep snow, it is better adapted than the Coyote for winter occupation of high-elevation areas and can avoid competition with and exposure to Coyotes in those areas. In the Bulkley Valley, I observed several cycles in which fox sightings became common for a year or two when Coyote sightings were not, and vice versa. A possible factor was weather, with foxes surviving better than Coyotes in severe, deep-snow winters. On the other hand, former Animal Control Officer Jack Lay advises that in the Lower Mainland (where snow is not a factor), foxes were common when he arrived there in the mid 1930s and became rare after the spread of Coyotes into the area.

During periods of inactivity, particularly during the day, foxes occupy a variety of day bed sites, using tree canopy or other structures to provide protection from sun, wind or precipitation when that is necessary, and resting in open areas when it is not. Dens used for birthing and rearing of young are often underground burrows dug into the sides of knolls, banks or steep slopes, but also in other protected sites such as rock caves, dense thickets, blowdown tangles, and the hollows under the roots of large trees. Foxes excavate their own burrows or modify burrows made by other animals. An active Grey Wolf den that Don Eastman and I located on Level Mountain in 1979 (figure 44) was occupied by a family of Red Foxes the following year. Although a particular pair may use the same den year after year, it is also common for adults to move their pups to a number of different dens in any one year.

Natural History

Feeding Ecology
Red Foxes eat a wide variety of animals and plants, taking whatever is available to them. Small rodents are important prey items in most areas, particularly voles and mice, but also shrews, moles, chipmunks and pocket gophers. Foxes also take larger mammals, including Snowshoe Hares, cottontails, marmots, ground squirrels, tree squirrels and Muskrats, and, much less frequently, other carnivores, including skunks, American Minks, weasels and Northern Raccoons. They prey on numerous birds, including the eggs and young of grouse, quail, waterfowl, shorebirds and many ground nesting passerines, as well as on various amphibians, reptiles and insects. Plant materials eaten by foxes include berries and other fruits in season, grains such as oats and corn, and nuts. In rural and urban areas, the menu of individuals may expand to include rats, pigeons, pets and livestock such as poultry, rabbits, cats and lambs, and garbage. Their superb senses of smell, sight and hearing make foxes efficient scavengers, readily locating carcasses of all sizes whether in the open or hidden by other species.

The opportunistic nature of the Red Fox was nicely demonstrated by Donald Jones and John Theberge in separate study areas in northwestern British Columbia and the Yukon. There, the diet of foxes occupying areas north of the treeline consisted primarily of Arctic Ground Squirrels, mice and voles, while that of foxes in the boreal forest below was more diverse, though mostly Snowshoe Hares. Other isolated incidents further illustrate fox opportunism: one took a Ringed Seal pup on the ice in the Beaufort sea (it was the only Red Fox seen by Canadian Wildlife Service in over 100,000 km of Polar Bear surveys over a 16-year period); another attacked a Beaver that had been forced to forage above ice in winter (it would have been successful had the observers not intervened); in North

Dakota some foxes ate sunflower seeds at a commercial sunflower operation; and others, in Newfoundland, climbed into small balsam trees to obtain seed cones. Porcupine remains have been found in fox digestive tracts and scats, and Horace Quick found quills in 4 of 12 fox carcasses examined in the Fort Nelson area; predation on Porcupines is likely opportunistic and focused on juveniles or unusually vulnerable adults.

Red Foxes hunt primarily by slow cat-like stalking and rapid pouncing, but may chase prey for short distances when circumstances require. To me, the most characteristic and enthralling hunting method is the fox's high, graceful pounce to capture a vole in thick vegetation or under snow. Detecting a vole, apparently often by sound, the fox leaps (seemingly floats) high into the air and comes to ground so that its front paws pin the prey to the ground. I do not have detailed data on the success of that method, but I have seen foxes succeed on more than half of their attempts, catching 8 to 10 voles in a single session. When opportunities arise, foxes will often kill more than they need, transporting and storing excess items at the home den when pups are present, or at other times burying them in caches. In one case, researchers found intact carcasses and partially eaten remains of 67 adult dabbling ducks cached at a den. While often wasteful, cached food is probably a significant survival factor in some cases. The foxes that climbed trees for balsam cones lived on a remote island in Newfoundland; their primary winter food, as reported by biologist B. Sklepkovych, was seabird carcasses, mostly storm-petrels, which they had caught and cached during the summer.

Home Range and Social Behaviour

Home-range sizes of Red Foxes vary widely, depending mostly upon the distribution and density of prey. In subalpine habitats studied by biologists Donald Jones and John Theberge in northwestern British Columbia, the summer ranges of eight adults ranged from 3 km^2 for a female with pups to 34 km^2 for each of two adult males, and averaged 16.1 km^2. In other areas, most studies have been in agricultural or urban areas where home ranges were smaller, averaging 0.9 km^2 in southern Ontario and just 0.1 km^2 in Bristol, England. Foxes maintain their home-range boundaries throughout the year, but their areas of intensive use may shift seasonally in response to changes in food supply and cover.

The Red Fox has been the subject of many studies over its enormous circumpolar range, and its social structure appears to vary in different areas and habitats. For non-urban foxes in North America, the primary social unit seems to be the family group, comprised of an adult pair and their pups. Generally the home ranges of such groups do not overlap, indicating some degree of territorial exclusion which is likely maintained by scent-marking (urine and feces) and vocalization, but only rarely by aggression. In the fall, after the pups disperse, the adults become more independent and solitary, usually travelling and foraging alone until the mating season in mid winter. The same pair may re-mate, but not always. Adult males appear to be less constrained by the boundaries of family territories, especially in winter. The overall picture of social structure is muddied by the presence of subordinate animals, probably mostly juveniles and yearlings, with some of the females remaining in their mother's territories and most of the males in transient status, only passing through. The distinctness and discreteness of fox territories probably varies with local population density and prey availability, being less defined where fox numbers are high or foraging areas are compressed and constrained by local habitat features or by the presence of competitors such as Coyotes.

Activity and Movements
Foxes are crepuscular and nocturnal, conducting most of their activity away from the den between dusk and dawn. They can also be active during daylight hours, particularly at the den when pups are present. Nightly activity is directed primarily to foraging and territory maintenance, and the animals move about extensively while so engaged. In one of the classic early studies of the species (mid 1950s), Michigan biologist Raymond Schofield and associates logged almost 1800 km snow-tracking foxes to assess food habits. They trailed 84 foxes to their day beds, averaging 5.9 km from the tracker's point of interception of the trail to the bed; thus the average total distance moved (adding the distance back to the previous day's bed that would have been obtained by backtracking) was farther than that. Radio-tracked foxes have moved up to 10 km overnight.

As with most carnivore species, the longest movements are characteristically undertaken by juveniles dispersing from the home ranges where they were born. The nature and extent of that dispersal is of particular interest in eastern North America as related to the potential for the spread of rabies. In a midwestern United States study, young foxes began to disperse in late September or early October, when they were about seven months old. Among those tagged as pups and subsequently recovered, 80 per cent of males and 37 per cent of females had dispersed more than 8 km from their natal ranges by their first birthday, and those numbers increased to 96 per cent and 58 per cent, respectively, for animals recovered in their second year. In addition, juvenile and subadult males generally travelled farther, up to 211 km and averaging 31 km, in comparison to their female counterparts (up to 108 km, average 11 km). Those figures represent straight-line distances between tagging and recovery locations. Actual distances travelled averaged 40 to 50 per cent farther for ten animals that were also radio-tracked. Red Fox dispersals of 100 km or more are common for both sexes; the longest recorded are 302 km for a female in North Dakota and 395 km for a male that moved from Wisconsin to Indiana. Long-distance dispersal by adults also occurs occasionally, usually in response to reduced availability of food on the home range due to climatic factors or a major population decline of an important prey species.

Reproduction

The Red Fox has a fairly high reproductive potential. Both sexes mature at ten months of age and females (vixens) often produce young in their first year. The proportion of vixens in a population that bring pregnancies to term and the number of pups that are born vary with food supply, age and factors associated with population density. Reproductive success generally increases with the vixen's age and that likely reflects her experience and the good nutrition associated with an established home range. High fox density results in both increased competition for food resources and decreased likelihood that yearlings, vixens in particular, will find a suitably secure and productive home range.

Red Foxes generally breed in late winter, probably February and March in British Columbia, and the vixens bear pups in April or May after a 51–53 day gestation period. Litter size ranges from 1 to 8 pups and averages 5 or 6 in most areas, but viable litters as large as 12 have been documented. The largest known litter was 17, in a den in Michigan; all of the pups were undersize and showed evidence of malnutrition, and probably would not have survived.

At birth, Red Fox pups weigh about 100 grams and are fully furred but blind and helpless. Their eyes open and they become more mobile at about three weeks, they begin to venture outside the den by four to five weeks, and are weaned at five to six weeks. The pelage of fox pups is mostly grey and woolly to about four weeks, when the guard hairs begin to appear, and by seven weeks the lighter-colour-phase animals appear tan to yellow-brown. The transition to adult pelage, including colour, is complete by about fourteen weeks. Juvenile foxes are nearly full grown and largely independent by autumn when they are six to seven months old.

As with other members of the dog family, both parents care for their pups, thereby increasing pup survival. A behavioural study by Valeria Vergera in southern Ontario demonstrated that the contribution of the male appears to vary with the individual. Some males in her study area visited the den frequently, grooming the pups and showing considerable attentiveness, and others showed up only occasionally and never interacted directly with their pups. Although both parents brought food and guarded the pups, vixens generally brought the most food. Males provided more vigilant protection from predators, including other foxes. Two of eight family groups observed in the Ontario study included an additional non-breeding adult, a female in one case and undetermined in the other. Those adults also fed and interacted with the pups, although to a lesser degree than the parents. Red Foxes apparently adopt readily, another behavioural trait that favours pup survival. In Sweden, T. von Schantz observed a supplemental vixen take over care of the pups when the mother died. Of 61 pups, age 4 to 10 weeks, that were experimentally transplanted to other dens in Iowa, nearly half survived to independence.

Health and Mortality

Red Foxes can have a variety of parasites, both internal (roundworms, tapeworms, flukes) and external (fleas, lice, ticks, mites), but most do not significantly affect fox populations. Mange caused by Mites that irritate the skin and cause hair loss may kill foxes, particularly in winter in northern areas. Mange is most prevalent when one (or more) of the local canid species (Grey Wolf, Coyote, Red Fox) attains a high population density, and is most serious for the foxes. The Red Fox carries a tapeworm that can transfer to humans and cause Hydatid Disease (see Grey Wolf, page 88).

Of the wild canids, the Red Fox appears to be the most susceptible to rabies. Outbreaks of rabies have killed up to 80 per cent of a fox population and have aroused considerable human health concerns in some areas of North America, but rabies has never been detected in BC foxes. Diseases that affect other canids (parvovirus, various forms of distemper) can also cause widespread mortality in foxes, particularly when populations are high and animals are more likely to come into contact with infected individuals.

Coyotes are the most important predators of Red Foxes in many areas, taking both adults and juveniles, but Grey Wolves, Bobcats, Canada Lynxes and Golden Eagles also prey on foxes. Bob Stephenson and associates in Alaska documented several cases of Lynx predation on foxes, all during periods of Snowshoe Hare scarcity when Lynx prey choices were limited. Foxes sometimes kill the pups of other foxes. During a study of eight family groups in southern Ontario, V. Vergera observed two instances of a "stranger" visiting a den when the parents were absent and killing a pup. In many areas, most mortality is human-caused: legal harvest by trapping and hunting, animal-control measures by land owners or government officials in response to depredations, road kills, accidents with farm machinery (mostly affecting pups) and incidents involving uncontrolled domestic dogs. Wild Red Foxes as old as ten years of age have been documented, but few live longer than five years.

Abundance
There have been no studies to determine the local or general abundance of Red Foxes in British Columbia, but my observations suggest that the highest numbers and densities are in the northern half of the province, especially in northern uplands where Coyotes are rare or absent. Density estimates from other areas range from 0.1 fox per km^2 in boreal forest habitats to 1 per km^2 in agricultural areas of southern Ontario, up to 30 per km^2 in a food-rich urban study area in Great Britain. Fox numbers have been shown to fluctuate with the ten-year Snowshoe Hare cycle in northern regions, and that is probably the case in northern BC; some fox populations in Sweden exhibit a three to four year cycle in concert with voles, their primary prey in that part of the world.

Human Uses
With its bright coloured and luxuriant pelage, the Red Fox has long been one of the most important species in the international fur industry. Its popularity in the first half of the 20th century was reflected in comparatively high prices paid for pelts, which led to the establishment of large-scale fox farming in both Europe and North America. Since that time the large market demand for fox pelts has been maintained and mostly supplied by the farmed product. In BC, annual harvests of wild Red Foxes were highest in the 1930s and early 1940s, peaking at almost 10,000 animals in 1944, but they declined sharply as pelt prices dropped with the

increasing supply from farms. The annual wild harvest in BC has not exceeded 1000 animals since 1946, and has mostly been in the range of 200 to 400 since the early 1990s.

Taxonomy

Ten subspecies are recognized in North America, with two occurring in British Columbia, but no modern genetic research has been done to assess their validity. The genetic integrity of the Red Fox has been obscured by introductions and escapes from fur farms. Keith Aubry concluded that the only indigenous Red Fox populations in Washington State are those inhabiting the Cascade Mountains; coastal lowland and northeastern populations in the state are derived from introductions. Likewise, Cowan and Guiguet noted that they could not determine the systematic status of Red Foxes in south-central BC because of fur-farm introductions in that region.

Vulpes vulpes abietorum Merriam – northern Saskatchewan, Alberta, the southern Northwest Territories and Yukon Territory, and northern and central BC. Although this subspecies is reportedly darker with a longer tail than *cascadensis*, it is a weakly defined subspecies that Phil Youngman found to differ little from *alascensis*, a northern subspecies found in Alaska and northwestern Canada.

Vulpes vulpes cascadensis Merriam – Oregon, the Cascade Mountains in Washington, and southern BC (lower Fraser River valley, the Cascade Mountains and southern Coast Mountains, to the Kootenays). The extent to which introductions have obscured the genetic structure of Red Foxes in the lower Fraser River valley is unknown.

Conservation Status and Management

The Red Fox is officially designated as both a furbearer and as a small game species in British Columbia; but there are no open seasons for hunting and so only licensed trappers can legally harvest the species in prescribed seasons and areas. The intelligence of foxes is legendary and that, together with its high reproductive potential and high dispersal capability, results in a low likelihood of overharvesting and, therefore, little conservation concern for the species.

Red Foxes occasionally come into conflict with rural and suburban British Columbians by killing pets, poultry and other small livestock. But they often compensate for their depredations to some extent by limiting rodent populations.

The species is a well known predator of game birds, particularly during the nesting season, eating eggs, young and adults of both waterfowl and upland species such as grouse and pheasants. The most research on that problem has been conducted in the American plains (the Dakotas, Minnesota, Nebraska and Iowa). Based on observations at 1432 fox dens over a five-year period, Alan Sargeant and his co-workers estimated that foxes killed 900,000 adult ducks annually, mostly hens apparently caught on nests, and were seriously reducing duck production in that area. There are also records of extensive fox predation on colony-nesting birds. On South Manitou Island in Lake Michigan, predation and disturbance by small numbers of foxes caused an 84 per cent reduction in breeding pairs among Ring-billed Gulls (from more than 5000 to less than 1000) and a 53 per cent decline in Herring Gulls over a nine-year period. As reported by William Southern and associates, those negative effects followed a reduction in fox-control efforts by local residents after the area was taken into the national park system.

Remarks

Although Red Foxes have three basic colour phases, there is considerable variation in each of them. Red-phase foxes vary from pale yellowish to intense red, cross-phase foxes can be greyish to golden tan, with the cross pattern ranging from large and distinct to essentially absent, and silver-phase animals show a considerable range of silver-tipping on the guard hairs, such that some appear completely black and others almost silvery grey. In addition, I have seen intermediate individuals, with characteristics of both the cross and silver phases, and there is a rare pelage mutation (the "samson" condition) in which there are no guard hairs and the animal has an overall woolly appearance. There are also variations in markings, of which those on the tail may be particularly noticeable. Once, on a summer aerial survey in the Tatshenshini area of northwestern BC, I glimpsed what appeared to be a white animal streaking through a patch of low shrubs. A closer look proved it to be a red-phase fox with a tail that was completely white from base to tip.

In a recent monograph on the Red Fox, J.D. Henry labels the species "the catlike canine", in reference to a number of features that are indeed catlike, including the animal's agility and gracefulness, stealthy stalk-and-rush hunting method, partially retractile front claws, vertical pupil and well-developed night vision, and proportionally large vibrissae (whiskers). It is also reputed to have an aversion to water, wading and swimming only reluctantly and greatly reducing its activity during periods of precipitation. In other respects, especially in intelligence and adaptability, the Red Fox is definitely canid. A fox's intelligence is the basis for its character in fables and legends, and has inspired phrases and words such as "cunning as a fox", "sly as a fox", "foxy" and "outfoxed". Consistent with its intelligence and perceptiveness, the Red Fox has shown curiosity and resourcefulness when encountering strange objects in its environment. Many golfers have had their golf balls carried off the fairway by a fox. In northern BC I was investigating a radio-collared Caribou that had been killed by wolves and, although I had located the Caribou remains, I had difficulty finding the collar. The location of the strongest radio signal was in a small, apparently undisturbed opening amid ground balsam, but the collar was not to be seen. Eventually, after several puzzling minutes, I discovered it under a moss mat; a small section of moss had been pulled back, the collar placed under it, and then the mat rolled back over – a fox scat on top of it told the rest of the story. On another occasion I collected several vials of blood samples from Caribou and placed them in a snowbank on a remote ridge in the Spatsizi Park area. When I returned a few hours later to retrieve the vials, all that remained were fox tracks.

Selected References: Ables 1975; Aubry 1984; Cowan 1938a; Dekker 1989; Hatler et al 2003b; Henry 1986; Jones and Theberge 1982, 1983; Kamler and Ballard 2002; Lariviere and Pasitschniak-Arts 1996; Lindstrom et al. 1994; MacDonald 1976; McGowan 1936; Sargeant et al. 1984; Schofield 1960; Vergara 2001; Voigt and Berg 1987.

American Black Bear   *Ursus americanus*

Other Common Names: American Bear, Black Bear, Cinnamon Bear, Glacier Bear, Kermode Bear; *Ours Noir*. Provincial Species Code: M-URAM.

Description
Most people are familiar with the general body shape of bears: the compact, stocky torso with thick legs, relatively large, flat-soled feet and a tail so small that it is rarely visible. The American Black Bear, referred to simply as "Black Bear" in the remainder of this account, is a large animal, adult males generally standing taller and weighing more than twice as much as the largest domestic dogs. It has a broad head, short but conspicuous rounded ears and small eyes. Black Bear pelage is heavy and thick, except during the summer moult (mid June through August), and despite the species name, comes in a variety of colours: black (in all regions), brown

(blonde to auburn to dark chocolate, generally referred to as a Cinnamon Bear), white (Kermode Bear) and blue-grey (Glacier Bear). Most dark-coloured Black Bears (black, brown, blue) have distinctly lighter coloured muzzles (tan to brown) and some also have irregular white markings on their chests. The non-black colour phases, all of which may have black litter mates, are not uniformly distributed in British Columbia. Brown-phase bears occur throughout the interior but most commonly in the central and southern regions, the white phase is found along the north coast and in the lower Skeena and Nass river drainages, and the blue phase is in the Haines Triangle in the extreme northwest.

Black Bears often appear larger than they really are, probably because of their bulky shape and their big reputation ("big" and "bear" seem to go together). In fact, the size of Black Bears is difficult to describe and characterize with terms and concepts such as average and normal. In addition to differences between the sexes, with mature males often twice as heavy as females, Black Bears take five years or more to attain full size. The rate and extent of growth varies regionally depending upon factors such as the availability of protein and the length of the growing season, and the weights of individual bears may also fluctuate seasonally by 25 per cent or more due to fat storage for winter denning. Because data from coastal populations were not available, the figures given below probably do not depict the full range that might be expected for Black Bears throughout BC. In general, a 115 kg female or a 180 kg male would be considered large anywhere, but there are extremes well beyond that. The largest Black Bear on record, a male killed in collision with a vehicle in Manitoba in 2001, was 235 cm long, had a girth of 196 cm, and weighed 402 kg.

Measurements:

total length (mm):
  male: 1603 (1300-1880) n=183
  female: 1413 (1160-1660) n=112

tail vertebrae (mm):
  male: 104 (65-130) n=9
  female: 98 (75-121) n=4

hind foot (mm):
  male: 258 (229-295) n=12
  female: 255 (242-270) n=3

ear (mm):
  male: 128 (120-135) n=5
  female: 117 (115-119) n=2

weight (kg):
  male: 93.4 (36-148) n=170
  female: 58.4 (31-118) n=107

Dental Formula:
  incisors: 3/3
  canines: 1/1
  premolars: 4/4 (1 or more often
                     missing)
  molars: 2/3

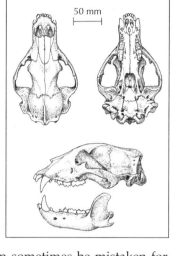

Identification:
Black wolves and large black dogs can sometimes be mistaken for Black Bears if seen only briefly or at a distance, but the presence of a tail (canids) or its absence (bears) suffices to make the distinction. The Black Bear is more likely to be confused with the Grizzly Bear, BC's only other bear species. Although useful as first clues, neither size nor colour are sufficient to confirm identification. A large Grizzly Bear will rarely be mistaken for anything else, but the great variation in size within species results in considerable overlap between them. For example, adult male Black Bears are larger than many female and young male Grizzly Bears. Some grizzlies may look dark when wet or in poor light, but few are truly black – thus, a black bear is very likely a Black Bear. However, as described above, many Black Bears are shades of brown or even blonde and cannot be clearly separated from Grizzly Bears by their colour. A Grizzly Bear's "frosting" or "silver-tipping" of the hairs on its head, neck, shoulders, and sometimes the back and sides is a strong clue for identification when it is present, but some individuals show little or none. Figures 13 and 14 (page 45) show several features that differ between the two species. Black Bears have more prominent ears than Grizzly Bears, lack the ruff of longer neck hairs on many Grizzly Bears and, when viewed from the side, the profile from a point between the eyes to the nose along the top of the snout is straight or slightly raised (convex) on Black Bears and tends to be dished (concave) on Grizzly Bears. Unfortunately, those distinctions are more evident when both species are present for comparison than when only one is being observed. More reliable

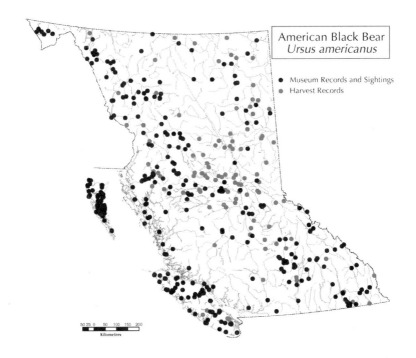

American Black Bear
Ursus americanus

● Museum Records and Sightings
● Harvest Records

50 25 0 50 100 150 200
Kilometres

distinguishing features are the Grizzly Bear's prominent hump over the shoulders and its longer front claws (up to 10 cm, figure 45); Black Bears have no hump and their claws are less than 4 cm and fairly inconspicuous. Black Bears are adept at climbing trees – their short, curved claws appear to be adapted primarily for that purpose.

Distribution and Habitat

The Black Bear occurs in forested areas over most of North America, from the northern treeline in Alaska, the Canadian territories and Labrador to the Sierra Madre of Mexico. It is absent from treeless plains, prairies and deserts, and occurs in a discontinuous distribution in the more developed areas of the eastern and southeastern United States. In Canada, it occurs in all provinces and territories except Prince Edward Island. It occurs in all parts of British Columbia, including the coastal islands; individuals can be seen everywhere in the province (including, occasionally, the large urban centres).

Black Bears are forest-dwelling animals, occurring in both coniferous and deciduous forests, but do much of their foraging in non-

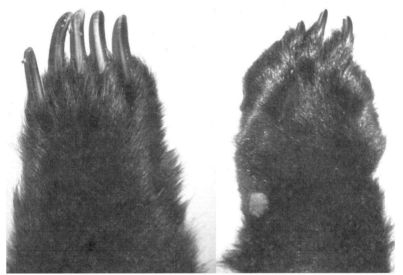

Figure 45. Bear claws: Grizzly Bear (left) and Black Bear.

forested patches: riparian areas, wetlands, meadows, estuaries, subalpine parkland, and natural openings of all sizes created by burns, insect infestations, floods, avalanches and falling trees in older forest. Black Bears also use man-made openings near forest cover, including logging cutblocks, pastures and croplands. Large tracts of dense, unbroken forest are not particularly good year-round bear habitats. Diversity is important, with the trees providing protective cover, the various openings in the canopy allowing increased penetration of sunlight to support growth of food plants, and the local moisture enhancing that growth wherever it occurs. Black Bears may pass through, but do not regularly occur in, extensive grassland or desert habitats, such as those in the southern interior, or large expanses of subalpine shrubland or alpine tundra.

In BC, a Black Bear goes through two distinct phases in a year: one in spring through fall, when it is active and intensely focused on foraging, and the other when it hibernates in a winter den. The duration of each phase varies regionally and individually, with the result that some bears in the south and most in the north must obtain their entire year's sustenance in just five or six months. During the active period, bears exploit a number of habitats where food resources are concentrated or at their most nutritious stage. For example, in mountain landscapes Black Bears commonly begin their post-hibernation activity in lowland or south-facing areas where vegetative green-up first occurs in the spring, and then gradually (over a period of weeks) move upslope following the spread of spring and summer conditions. Many bears will be in high subalpine habitats by late summer or early fall, but most move to lower elevation as growing conditions diminish with the change of seasons.

Given that Black Bears spend the most inhospitable part of the year in hibernation (more than six months in some cases), and that their young are born during that time, it follows that the winter den is an important component of habitat for this species. In the only intensive study of dens in BC, undertaken by Helen Davis in the Nimpkish River area (northern Vancouver Island), the hibernation period varied among individuals and by sex. Most males entered dens after November 21 and emerged before April 1, while most females entered during the first three weeks of November and emerged by April 30. Females with cubs remained in dens the longest, up to 190 days, with an average of 160 days. Based on my observations, most Black Bears in the central and northern interior are denned by mid October and emerge in early to mid May while those denning at lower elevations in the southern interior may stretch the active period a month or more longer, entering dens in late October and emerging in early to mid April.

In Davis's Nimpkish study, all of 67 dens located over a four-year period were in or beneath large-diameter (average 143 cm) trees, logs or stumps (figure 46). She concluded that tree-related structures were essential for keeping bears warm, dry and secure in the persistent rain typical of that area in winter. In northeastern Oregon, nearly two-thirds of 165 dens examined by Evelyn Bull and her co-workers were also associated with large-diameter tree structures, while the rest were in rock caves (22%), in burrow-like excavations in soil (12%) or (in one case) on the soil surface at the bottom of a tree

Figure 46. Black Bear den on Vancouver Island.

well (a hollow formed by build-up of snow around the base of a tree). In the Oregon study, 30 of the dens were in large trees that had broken off entirely or at the insertion point of large branches, exposing cavities or areas of internal decay that could easily be excavated. Those top entry dens averaged 16 metres above the ground, in trees averaging 114 cm in diameter at breast height (dbh). As usually occurs under intensive forest management, the reduction in numbers and dispersion of trees large enough for bear denning was considered cause for concern in both the Vancouver Island and Oregon study areas.

In areas with regular rain or frequent thaws in winter, underground dens, including those excavated under logs and at the bases of trees and stumps, are more subject to flooding and are probably somewhat less secure than are elevated tree structures. But in areas with a more continental climate and/or where large trees are rare, underground dens appear to be the rule (35 of 37 in a study by William Tietje and Robert Ruff in the boreal forest of northern Alberta, and 23 of 25 in Walt Klenner's and Darryl Kroeker's study area in west central Manitoba). The nest chambers of all but one of the excavated dens in those two studies were lined with grasses and litter from the surrounding area. A female Black Bear in Montana had lined her nest chamber with a thick layer of cedar bark she had stripped from nearby trees. The warmth and security of ground dens is enhanced in northern regions by snow cover. Taking advantage of the insulative value of snow, a technique regularly recommended by instructors of human survival training, has been practiced by hibernating bears for eons.

Eleven radio-collared bears monitored by Tietje and Ruff took 5 to 10 days (average 7) to construct their dens. Most of the excavated dens (24 of 34) consisted of a distinct tunnel (average length 81 cm, maximum 206 cm) leading to the nest chamber. Entry tunnels averaged about 68 cm wide by 47 cm high, and the average dimensions for nest chambers were 117 cm long by 105 cm wide by 70 cm high. The dens of large bears were larger than those of small bears, suggesting that individuals stopped digging once they felt comfortable with the space. Open-air space appeared to be kept to a minimum, probably to maximize retention of body heat. Bears that excavated dens under woody structures, such as the root-masses of fallen trees, dug the floor to the depth required and let the woody structure determine the nature and thickness of the ceiling. Dens excavated in more open soil had ceilings that averaged 41 cm thick and floors an average of 111 cm below ground. Reuse of dens in subsequent years by the same or different bears was fairly common in the Oregon and Vancouver Island studies (about 25%), particularly those in or under trees, but was extremely rare for the excavated dens in Alberta (1 of 34 dens monitored).

Natural History

Feeding Ecology

Although all animals require food, the phrase "hungry as a bear" characterizes the particularly voracious nature of Ursine appetite. Because they need to meet all of their annual nutritional requirements in a little more than half a year, while active outside the winter den, Black Bears are veritable eating machines. They readily and enthusiastically eat animal protein whenever they can obtain it, but most of their food is plant material. Based on chemical analysis of hairs, Canadian Wildlife Service biologist K. Hobson and his co-workers calculated that plants contributed over 90 per cent to the diets of Black Bears in southeastern British Columbia. Although bears have physical adaptations for such a diet, including flat molars for grinding and a longer digestive tract than other carnivores, they do not digest vegetable matter efficiently and, to compensate, they must eat prodigious quantities.

In most areas of western North America where Black Bear food habits studies have been undertaken, green vegetation was the primary plant component in the spring and early summer and berries predominated in late summer and fall. The list of plants consumed is large, but some of the most common items in most areas are horsetails, various grasses and sedges, legumes such as peavines, vetches and clovers, and forbs such as Cow-parsnip, Skunk Cabbage and Common Dandelion. The stomach of a large male killed near Houston, BC, contained nearly four litres of Common Dandelion, mostly the flowers. In some areas, the catkins and new leaves of poplars are also regular spring foods. From a single vantage point one spring morning in the Bulkley Valley, I saw three bears feeding high in Black Cottonwoods, hundreds of metres apart; claw marks left by bears climbing Trembling Aspen trees, presumably for feeding, are common in some areas (figure 47). Another plant food consumed by Black Bears in spring and early summer is the sapwood (phloem layer) of trees, mostly conifers and especially Douglas-fir, Western Hemlock, Western Redcedar and Sitka Spruce. The bears use their teeth and claws to strip the bark from the trees, then feed on the exposed sugar-rich sapwood.

Despite its predominance in the diet for the first 2 to 2.5 months of a bear's active season, green vegetation does not appear to contribute to the fat deposits that the Black Bears require for overwintering. Most of the annual weight gain and fat storage is obtained from the fall diet, which is primarily berries in most inland areas. Of the berries regularly consumed by Black Bears, the *Vaccinnium* species (huckleberries, blueberries, Lingonberries) and Soopolallie are among the most nutritious and digestible, and often occur locally in the large quantities bears require. For perspective on the issue of quantity, the stomach of a medium-sized female I examined in Alaska contained 3070 cc of material that, based on detailed sampling, included about 12,900 blueberries. Berries pass through a bear's digestive tract quickly, and it is likely that total represented the results of just one feeding session. Other fruits consumed by western Black Bears in late summer and fall include raspberries, elderberries, currants, rosehips, and the fruits of Highbush-cranberry, Saskatoon, Choke Cherry, Devil's Club, twistedstalks, Crowberry and Kinnikinnick. The last two in that list overwinter well and are also eaten by bears in the spring, although quantities available at that time are usually limited.

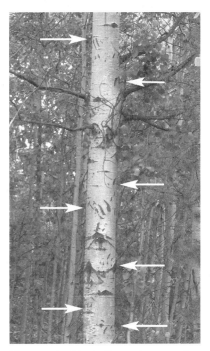

Figure 47. Black Bear claw marks on a poplar trunk.

The consumption of animal protein by Black Bears in most areas is largely opportunistic, taken as and when encountered rather than specifically hunted. The list is long but some of the most regularly consumed items are ungulate carrion, newborn ungulates (calves, fawns), the eggs and young of various birds, rodents (especially voles), insects (ants, bees, wasps, beetle larvae) and fish. On the Kenai Peninsula, Alaska, biologists Charles Schwartz and Albert Franzmann found Moose-calf remains in almost one-third of 241 spring Black Bear scats examined, and calculated annual kill rates of 1.4

calves per bear in one portion of the study area and 5.4 calves per bear in another. They further calculated that a single calf provided an average of 7.5 times as much digestible energy as an equivalent amount of spring plant foods.

There are also records of Black Bears killing adult ungulates, but usually in exceptional circumstances. Yellowstone Park officers William Barmore and David Stradley reported that an adult bear killed two bull Elk on successive days, catching them after they became bogged down in deep snow. In Ontario, biologist Matt Austin and associates found evidence of an old cow Moose killed by a large male Black Bear; they speculated that the bear had happened upon the Moose while it slept.

Black Bears prey on smaller animals as opportunities arise. A Colorado study reported the loss of seven radio-tagged Yellow-bellied Marmots to one or more Black Bears. In a study on woodpeckers in the Hat Creek valley of south-central BC, University of Victoria researchers Eric Walters and Edward Miller recorded four cases of Black Bear predation on nests (three Northern Flicker, one Williamson's Sapsucker) and eight unsuccessful attempts. Alaskan biologist Peter Shepherd told me in the 1960s that as duck egg predators, Black Bears were second only to Mew Gulls in the interior wetlands of that state; diving ducks, such as scaups, were particularly vulnerable, since they nest on floating mats near the water's edge where bears regularly travelled while foraging on succulent emergent water plants.

The consumption of insects by such a large animal may seem incongruous at first glance, but it has been reported in most major Black Bear food-habits studies. Bears eat insects that congregate in great numbers and so attack the nests of colonial species, especially ants and wasps. In Alaska I found evidence of bears consuming entire nests of wasps, and obviously the many hundreds of larvae packed into the combs constituted a significant source of animal protein. In northeastern Oregon, Evelyn Bull and her associates found insects, mostly ants, in 434 (70%) of 621 scats collected over a two year period, and estimated that they composed 24 per cent of food volume during that time. A radio-collared adult female monitored by Lynn Rogers in Minnesota spent more than three quarters of her active time foraging for ants one fall when nuts and berries were scarce. In Black Bear country, one may find many locations where bears have ripped apart logs, turned over stones and excavated anthills in their foraging efforts. Naturalist William Wright describes a bear's raid on an ant colony:

*Out rush the ants by companies and regiments and
brigades; mad as hornets, brave as lions, smelling like a
spoiled vinegar mill, and looking for trouble. They get it
almost immediately. They discover the bear's furry paws
and, struggling and tumbling in the hair like angry and
hurrying warriors in a jungle, they begin to swarm over
them. And as fast as they come the bear licks them up.
When the excitement dies down, he gives (the nest)
another poke.*

Many of the swarming ants in a disturbed nest carry eggs and larvae,
adding to the protein benefit of the bear's raid.

The extent to which coastal Black Bears forage in the intertidal
zone has not been described, but I have seen scats packed with the
remains of Purple Shore Crabs and suspect that some individuals
learn to exploit those and other animals that hide under boulders
at low tide, hunting them much as they do ants in upland habitats.

Where spawning fish concentrate, Black Bears gather and consume little else for weeks, obtaining a nutritional boost not available to bears in other areas. In a study on the Queen Charlotte
Islands, Tom Reimchen counted 4281 Chum Salmon carcasses
taken by Black Bears in autumn 1993. He estimated that each bear
caught 13 salmon per day, as many as 74 per cent of the fish that
entered the stream (he believed that most had already spawned, so
the bears had little impact on the salmon population). When
salmon numbers are low, bears consume most of each carcass, but
when numbers are high, they maximize their nutritional gains by
eating only certain energy-rich body parts, such as brains and roe.

Although the popular portrayal of bear fishing is the use of a
front paw to swat the fish out of the water and onto shore, that
technique appears to be used only rarely if at all. Reimchen
described the two most common fishing methods in his study area:
the bear leaped at a target fish and trapped it with forepaws and
mouth, and the bear, standing motionless, suddenly struck into the
water and caught the fish in its mouth.

Black Bears also scavenge spawned-out salmon that are dead, or
nearly so. Some bears seem to prefer them. A large male that I
observed for almost an hour at close range in the Megin River
(west coast of Vancouver Island) caught a large, fresh looking
Chum Salmon by pinning it against the bank with his forepaws
and grabbing it in his jaws, then sat back and released it ... watching it struggle in shallow water with no attempt to recapture it. He

then spent some time looking into the water and groping with a front paw at the upstream side of woody debris at the river edge, sometimes completely submerging himself. He brought seven dead salmon to the surface and, holding them in his front paws, sniffed them from head to tail and back again. Following inspection, he rejected five of the fish, letting them slip back into the water, and took the other two to shore. He ate only a bite or two from the first, but spent more than five minutes devouring the other, which was completely covered with mould and, to my eye, the most rotten of all the fish he tested.

Like most carnivores, Black Bears make great efforts to obtain food. For example, in the 1960s, when I was working as a fisheries technician in Alaska I put about 20 small Arctic Grayling in a bucket of a strong formaldehyde solution to preserve them for future study. I left the bucket outside for the night and the next morning I found that a small Black Bear had visited, recognized the bucket of fish as a potential meal and given it his best shot. He left footprints around the bucket and one-half of what must have been a very bitter-tasting grayling on my doorstep.

Home Range and Social Behaviour

Black Bears appear to have a less rigid social system than most of the other carnivores. As elaborated by University of Calgary biologist Steve Herrero, they do not require cooperation in foraging and live a solitary existence over most of the year. The only exceptions are brief male-female pairings during the mating season, and family groups consisting of adult females with dependent young. Adult females appear to be more territorial than males, tending to maintain separate home ranges most of the time, though they freely overlap with the ranges of local adult males and subadults of both sexes. Adult males occupy distinct home ranges for long periods, but the ranges of adjacent males may overlap almost completely and there is no evidence that they defend boundaries or attempt to exclude other bears from particular areas. The marking of "bear trees" by rubbing, biting and clawing, mostly involving adult males, may have a social (territorial) function, but that has not yet been clearly demonstrated. Despite the lack of hard evidence for male territoriality, experimental removal of several adult males in an Alberta study area was followed by increased immigration and survival of young bears and greatly increased local bear density. Whether that reflected territorial or predatory behaviour of the adult males remains unclear. One result of the apparently flexible social system of Black Bears is that numbers may gather in relatively small areas to exploit concentrated food sources (e.g., berry patches, salmon runs, garbage dumps). In those cases, interactions are hierarchical, with subordinate bears avoiding conflict by giving ground to approaching dominant individuals.

As with most species, home-range sizes vary considerably depending upon habitat conditions (especially food supply), and male home ranges are larger than those of females. Average home-range sizes of 495 km^2 for males and 295 km^2 for females, in a study area in Manitoba, are among the largest recorded. Average home-range sizes for males and females, respectively, in some western study areas include 119 km^2 and 20 km^2 (Alberta), 112 km^2 and 49 km^2 (Idaho), and 84 km^2 and 21 km^2 (Washington). In general, home ranges are smaller in areas or years when food is abundant and larger when food is scarce.

Activity and Movements

Activity and movement patterns of Black Bears are highly variable, differing by sex, age, season, habitat and local circumstances. A radio-collared adult female intensively monitored by biologist Lynn Rogers in Minnesota foraged primarily during daylight hours. She consistently bedded and remained inactive from about two hours after sunset to an hour before sunrise, and for up to two hours during the day. On average she was active for about sixteen hours a day, covering her home range in circuits that took one to four days to complete. I have found the most predictable times for observing Black Bears to be morning and evening, particularly the latter. The extent of midday activity probably depends mostly upon food availability, with the bears needing to forage for longer periods when food is scarce or, in the case of young bears, foraging outside of a resident's prime time to avoid conflicts. Although there is evidence that Black Bears do not see well in the dark, that does not preclude nighttime activity. I occasionally encountered bears while doing night surveys for American Minks and Northern Raccoons along beaches on the west coast of Vancouver Island, and University of Victoria biologist Tom Reimchen recently used night-vision goggles to study Black Bears along salmon streams on the Queen Charlotte Islands.

Based on straight-line distances between locations, Barry Young and Robert Ruff calculated that radio-collared Black Bears in eastern Alberta moved an average of 1.7 km per day while foraging. But an individual can move farther than that on any given day between feeding locations or, in the fall, to den a site. When a bear has located a concentrated food source it may bed nearby and move very little for several days. As with most carnivores, the Black Bear travels farthest when it is young and dispersing from its natal range. Young males make most of the long movements, up to 219 km documented in a Minnesota study. During years of food scarcity or failure, some adults may abandon home ranges and make similar movements.

Black Bears that have been captured alive in nuisance situations and translocated to distant areas have shown a remarkable ability to return to their original locations. In an analysis of 77 cases from seven states and provinces, Lynn Rogers found that 34 (44%) of the bears that had been moved more than 64 km (longest 229 km) returned to their previous home ranges and 18 others (24%) were heading in the right direction. Those that did not return or moved in random directions were probably non-resident where captured and did not have an established home area that compelled them to return. In southeastern BC, Conservation Officers R. Rutherglen and B. Herbison live-trapped, tagged and translocated 236 nuisance Black Bears between 1968 and 1973. Of 54 that were subsequently recovered, 37 had returned to the original capture location one or more times. One adult female returned to Champion Lakes Park, south of Castlegar, from three different release sites 38, 88 and 99 km away. The means by which such bears navigate home is not known, but previous experience in or near the release area can be ruled out in most cases.

Reproduction

Black Bears mate in the summer, mostly in June through mid July. The mating pair stays together for only a short time, (a few hours to two or three days). The short mating association allows males, which travel widely during the mating season, to locate and breed with more than one female each year. It also results, in some cases, in females producing litters sired by more than one male. The age at which females become sexually mature varies both regionally and individually, and relates mostly to growth and development as affected by the level of nutrition. Well-fed females mature earlier, as young as two years old in some populations, while those in areas where food is less abundant or nutritious may not mature until they are six to eight years old.

Most males mature at two to three years, but are not yet full grown and usually cannot compete with other males for mating privileges until they are six years old or more. In a genetics-based study in North Carolina, Adrienne Kovach and Roger Powell found that all three large resident males in their study population reproduced successfully, siring 20 of 22 cubs that were tested over a two-year period. In contrast, only two of nineteen smaller males known to the researchers were successful, each siring one cub. There was evidence that the breeding dominance of the larger, older males was enforced in part by fighting.

The young are born in the winter den in January or February after a gestation period lengthened by delayed implantation. The actual length of pregnancy after implantation is 60 to 70 days. I have always found it remarkable that female Black Bears go through all stages of pregnancy and then produce milk for growing cubs for four to five months, all while fasting in hibernation. Studies throughout the species' range have highlighted the importance of fall nutrition, mostly as related to climatic factors, in reproductive success. During a year of poor berry production, for example, a pregnant female may enter the den with inadequate fat stores and either fail to support the pregnancy or be unable to feed some or all of the litter. A delayed spring, which keeps them in the den longer, may exacerbate those effects.

Newborn Black Bears are small and undeveloped, no doubt an adaptation to minimize the energy demands on the denned mother. Each weighing 200 to 300 grams, about the same size as a newborn Coyote, they are sparsely-haired, blind and able to move only feebly. Their eyes open at about 40 days, they can walk in a coordinated way by about 60 days, and are able climbers when they emerge from the den at about 120 days. The number born in the den may be higher, but usually 1 to 3 cubs emerge in the spring (average 2 to 2.5 cubs in most areas). Litters of 4 cubs are not uncommon, especially following good food years (and litters as high as 6 have been reported, though rarely and only in the east where various energy-rich nuts form a major part of the fall diet).

The mother provides all parental care. Although the cubs continue to nurse for several months after emerging from the den, they also begin to take solid food fairly early in their new life outdoors. My observations of small cubs in the spring indicate that they consider virtually everything as potential food, probably learning by trial and error as they go. In most cases, cubs-of-the-year den with their mother during the first winter and will continue to travel with her the following spring until family break-up during the mating season, usually between the last week of May and early July. Some yearling males begin to disperse at that time while others may linger in or near their mother's home range and delay dispersal until the following year. After separating from the family group, most yearling females remain in small home ranges within their mothers' territories.

Females with young-of-the-year usually either do not breed or are unable to produce young after lactating all summer, thus putting them on an alternate-year breeding schedule. Some may skip additional years between litters, probably depending upon their condition and food availability. Female Black Bears may breed in consecutive years if they lose or abandon cubs before the breeding season. Some adult females have been seen with cubs of different ages (young-of-the-year and yearlings), possibly the result of consecutive-year breeding, but also possibly from adoption.

Health and Mortality

Black Bears are host to a few ectoparasites (fleas, lice and ticks) and, across their wide continental range, to a variety of internal parasites, including at least 20 species of roundworms, 10 tapeworms, 3 protozoans, 1 fluke and 1 spiny-headed worm. Rates of occurrence are usually low and infestations light, and none of those parasites and no diseases are known to be significant to bear populations. Black Bears seem to be particularly susceptible to tooth problems, including both dental caries and periodontal disease, possibly resulting from the large amounts of sugar they consume while eating berries. I have seen a number of Black Bears with tooth cavities large enough to cause severe toothaches, and I wondered about the temperament of those individuals. Among the roundworms carried by some Black Bears is the larval form of the Trichina Worm, which can cause serious illness (trichinosis) to humans who eat inadequately cooked bear meat. Outside of concentrations at artificial feeding sites, such as garbage dumps, the Black Bear's solitary, mobile existence at relatively low density is not conducive to the spread of disease.

Factors contributing to natural mortality of Black Bears are not well known, but several studies have shown that the effects of inadequate nutrition on young bears extends beyond the natal denning period. In Minnesota, the cubs and yearlings of food-deficient females were significantly smaller than those of well-fed mothers, and were four times more likely to perish before family break-up. Whether they died directly from starvation or were predisposed by poor condition to other causes, such as predation or disease, could not be determined. Results from several studies have highlighted the even more precarious existence of young bears after separation from their mothers. They must obtain adequate food while avoiding older bears, and a large portion of them fail, especially males. Among a sample of radio-collared subadults in Alaska, 7 of 24 females (29%) and 17 of 21 males (81%) died between the ages of two and four years. Cause of death was not recorded for most, but many were human-caused relating to the animals being particularly vulnerable to hunting or being shot in nuisance situations.

Black Bear cubs and yearlings have been preyed upon by Bobcats, Coyotes, Wolves, Cougars, Grizzly Bears and, in at least one case, a Golden Eagle, but their most commonly documented predators are other Black Bears. Of nine cases of cannibalism reported by Lynn Rogers (including three from Alberta), most involved large males taking cubs or yearlings, but in one case a mother killed and ate a cub from another litter and in another case a male killed and ate an adult female and her two cubs at their den. Helen Davis and Alton Harestad recorded similar observations on Vancouver Island in the early 1990s. Biologists debate about whether cannibalism is an extension of foraging behaviour, a strategy for reducing competition for food or space, or a strategy to gain genetic advantage (individual bears maximizing their own genes in the population by reducing the reproductive contributions of others). Davis and Harestad speculated that cannibalized adult females probably die trying to protect their cubs from predatory males; once dead, they become food.

Predation on subadult or older Black Bears by other species is largely limited to Grey Wolves and Grizzly Bears, and most documented cases have been at winter dens. Bears killed and eaten by wolves at dens include an adult female and her newborn cubs in Minnesota (nine wolves involved), an adult female in Alberta (eight wolves), and three separate bears of undetermined sex in Manitoba (one subadult killed by five wolves and another by up to 10, and one adult killed by two wolves). Predation on denning Black Bears by Grizzly Bears has been documented in Alaska (adult female killed by apparent adult male Grizzly), Montana (yearling killed by adult female or young male Grizzly), and Alberta (two cubs killed by adult female Grizzly accompanied by two yearlings). The cases in Alaska and Montana appeared to be opportunistic, the predators happening to pass by close enough to detect the dens and occupants. As reported by Ian Ross and his associates, the Alberta incident involved the Grizzly Bear family tracking the Black Bear family in the snow to their den where they killed and ate the cubs; the mother apparently escaped.

Black Bears also die from natural accidents. While observing migrating salmon at Moricetown Canyon, on the Bulkley River, Mason Stucklberger of Smithers noticed a Black Bear swimming across the river above a falls. Apparently, the bear miscalculated the current and was swept over the falls into the turbulent waters below. Several minutes later, Stucklberger saw the bear's body floating under a bridge about 100 metres downstream from the falls.

The most common human-caused death of Black Bears throughout their North American range are from vehicle collisions on roadways and railways, problem wildlife control activities, and hunter harvest. Those killed by trains were often attracted to the rail right-of-way by food sources such as train-killed ungulates or spilled grain.

Once a Black Bear has reached adulthood, it has a good chance of surviving for another 15 to 20 years, especially if it obtains a stable home range in good habitat. Among 167 Black Bears captured by biologist John Beecham on a study area in Idaho, the oldest female was 24 years and the oldest male was 20 years. The maximum known age of a Black Bear in captivity is also 24 years.

Abundance

The Black Bear has successfully maintained numbers in the face of the expanding human population in North America; indeed, it has increased in many areas because of human-caused habitat changes that result in young forest stages and forage-producing openings. As reported by Ray Demarchi and Carol Hartwig, most of British Columbia has habitats of at least moderate capability for the species, and the currently stable or increasing provincial population is estimated by the Ministry of the Environment at between 120,000 and 160,000 animals.

The highest density recorded in western North America, 12 to 15 bears per km² over a three-year period, was on a small coastal island in Washington. As described by biologists Fred Lindzey and Charles Meslow, the lush habitat on that island had been enhanced by logging ten or more years earlier and supported the year-round needs and continuing reproduction of the local bear population during the study period. In contrast, intensively studied (larger) inland areas with good bear habitat have generally supported less than 1 bear per km² (0.2 in south-central Alaska, 0.4 in western Alberta, 0.4 to 0.5 in northern Montana, 0.8 in central Idaho).

In some years Black Bear sightings increase dramatically and people tend to interpret that as a population increase. But those sighting increases are usually due to widespead failure of important food supplies, usually berry crops in the fall, forcing bears to find alternative foods, often near human settlements (farms, orchards, suburban areas, garbage dumps) where they are more likely to be seen. In years of natural food shortage, they may also forage longer during the day, making them more conspicuous.

Human Uses

The Black Bear's flesh is quite palatable, and the species has long been hunted both for food and for its thick, warm pelt. Over the course of history, the hides have been used for rugs, sleeping robes, winter clothing, as covering for furniture, and as lining for sleighs. Most current use is in taxidermy but Black Bear fur is still used to make the distinctive hats (busbies) worn by members of the Queen's Royal Guard at Buckingham Palace.

First Peoples traditionally used the fat from fall and early winter bear carcasses as an ingredient in pemmican, a base for paint and cosmetics when mixed with natural pigments, and a skin-softening lotion. North American pioneers used rendered bear fat for a variety of purposes, including soap and candle-making, waterproofing of leather boots, lubrication of machine parts, rust-proofing metal, hair pomade ("greasy kid stuff"), hair restorer (akin to snake oil) and making pie-crusts (to this day some rural chefs consider bear fat the best shortening). In recent decades, the Asian market for bear parts such as paws and gall bladders has attracted international attention. Although the concern was for conservation of the Asian species of Black Bear, the (legal) sale of such parts from look-alike species (all other bears) was banned in many jurisdictions, including British Columbia.

Because these fascinating animals are large, conspicuous, and often active during daylight hours, they are popular subjects for nature viewing. Many people, myself included, know of locations frequented by bears in certain seasons and make a practice of periodically visiting those areas to observe and photograph the animals.

Taxonomy

Based mostly on skull morphology, 16 Black Bear subspecies are recognized and 5 occur in British Columbia. Although pelage colour was used as a taxonomic trait in the original descriptions of western races such as *kermodei* (the white Kermode Bear) and *emmonsi* (Blue or Glacier Bear), E. Raymond Hall affirmed that the various colour phases are not useful for defining subspecies. Recent genetic studies by Karen Stone, Joe Cook, and Ashley Byun and her colleagues have shown little support for separating BC Black Bears into five subspecies. They found two distinct genetic groups, one continental and the other coastal. The coastal group includes the currently designated subspecies *altifrontalis*, *carlottae*, *kermodei* and *vancouverensis*. The continental group includes the BC subspecies *cinnamomum*. But there are inconsistencies in distribution with Black Bears of the coastal lineage found as far east as the Rocky Mountains and animals of the continental group ranging as far west as the Coast Mountains. Further taxonomic work is clearly needed but, for the present, the five currently accepted Black Bear subspecies in BC are as follows:

Ursus americanus altifrontalis Elliot – a coastal form, characterized by a dish-shaped forehead and large teeth, ranging from Oregon to Bella Coola and Tweedsmuir Provincial Park.

Ursus americanus carlottae Osgood – restricted to the Queen Charlotte Islands; with its geographic isolation and distinctive structure (large size, massive skull and heavy teeth) this population probably has the most legitimate claim to being a distinct subspecies of the five listed.

Ursus americanus cinnamomum Audubon and Bachman – Alberta, Montana, Wyoming, Idaho, eastern Washington and Oregon, and most of mainland BC east of the coastal mountain ranges. The precise boundary that separates the range of this subspecies from *altifrontalis* and *kermodei* is not clear. It is distinguished from *altifrontalis* by its larger skull and less dish-shaped forehead.

Ursus americanus kermodei Hornaday – restricted to the mainland coast of BC, where it occurs from Burke Channel to the Nass River and on many adjacent coastal islands (e.g., Gribbell, Hawkesbury, Pitt, Pooley, Princess Royal, Roderick and Yeo). Because of the high incidence of the charismatic white colour phase (up to 12% in some populations), it has attracted considerable attention. It is a weakly defined subspecies based on skull traits, but a genetics study by H.D. Marshall and K. Ritland showed that it is strongly differentiated from *vancouveri*. Its relationship to the adjacent mainland subspecies (*altifrontalis* and *cinnamomum*) was not studied.

Ursus americanus vancouveri Hall – restricted to Vancouver Island and larger adjacent islands, this subspecies is distinguished from adjacent mainland forms by its larger, more robust skull. The precise boundary that separates *vancouveri* from island forms of *altifrontalis* near the mainland has not been determined.

Conservation Status and Management

As in most areas throughout its range, the Black Bear is of no conservation concern in BC. The primary concerns and management issues relating to the species in this province, as in many other jurisdictions, involve human-wildlife conflicts. Most of those conflicts centre around the bears' choices of foods, which often include items belonging to or desired by humans, and their means of obtaining those foods, which often result in damage to human property and sometimes compromise human safety. The list of documented transgressions is long, including the killing of livestock, a variety of agricultural, horticultural and apicultural depredations, raids at campsites that damage vehicles, boats and other equipment, and cottage/home offenses such as breaking-and-entering and pantry larceny. BC Conservation Officers commonly log 8000 to 10,000 complaints annually, attending an average of about 4000 of those and destroying or relocating about 1200 bears in the process (to a peak of 1728 in 1998). Many communities have programs (Bear Aware, Bear Smart) directed at reducing bear problems in urban and suburban areas by minimizing the availability of attractants. Those programs have merit, but they will not completely solve the problem as long as high bear populations continue to produce non-resident and dispersing individuals, particularly during autumn in years of poor berry production or low fish runs.

Damage to commercial trees by Black Bears feeding on sapwood in the spring has been reported in Montana, California, Oregon, Washington and BC. Authorities in Washington have expressed the greatest concern, stating that one bear can girdle 60 to 70 trees per day, resulting in millions of dollars in commercial losses annually. As reported by researcher G. Ziegltrum, the use of supplemental feeding to lure bears away from trees may be a cost-effective and publicly favoured solution. That conclusion is based on preliminary work involving the distribution of 450,000 tons of commercial feed pellets at 900 feed stations over a seven-year period. In BC, Tom Sullivan evaluated damage to second-growth forest stands on the lower mainland coast. He found that the highest incidence of damage was to Western Redcedar, which was a minor component of the new forest in that area. Those findings apparently did not stimulate any major concerns or protection initiatives in that area, and I am not aware of any in other areas of the province.

In some areas, Black Bears are of concern to wildlife managers because of depredations on young ungulates and the resulting impact on populations. This is a large and controversial subject that I can only touch upon here. For example, biologists have long known that Black Bears kill Moose calves but the extent was not possible to determine until the advent of radio-tracking. On the Yukon Flats in interior Alaska, only 17 of 62 radio-collared calves (28%) survived to 14 weeks of age and 92 per cent of the losses were due to predation, including 45 per cent to Black Bears and 39 per cent to Grizzly Bears. In another Alaska study, on the Kenai Peninsula, only 42 per cent of radio-collared calves survived to the end of July (8 to 9 weeks old). Most of the losses (23 of 27) were to predation, mostly by Black Bears (16 of 23, or 70%). There have been no comparable studies in BC. Provincial ungulate specialist Ian Hatter noted that of 28 radio-collared Black-tailed Deer fawns in a Vancouver Island study, only one was killed by a Black Bear.

It is sometimes difficult to reconcile that the "cute" animal one observes calmly grazing on dandelions in the roadside ditch or comically scratching its behind on a stump is a powerful and dangerous predator. BC government records, provided by Matt Austin, show 64 incidents of Black Bear attacks on humans between 1985 and 2003 (average of 3.4 per year); the bears injured 56 people and killed 8. University of Calgary professor Steve Herrero has done considerable research on this subject and has written an excellent book, *Bear Attacks: Their Causes and Avoidance*. According to Dr Herrero, Black Bear attacks on humans are usually predatory rather than defensive or territorial, and often involve bears that have become habituated to humans in parks and campsite areas. Of the thousands of Black Bear - Human encounters that occur each year throughout North America, the vast majority are completely benign (usually giving a measure of enjoyment to the humans involved). Nevertheless, people must be aware and respectful of the potential danger, and heed the time-honoured wisdom of experienced outdoors people who say, "You never know about bears...".

Remarks

Some researchers have been reluctant to label the winter denning of Black Bears as hibernation, preferring terms such as "torpor" or "dormancy". But based on evidence such as reduced metabolic rate and body temperature and greatly reduced heart rate (as low as 8 to 10 beats per minute), physiologist Edgar Folk and associates have affirmed that hibernation is fully applicable. One of the features of hibernating bears that has confused the issue is that they are easily woken up and can become active immediately. That feature has inspired my admiration for the people whose scientific curiosity compels them to crawl into dens and take rectal temperatures of sleeping bears.

Whatever it may be called, the bear winter denning phenomenon verges on miraculous. For the duration of the denning period – up to seven months – Black Bears do not eat, drink, urinate or defecate, yet they suffer no deleterious changes in blood chemistry and related life function, and do not accumulate any toxic waste. Accordingly, denning bears have been the subjects of medical research in some of the most prestigious institutions such as the Mayo Clinic in Minnesota, with particular reference to problems with obesity, kidney diseases and patient care both during and after surgery.

The importance of Black Bears in local ecosystems extends beyond the plus/minus aspect of their consumption of plants and other animals. They are well known dispersers of berry seeds (and probably the seeds of other plants), and one study found that seeds passed through a bear's digestive tract germinated at significantly higher rates than those that didn't. On the coast, Tom Reimchen and others have produced evidence that the nutrients transported to upland habitats by bears in feces and salmon carcasses significantly increases plant growth and general productivity in those areas.

A Black Bear given to the London Zoo by a Canadian soldier, and named "Winnie" after the soldier's home town of Winnipeg, was the inspiration for A.A. Milne's beloved storybook character, Winnie the Pooh.

Selected References: Bull et al. 2001; Byun et al. 1997; Cowan 1938a; Davis 1996; Davis and Harestad 1996; Demarchi and Hartwig 2000; Hatler 1972; Herrero 1972, 1985; Hobson et al. 2000; Holcraft and Herrero 1991; Jacoby et al. 1999; Jonkel and Cowan 1971; Lariviere 2001; Marshall and Ritland 2002; Nelson et al. 1973; Reimchen 1998, 2000; Rogers 1987; Rutherglen and Herbison 1977; Schwartz and Franzmann 1991; Stone and Cook 2000; Sullivan 1993; Wright 1910.

Grizzly Bear   *Ursus arctos*

Other Common Names: Barren Ground Grizzly, Big Brown Bear, Brown Bear, Grizzly, Interior Grizzly, Kodiak Bear, Moccasin Joe, Old Ephraim, Silvertip; *Ours Brun*. Provincial Species Code: M-URAR.

Description
The Grizzly Bear is the largest carnivore in British Columbia with some individuals as bulky as domestic cattle. As is typical of bears, the Grizzly has a compact, stocky torso with thick legs, large flat-soled feet, and a tail so small that it is not apparent. It has a broad, massive head, short rounded ears and small eyes. A ruff of longer hairs around the neck often makes the head look even larger than it is. The facial profile, from forehead to snout, tends to be concave

(dished). The pelage is thick, heavy and variable in colour. Most Grizzly Bears are some shade of brown, but they range from pale blonde to black (the latter rare, especially in BC). The legs and belly are usually darker than the rest of the body, and most individuals have a lighter wash (frosting or silver-tipping) across the head, neck and shoulders. The grizzled effect, from the pale-tipped guard hairs, is the basis for the species' common name. Nevertheless, it varies greatly in extent among Grizzly Bears and is absent on some, especially along the coast and among large males in the interior. Adapted for digging, the Grizzly Bear has long, conspicuous front claws and its shoulder musculature forms a prominent hump on the back. The claws are longest in the early spring, (wearing down from digging over the summer), and may be very light coloured (ivory to tan, see figure 48).

Grizzly Bear size is difficult to characterize in terms of averages because there is great variation individually, locally and regionally. Adult males are often more than twice as large as adult females in all areas, and both sexes take six years or more to attain full size. The weights of individual bears may also fluctuate seasonally by 25 per cent or more, as related to deposition of fat for winter denning. Regional size differences probably have some genetic basis, but mostly reflect differences in nutritional factors, such as the availability of protein and the length of the plant growing season. The largest Grizzly Bears live in coastal habitats where salmon are available in the fall, and the smallest inhabit northern interior areas with short growing seasons and no regularly available source of autumn protein. Some average adult weights from representative areas at those two geographic extremes are: Alaska Peninsula, males 389 kg and females 207 kg; Southwest Yukon males 139 kg, females 95 kg. There are unconfirmed records of coastal Grizzly Bears weighing more than 600 kg, but the heaviest living specimen weighed by researchers was 442 kg (in Alaska). Occasional individuals grow much larger than the local average – a cattle-

Figure 48. Grizzly Bear in Glacier National Park, BC, showing light claws in the spring.

killing Grizzly Bear from BC's central interior reportedly weighed 459 kg when killed by Conservation Officers in October 2001; he now occupies a glass case as a mounted specimen in the Smithers airport.

Measurements:

total length (mm):
male:	1876 (1530-2280) n=84
female:	1659 (1400-1850) n=82

tail vertebrae (mm):
male:	106 (65-160) n=7
female:	100 (80-130) n=3

hind foot (mm):*
male:	239 (215-263) n=8
female:	208 (190-245) n=7

ear (mm):
male:	130 (130-130) n=2
female:	127 (120-130) n=6

weight (kg):
male:	168.6 (73-329) n=82
female:	108.8 (59-175) n=85

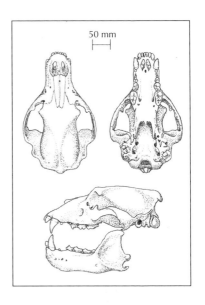

50 mm

Dental Formula:
incisors:	3/3
canines:	1/1
premolars:	4/4 (1 or more often missing)
molars:	2/3

Identification

The most common difficulty in identifying a Grizzly Bear is in distinguishing it from the American Black Bear. Although useful as first clues, neither size nor colour are sufficient for that determination. A large male Grizzly Bear will rarely be mistaken for anything

---

* These hind foot measurements are not comparable to those for other species, including Black Bears. All were obtained from Grizzly Bear researchers who use a modified hind-foot measurement extending from the rear of the hind pad to the front of the longest toe pad, not including claw or heel.

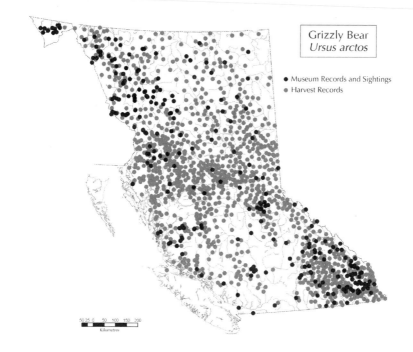

Grizzly Bear
Ursus arctos

● Museum Records and Sightings
● Harvest Records

50 25 0   50   100   150   200
Kilometres

else, but the variation in size within both species results in overlap between them. Adult male Black Bears are larger than many female and young male Grizzly Bears. A black bear is very likely a Black Bear, since few Grizzly Bears appear that dark, even when wet, but both species exhibit a full range of base colours, from pale blonde to dark brown. Grizzled or silver-tipped hairs on the head, neck, shoulders and sometimes the back and sides are strong clues, but some Grizzly Bears show few or none. Figures 13 and 14, (page 45) show several features that differ between the two species. Grizzly Bears have less prominent ears, often have a ruff of longer neck hairs that does not occur on Black Bears and, when viewed from the side, the profile from between the eyes to the nose along the top of the snout tends to be dished (concave) on Grizzly Bears and straight or slightly raised (convex) on Black Bears. Unfortunately, those distinctions are more evident when both species are present for comparison than when only one is being observed. More reliable distinguishing features are the Grizzly Bear's prominent hump over the shoulders and its longer front claws (up to 10 cm); Black Bears have no hump and fairly inconspicuous claws (less than 4 cm) (see figure 45).

Distribution and Habitat

The Grizzly Bear ranges across all of northern Eurasia, from scattered small sub-populations in the mountains of the west (Scandinavia, Italy, Spain, France) to an essentially continuous distribution in the vast boreal forest and tundra habitats of Siberia and onward to coastal mountains and islands along the Bering Sea and Sea of Okhotsk. The Eurasian distribution includes the Himalayas in Tibet, where wandering bears likely left tracks attributed to the yeti (abominable snowman), and extends as far southeast as the island of Hokkaido in Japan. In North America, the Grizzly Bear has never occurred in the eastern half of the continent. Its original range extended from Alaska to central Mexico and included the western prairies to about the Red River valley. The demise of the Grizzly Bear on the prairies likely parallelled that of the Bison, and its disappearance over much of the rest of its former range south of the Canadian border was largely an adjunct to human expansion and settlement in those areas.

The Grizzly Bear's distribution in North America now includes Alaska, Yukon Territory, northern Northwest Territory and Nunavut to the Keewatin District in the north, most of British Columbia, western Alberta, and southern extensions of the Rocky and Columbia mountains in Montana, Idaho and Washington. It also occurs in a few semi-isolated areas, including the Swan Hills of Alberta and Yellowstone National Park in Montana and Wyoming. In BC, Grizzly Bears still occupy 85 per cent of their pristine range, regularly occurring in suitable habitats over all but the most heavily developed and human-settled areas (Lower Mainland, south Okanagan). They have never occurred on the Queen Charlotte Islands and, although there are records from several locations along the mainland coast, there are few from adjacent near-shore islands. The Grizzly Bear is not considered resident on Vancouver Island, although a marauding male was recently (August 2003) shot in a human conflict situation near Port Hardy. Wandering individuals may occasionally swim across Johnstone Strait, a distance of little more than 3 km with selective island-hopping.

The Grizzly Bear in BC is typically an animal of the mountains, regularly occurring on the highest vegetated ridges and peaks. Most individuals use a number of different habitats during the year, moving from one to the next (often up and down slope) in response to seasonal changes in the quality and quantity of available foods. Biologists Bruce McLellan and Fred Hovey identified two distinct habitat-use patterns among radio-collared Grizzly

Bears they monitored in the Flathead River drainage of southeastern BC, characterizing one as "elevational migrant" and the other as "mountain resident". The elevational migrants emerged from high-elevation dens and descended to forage primarily in valley-bottom riparian habitats for several weeks in the spring, then moved upslope in summer to feed on berries in shrub and open forest habitats produced by wildfires (burns). At the end of the berry season, they returned to lower elevation, including riparian habitats, before again moving upslope for denning. In contrast, the mountain resident bears did not move down to the valley bottom, foraging in spring through fall in various upland habitats, particularly avalanche chutes and burns.

Grizzly Bears use forest cover for protection from weather and to some extent for security, to keep out of sight of potential predators and competitors, such as humans and other bears. But they are poor tree climbers and not as strongly associated with forest cover as are the more arboreal Black Bears, which find security and do some of their foraging high in trees. As outlined by University of Calgary professor Steve Herrero, Grizzly Bears evolved to exploit the open (treeless) landscapes left behind by the receding Pleistocene glaciers, and may often be found in open country such as the areas near present-day alpine glaciers far above timberline. When below timberline, they forage mostly in non-treed habitat patches within the overall forest matrix. Riparian habitats are among the most important of those in areas where they have not been occupied or extensively disturbed by humans; and avalanche chutes are extremely important during spring and early summer in most mountainous areas, because they shed snow and green-up early, usually retain sufficient moisture to support the growth of favoured food plants, and prevent the establishment of trees that would eventually shade out those plants.

Other open or semi-open habitats used by Grizzly Bears for foraging in or adjacent to forested areas include burns, wetlands, meadows, estuaries, subalpine parkland, and areas with gaps in the forest canopy caused by tree mortality from factors such as insect infestations, floods and disease. They also forage in logging cutblocks, pastures, croplands, roadside ditches and berms, and other man-made openings when those areas provide food of sufficient quality, or when natural foods are in short supply. Grizzlies rarely forage in large tracts of dense forest – from the pole/sapling stage in regenerating cutblocks and burns to the mature stage before trees start to die – because they produce very little food.

A Grizzly Bear's year is divided into two distinct phases, a spring-through-fall activity period, when the animal is focused on foraging, and a winter-dormant (hibernation) period, when foraging is no longer possible and the animal remains in a den. The duration of each varies regionally and individually, but bears in most areas of BC obtain their entire year's sustenance in as little as six to seven months, emerging from dens as early as mid April and returning to them by late October in the north and as late as the second half of November in the south or in mild years.

The winter den is an important component of the Grizzly Bear's habitat, providing cover and protection for half the year or more, and as the location where the young are born. Since the advent of radio-tracking in the early 1970s, biologists have found hundreds of Grizzly Bear dens. Most have been at relatively high elevation in local terms (near timberline) and were excavated by the bears, although some den in natural rock caves. Excavated dens are often at the bases of trees or shrubs with the roots providing structural support to the ceiling (figure 49), but I have seen dens in northern BC that were on alpine slopes or ridges some distance from the nearest woody vegetation.

The use of high-elevation sites likely ensures a stable snow cover for den insulation while minimizing the risk of being covered

by avalanches or flooded by mid-winter thaws. But bears adapt well to the local environment, so high-elevation denning is not universal. A young adult female, radio-collared and monitored by biologists Mike Demarchi and Steve Johnson in the Nass Ranges of northwest-

Figure 49. David Hatler emerging from a Grizzly Bear den at the Nass River drainage in northern BC. The entrance tunnel is 56 cm wide, 66 cm high and 40 cm long; inside, the roomy nest chamber is 150 x 114 x 147 cm. Smaller excavations nearby suggest that the bear tested other sites before settling on this one.

ern BC, denned in small timber patches at low elevation (306 and 311 metres) in two successive winters. Farther north, a subadult male that I radio-collared in Spatsizi Provincial Park denned in an aspen copse on a sidehill about 20 metres above the Stikine River. Denning at mid elevations also occurs, as in an interesting case near Prince George described by biologist Grant Hazelwood: on April 15, 1980, a forest worker was treed for five hours by a female Grizzly Bear after unknowingly passing within 40 metres of its den containing cubs; the den was in thick spruce-balsam forest at 975 metres.

An excavated den generally consists of a relatively narrow entrance tunnel leading to a larger nest chamber. Dimensions vary considerably, probably depending upon the size of the bear as well as the ease of digging, but the depth from the entrance to the back of the chamber is usually about 2 metres; one in the Nass study area was 3.1 metres. Grizzly Bears expend considerable energy constructing a den, removing a large volume of soil, rocks and roots. At a high-elevation den inspected by helicopter pilot Darryl Hodson in the Ogilvie Mountains, Yukon, the excavated debris at the entrance included boulders weighing 10 to 15 kg, and the bear constructing the den near Prince George described by Hazelwood, had chewed off 12 roots up to 9.5 cm in diameter to enlarge the entrance. Denning Grizzly Bears usually line the nest chamber with vegetation, especially conifer branches (the original "bough bed"), and other materials. The bedding in the large Nass den mentioned above consisted of hemlock branches, small shrubs and Bunchberry, while the den described by Hazelwood consisted entirely of spruce branches. Some of the spruce branches had been chewed off a small tree at a height of 3 metres, suggesting that the bear had emerged and added more bedding in winter when the snow was deep enough to allow that reach.

Known Grizzly Bear dens have rarely been used in more than one year, suggesting that at least in those study areas, denning habitat is not limiting.

Natural History

Feeding Ecology
Provincial bear biologist Tony Hamilton once remarked that there are only two kinds of organisms – Grizzly Bears and Grizzly Bear food. Decades earlier, the naturalist John Muir said it somewhat

differently: "To him [the Grizzly Bear] almost everything is food except granite."

To meet their nutritional requirements in the half year or so when they are outside the winter den, Grizzly Bears consume a large variety of plants and animals. They eat animal protein whenever possible, but the bulk of their food throughout the year is plant material. Based on chemical analysis of hairs, Canadian Wildlife Service biologist K. Hobson and his co-workers calculated that plants comprised 91 per cent of Grizzly Bears' diets in southeastern BC. Although bears have physical adaptations for such a diet, including flat molars for grinding and a longer digestive tract than other carnivores, they do not digest vegetable matter efficiently and, to compensate, they must eat large quantities.

Food habits studies from a number of areas indicate that green vegetation is the primary plant component consumed by Grizzly Bears in the spring and early summer, and berries predominate in late summer and fall. A partial list of bear foods in British Columbia, compiled by Brian Fuhr and Dennis Demarchi, included 36 species and 26 genera of plants. The most commonly eaten vegetation in spring and summer are the leaves and stems of plants that grow in profusion (horsetails, Arrow-leaved Groundsel, grasses, sedges, dandelions, legumes such as pea-vines, vetches and clovers), large succulent forbs such as Cow-parsnip and Skunk Cabbage, the corms of Glacier Lily, and the starchy roots of sweet-vetches. Possibly after watching Grizzly Bears dig in sweet-vetch patches, First Nations Peoples harvested Spring Beauty corms, which has been referred to as "Indian potato". A bear's diet varies with what is available at the time. In the Sheslay River area of northwestern BC in July 1987, I saw several scats that consisted entirely of willow catkins. Although I had not seen that elsewhere, Canadian Wildlife Service biologist Art Pearson found willow catkins to be a regular spring food item for Grizzly Bears in the southern Yukon.

Despite their predominance in the diet for the first 2 to 2.5 months of a bear's activity season, the above plant items apparently contribute little to the fat deposits that the bear requires for successful overwintering. Most of the annual weight gain and fat storage is obtained from the fall diet, which is primarily berries in many inland areas. Berries of the *Vaccinnium* species (huckleberries, blueberries, Lingonberries), Soopolallie, Devil's Club and Highbush-cranberry are nutritious and digestible, and often occur in the quantities that bears require. In terms of quantity, the berry-laden scats of Grizzly Bears sometimes inspire awe (figure 50). Keith Koontz, a graduate student at the University of Alaska, counted 10,000 Soopolallie berries in a single Grizzly Bear scat. Grizzly Bears also regularly eat raspberries, elderberries, Mountain Ash berries, currants, Saskatoon berries, Choke Cherries, Crowberries and Kinnikinnick berries. In

Figure 50. A Grizzly Bear scat compared to a man's size-11 boot.

some areas, the seeds of Whitebark Pine are also a staple fall food.

My most memorable observation of a bear eating berries took place on a sunny day in mid October high on the Hudson Bay Mountain Block, overlooking Smithers. For 20 minutes I watched as a mature male Grizzly Bear ate cluster after cluster of Mountain Ash berries. The bear wrapped his foreleg around a clump of branches, pulled it toward him in a motion that compressed the branches into a smaller package, then shovelled the berries into his mouth with the other paw. He took no more than 20 to 30 seconds to strip each bush before moving on to the next one, until he had finished off most of the berries in view.

Although animal matter constitutes only a small portion of the annual food volume consumed by Grizzly Bears, there is evidence that it is of considerable importance. Biologist Grant Hilderbrand and associates demonstrated that the populations featuring the largest bears, largest litters and highest densities are those with regular access to animal protein, especially salmon. Although Grizzly Bears catch live salmon, they also scavenge dead ones. In a study in Alaska, 70 per cent of dead salmon marked by researchers were fed on by bears within a week. The impact of Grizzly Bear predation on salmon varies annually with salmon abundance. On a small stream near Bristol Bay, Alaska, Gregory Ruggerone and his co-workers found that bears ate 92 per cent (466 of 505) of Sockeye Salmon in a year with a poor run, compared to only 16 per cent (2569 of 15,631) during the year with the largest run recorded. Consumption patterns also vary: bears consume most of each carcass during years of low abundance, but often feed selectively when numbers are high, maximizing nutritional gains by eating only the energy-rich body parts such as brains and roe. In some areas, Grizzly Bears dine heavily on fishes other than salmon, but almost always in areas of concentration associated with spawning (e.g., Cutthroat Trout in Yellowstone National Park).

Grizzly Bears also eat ungulates (taken both by predation and as carrion), rodents (especially marmots and ground squirrels) and insects. Of ungulates, bears most often prey on calves or fawns in the first 4 to 6 weeks of life, but I documented three cases of Grizzlies killing adult Moose (two males and one female) in the Spatsizi Wilderness Park area. Both male Moose appeared to have been surprised, the bears lying in wait along a game trail, and the female (a radio-collared animal I was monitoring) was either disadvantaged by the late stages of pregnancy or died defending a new calf.

Grizzly Bears are not always successful in digging marmots and ground squirrels from their burrows, but the number of attempts and the effort expended in some areas suggests that the occasional rewards are at least satisfying and probably nutritionally important. A large male I watched on a steep slope in Alaska excavated three large holes, removing half a truck-load of soil and rock in the process, with no success. He dug a fourth hole and an Arctic Ground Squirrel emerged and ran down the slope. The bear turned and started to give chase, at which point the squirrel doubled back between the bear's front legs. Snapping at the squirrel as it disappeared under his chest, the bear lost balance and did three complete forward-rolls down the slope before regaining his feet. He returned to dig out more of the burrow but caught nothing while I watched.

Insects are small but rich in protein and some species live in sufficient concentrations to provide nutritional benefits to Grizzly Bears, especially early in the season or in years when other protein sources may be scarce. Among the most commonly consumed insects in all areas are colonial species such as ants and wasps, but other insect concentrations may also be exploited. In the southern Rocky Mountains, from the Absaroka Range in Wyoming to at least Glacier National Park, Montana, Grizzly Bears feast on Army Cutworm Moths, which aggregate every summer in rock talus in particular alpine areas. Bears have been seen foraging at those sites for 20 to 30 days consecutively, and the contents of bear scats from one Glacier Park site were 95 per cent moth remains, by volume.

While ungulates, salmon, rodents and insects constitute the bulk of protein consumption in most areas, Grizzly Bears opportunistically take whatever else they come across. They sometimes prey on denning Black Bears (see page 125), and Yukon biologists Bob Hayes and A. Baer reported that a Grizzly Bear had dug out a wolf den and eaten all but the heads of four pups. Grizzlies have also been implicated as significant predators of Canada Goose nests on the Copper River Delta, Alaska, and biologists Paul Henson and Todd Grant observed a Grizzly Bear preying upon a Trumpeter Swan nest despite the valiant defence of one of the adult swans. At Katmai National Monument, Alaska, biologist Tom Smith and Steven Partridge documented bears regularly digging for Pacific Razor Clams and Soft-shell Clams, and feeding incidentally on other clams, mussels, barnacles, marine worms and intertidal fishes. They judged that the clamming success rate (about one clam per minute) was nutritionally significant (better than plant feeding) for small-bodied bears, such as adult females, cubs and young males, but not for adult males.

As a final demonstration of foraging opportunism, David Mattson reported that Grizzly Bears digging for Northern Pocket Gophers in Yellowstone National Park often ate both the gopher and its food cache of starchy roots and other plant matter – clearly the Ursine equivalent of a "meat and potatoes" meal.

Home Range and Social Behaviour

Grizzly Bears are generally solitary, except for brief male-female pairings (up to a week) during the mating season, and females with dependent young. Individuals carry out their annual activities within distinct home ranges, that often overlap the home ranges of others regardless of sex or age. With only half a year to satisfy their annual nutritional needs, a bear's time is better spent foraging than fighting and since home ranges are large and feeding sites within them usually dispersed widely, attempting to defend them would be impractical. The marking of "bear trees" by rubbing, biting and clawing, mostly undertaken by adult males, may have a social (territorial warning) function, but that has not yet been clearly demonstrated. The social system of Grizzly Bears appears to involve dominant animals (mostly adult males) foraging wherever and whenever they want, and other bears avoiding potentially dangerous contact with them by foraging in other places or at different times. In some cases the avoidance distances may be small, enabling several bears to gather in relatively small areas to exploit concentrated food sources such as productive berry patches, salmon runs or garbage dumps.

Home-range sizes vary by sex, age, area and habitat conditions (especially food supply); they probably also relate to population density. The home ranges of males are generally larger than those of females and can be enormous, 2000 km^2 or more for adults in some arctic and interior temperate ecosystems, but averages in most areas are usually on the order of 500 to 900 km^2. In or near British Columbia, the following are some average annual home-range sizes (km^2) calculated for adult males and adult females, respectively: 446 and 200 (Flathead); 297 and 79 (Revelstoke area); 768 and 125 (northwestern Montana); 916 and 224 (Jasper National Park). The impact of food supply on home ranges can be inferred from experience in the Yellowstone National Park area where, for many years, Grizzly Bears had open access to and regularly congregated in several garbage dumps. The home ranges of radio-collared adult males and females, respectively, averaged 233 km^2 and 73 km^2 in the decade prior to dump closure, but increased to 828 km^2 and 324 km^2 in the decade afterward. A number of the dump-habituated bears were killed in nuisance situations after dump closure, thus the increase in home-range size may also have reflected decreased population density.

Individual bears do not remain in the same area year after year, so the total area covered over a period of years (total home range) is much larger than the figures given above. Four radio-collared males monitored by Bonnie Blanchard and Richard Knight in the Yellowstone Ecosystem averaged 3757 km^2 (maximum 5374 km^2) in total home range over four to five years. Females showed greater home-range fidelity in successive years, but six monitored for five years or more had average total home ranges of 897 km^2 (maximum 1391 km^2).

Activity and Movements

Grizzly Bears tend to be most active from an hour or two before dawn to mid-morning, and from late afternoon until a few hours after dark. But that pattern may vary seasonally and annually, depending mostly upon food supply and individually upon social status. In times when food is scarce all bears may need to extend their foraging periods, the smaller ones adopting non-crepuscular feeding patterns to avoid conflicts with dominant adults. Grizzly Bears spend 8 to 12 hours daily in foraging and associated travel, but some individuals are active for up to 20 hours in the fall.

In south-central Alaska, Warren Ballard and his co-workers documented radio-collared bears travelling up to 43 km in a day, but averaging 7 to 8 km for both sexes. Grizzly Bears are highly mobile and capable of covering great distances. But when a bear has located a concentrated food source, such as a particularly productive berry patch or an ungulate carcass, it may bed nearby and move little for several days. Young bears, especially males dispersing from their natal ranges make some of the longest natural movements. Robert Gau and his associates radio-tracked eight subadult males in the central Canadian Arctic (Northwest Territories) that travelled 201 to 779 km (average 431 km), in an average of 39 days. In contrast, dispersals of radio-collared Grizzlies monitored by Bruce McLellan and Fred Hovey in the Flathead (BC) study area were much shorter and more leisurely: 18 males dispersed an average of 30 km (maximum 67 km), and for most that involved a gradual expansion of range over months or even years rather than short-term directional movements; 12 females averaged 9.8 km (maximum 20 km), and one took four years. The authors concluded that the Grizzly Bear social system, which involves numerous overlapping home ranges for both sexes, may not require long-distance dispersal to avoid inbreeding or competition with relatives.

When a Grizzly Bear conflicts with humans, the usual public preference and expectation is that the bear be translocated to another area. In a study assessing that practice, carried out by Sterling Miller and Warren Ballard in south-central Alaska, at least 5 of 9 adult males and 7 of 11 adult females returned from remote release locations to their original home ranges. The distances of those returns averaged 211 km for males and 189 km for females, with a maximum of 258 km. One male was translocated and returned twice, the first time from 145 km and the second from 215 km. Of nine cubs moved with their mothers, six disappeared within 36 days of release, probably killed by resident males in the new locations. Among nine bears (eight adults) not known to have returned, at least six were headed in the right direction when last contacted (before their radio transmitters failed), and some may have returned undetected. Miller and Ballard could not explain the bears' homing method but they concluded that translocation of problem Grizzly Bears has a high probability of failure.

Reproduction

Based on observations of courting pairs (adult males and females interacting or travelling together), the breeding season of Grizzly Bears appears to be from mid May to mid July, but most actual mating occurs in June. Grizzly Bears court in the most basic sense of the word. A male does not attempt to win the affection of a female, just her consent. In the courting I have observed the attraction is very much one-sided. The male follows the female wherever she goes, lies down nearby when she lies down, gets up when she gets up, and follows again. Occasionally, he approaches too close and she rebuffs him, often vigorously with roars, cuffs and bites. Eventually she may become receptive, but none of my observations have ended with a successful mating. Other observers have reported a more abbreviated courtship with males taking a more assertive and dominating approach, particularly where bears concentrate at food sources such as salmon streams or garbage dumps. Males may mate with several females in one season and, particularly in areas of high density, females may breed with more than one male, and so produce a litter in which siblings have different fathers.

The age at which female Grizzly Bears become sexually mature varies both regionally and individually, and relates mostly to their growth and development as affected by the level of nutrition. In areas where a protein-rich food supply is available, such as in coastal areas where salmon spawning occurs, females may be ready to breed as young as 3.5 years, while those elsewhere may not become mature until they are 6 to 8 years old. In the Yellowstone Park area, the average age of first breeding changed from 4.5 years to 5.5 years after garbage dumps were closed to bear use. In or near BC, the average age at first breeding for females monitored by Bruce McLellan in the Flathead area was 5.5 years, as compared to 6.5 years in a study by Rob Wielgus and Fred Bunnell in the Selkirk Mountains of BC, Washington and Idaho. Males become sexually mature at 4 to 5 years, but are not full grown and cannot compete with other males for mating privileges for another 2 to 3 years.

Females give birth to cubs in the winter den in January or February after a gestation period lengthened by delayed implantation. Implantation occurs shortly after the female has denned, in about November, and the actual length of pregnancy is approximately 8 weeks. Evidence from several studies indicates that reproductive success (the number of cubs that emerge from the den in the spring) is strongly influenced by the level of fall nutrition. During a year of poor berry production or a failed salmon run, a pregnant female may enter the den with inadequate fat stores and fail to support the pregnancy or be unable to feed some or all of the litter for the full denning period. A delayed spring that keeps mother and cubs in the den longer can make survival even more difficult for infant bears.

Newborn Grizzly Bears are small and undeveloped; weighing about 400 grams, they are sparsely-haired, blind and able to move only feebly. The number of young born may often be higher, but the size of most emerging litters is 1 to 3, with averages that range from about 2 to 2.5 in most areas (long-term averages documented in BC include 2.3 for 31 litters in the Flathead area and 2.2 for 10 litters in the Selkirks). Litters larger than three are rare, but are more common in coastal areas where salmon are available. A female with six cubs was observed during a 1984 survey on the Alaska Peninsula and, in testimony to her maternal ability, all six were seen still alive and with her again a year later. I observed a female with four cubs on Level Mountain, northwestern BC, in October 1978, and am not aware of any larger BC litters.

All parental care is provided by the mother, and for Grizzly Bears that is an extended process. The cubs travel and forage with the mother for two full activity seasons, denning with her twice more after the winter of their birth. Families usually break up just before or during the following mating season (May or June), when the cubs are 2.5 years old; but some break up a year sooner or later, the cubs becoming independent at 1.5 or 3.5 years of age. Cubs nurse for several months after first spring emergence, but take more solid food as the summer progresses. A consequence of long maternal care is a correspondingly long breeding interval. During the first two summers, females with cubs apparently do not come into estrus, and therefore the time between pregnancies is usually three years or more. But if a female loses her litter before a mating season, she may breed sooner. Grizzly Bear mothers do not protect their cubs as Black Bears do, by sending them up a tree. Instead, they defend their cubs by aggression, at first using bluffs and threats but able and willing to follow through if required. Steve Herrero reported a mother Grizzly defending her yearling cub by attacking and eventually killing a large adult male, but then she died from injuries sustained during the fight.

Health and Mortality

Grizzly Bears are host to a variety of parasites, a few fleas and ticks externally, and mostly roundworms and tapeworms internally. Rates of occurrence are usually low and infestations light, and most parasites and diseases that have been identified in bears do not appear to be significant to populations. A possible exception is the roundworm Trichina Worm, which has been found at infestation rates of over 50 per cent in some Grizzly Bear populations. Researchers have documented larval occurrence in bear muscle tissue at densities considered lethal to humans in some cases, and have speculated (but not yet demonstrated) that effects on the nervous system could result in reduced survival or at least abnormal behaviour in host bears. Humans who eat inadequately cooked meat of infected bears can develop serious illness (trichinosis). The Grizzly Bear's solitary existence at a relatively low density is not conducive to the spread of disease, although that benefit may not be realized in areas of regular, long-term concentrations such as at garbage dumps.

Despite the level of maternal protection, cubs-of-the-year and yearlings have the highest mortality rates. Some die from inadequate nutrition (especially in years of poor or late food production), some from accidents such as drowning and others from predation by other Grizzly Bears. Subadults (ages 2 to 5), mostly males, suffer the next highest level of mortality. After separation from their mothers they must locate adequate food while avoiding older bears and staying out of trouble with humans, and some are not successful. Using data collected from more than 400 radio-collared bears in 13 study areas in the southern Rocky and Columbia mountains since the mid 1970s, Bruce McLellan and associates calculated the following overall annual survival rates, out of theoretical populations of 1000: 926 adult females, 877 adult males, 923 subadult females and 801 subadult males.

From all accounts, natural mortality among adult Grizzly Bears is low. Of nine such deaths documented for females in the interior mountains compilation described above, one was caused by a collapsed den, three were from avalanches or rock slides, and five were believed due to attacks by other Grizzly Bears. It is likely that most adult females killed by other bears died defending their cubs from larger Grizzlies, probably males. There are few specific records for natural mortality among adult males (which are difficult to keep radio-collars on), but pioneer bear biologist Frank Craighead recorded deaths from fights with other bears, gorings by male ungulates (Elk, Bison) during predation attempts and, in one case, apparent old age.

For the areas that have been intensively studied, most documented Grizzly Bear mortality has been human-caused. In the interior mountains studies, at the southern extremity of the species' continuous range and where human settlement and visitation is relatively high, the deaths of 99 bears over 22 years were known or suspected to be due to the following causes: natural (14), unknown (5), legal hunting (16), conflict management or self defense (22), illegal hunting (12), road accidents (2), unknown by humans (21) and research related (7). Interestingly, bears in parks and protected areas did not have lower mortality from human causes than in surrounding multiple use lands.

Based on evidence from a number of studies, Grizzly Bears that reach an age of 5 or 6 years and settle in good habitat have a good chance of living for another 10 years or more. The oldest reported wild Grizzly Bears, both still alive when last contacted, were a 28.5-year-old female from the Brooks Range in Alaska and a 27-year-old female in the Banff area, Alberta. Both of them, and others, have produced and raised cubs in their mid 20s.

Abundance

The Grizzly Bear is the most studied of BC's carnivores, and is the only one for which the provincial population estimate is based on detailed, region-by-region assessment rather than by broad-brush extrapolation. The general pattern for most local studies has been documentation of larger numbers than had been anticipated, with the result that the provincial estimate has been steadily growing over the past decade. Summarizing data available through 1999, Ray Demarchi and Carol Hartwig reported that the BC Grizzly Bear population was then estimated at 10,000 to 13,000 animals. The official estimate of 17,000 in 2005 may reflect a population increase since the 1990s, but is probably due mostly to better information. Grizzly Bear populations appear to be stable or increasing in most management units in the southeastern, central, and northern regions of the province, but are low and possibly decreasing in some areas of the south and southwest, where human populations are the highest.

The series of studies undertaken or led by forest-research biologist Bruce McLellan in the Flathead River drainage constitute what is arguably the most intensive and complete local population record for the species anywhere. Despite continuing hunter harvest and a high level of resource extraction activities (logging, oil and gas exploration), the Flathead Grizzly Bear population increased from 5.7 to 8.0 bears per 100 km^2 between 1981 and 1986, as compared to an average of 1.7 bears per 100 km^2 for several other interior study areas. The highest densities recorded for the species, from 25 to 30 bears per 100 km^2, have been on coastal islands in Alaska (Kodiak, Admiralty) where large annual salmon runs occur. Grizzly Bear densities (bears per 100 km^2) calculated from field studies elsewhere in BC include 2.3 in the Selkirk Mountains (south), 2.1 in the Nass River watershed (west central), 2.9 in the

Northern Boreal Mountains ecoprovince (north), and 1.0 in the Taiga Plains Ecoprovince (northeast). In the Nass study, Mike Demarchi and his associates directly observed 57 Grizzly Bears, including 42 in four consecutive days, while conducting an extensive aerial survey of Mountain Goats.

Human Uses

Although Grizzly Bear flesh is palatable, its pelt luxurious, and the fat on a fall-killed specimen copious, the species has never been sought for those products to the same extent as the Black Bear. That is probably true both because Grizzly Bears have always been less abundant than Black Bears and because they are much more difficult to kill and dangerous to hunt, particularly with weapons available in historic times. First Peoples traditionally feared and respected the Grizzly Bear, and featured its skull, teeth and claws in cultural and ceremonial contexts, though there is evidence that some societies did not use the whole carcass. A 1690 journal entry by Henry Kelsey, written while exploring the Manitoba prairies for the Hudson's Bay Company, describes the good meat of a Grizzly Bear his party had killed, but notes that the local First Peoples were troubled that he had eaten it and discouraged him from keeping the hide because "they said it was God".

It is likely that when pioneers killed a Grizzly Bear they used most of its parts – the meat for food, the hide for rugs and sleeping robes, and the fat for a variety of lubricating, medicinal and culinary purposes. Currently, Grizzly Bears in British Columbia are hunted, viewed and photographed by both residents and guided tourists, providing local economic and overall social benefits.

Taxonomy

Because they vary so much in size and colour, the Grizzly Bears of North America were classified by early taxonomists into a bewildering number of species and subspecies. Even as recently as 1984, the eminent mammalogist E. Raymond Hall recognized nine North American subspecies. However, based on studies of skull variation by Robert Rausch and Bjorn Kurten, most taxonomists now recognize only two or three subspecies in North America, one or two on the coastal islands of Alaska and *horribilis* on the Alaska mainland and the rest of North America.

In DNA studies, Lisette Waits and her colleagues found four distinct genetic groups of Grizzly Bears, two in northwestern Canada and Alaska, one restricted to the Alexander Archipelago of Alaska, and one in the southern Rocky Mountains of Canada and the United States. Those groups showed little relationship to the subspecies recognized from skull structure, suggesting that more work is needed.

Ursus arctos horribilis Ord – historically this subspecies had a large distribution, from northern Mexico to the Yukon Territory in western North America. It currently includes all extant North American populations except for those on some Alaskan islands.

Conservation Status and Management

The Grizzly Bear is a Blue-Listed species in British Columbia, meaning that, despite its relatively high numbers, stable or growing populations, and wide distribution throughout the province, it is considered "vulnerable". The concern relates largely to the historic (pre-management) pattern of its overall North American occurrence, in which it was extirpated from most of the southern and eastern portions of its original continental range. In response to that concern, and in recognition of BC's position as the centre of viable Grizzly Bear populations in the southern half of the species' current range, it is the most widely researched and intensively managed of the province's carnivores. A Grizzly Bear Conservation Strategy, initiated in 1995, has followed through with conservative hunting policies and regulations, local hunting closures, establishment of habitat reserves, development and application of forest management guidelines and input to high-level land-management plans as related to habitat, closing of numerous garbage dumps, development of viewing guidelines, increased census effort (together with development and improvement of census methodologies), and development of recovery plans for local areas (such as the North Cascades), where particular sub-populations are considered "threatened".

Notwithstanding that high level of interest, commitment and management effort, there are issues that will continue to challenge the process. One that has received considerable attention elsewhere, especially Alaska, is the extent and impact of Grizzly Bear predation on wild ungulates. Although such predation is unlikely to be an actual issue under current BC government policy, which is against control of predators for most wildlife management purposes, a description of some of the Alaskan research results is of interest to provide perspective. As described in the Black Bear account (see pages 129-30), predation on Moose calves by bears (both species) was the primary contributor to low calf survival

(28%) in the Yukon Flats area. In a south-central Alaskan study by Warren Ballard and associates, 52 of 66 radio-collared Moose calves (79%) that died of natural causes were killed by Grizzly Bears. Finally, in an examination of the situation from the other side, researchers Rodney Boertje and his co-workers observed 22 radio-collared Grizzly Bears daily during three seasons (spring, summer and fall). They found, in addition to the expected predation on Moose calves, that the bears also preyed upon adult Moose and Caribou in that area. From those observations, they estimated that each Grizzly Bear in the area killed five or six Moose calves annually and that, in addition, adult male bears also killed three or four adult Moose (mostly females during the calving season) and adult female bears killed about one adult Moose and one adult Caribou per year.

As with Black Bears, a recurring management concern for Grizzly Bears in British Columbia relates to conflict situations that compromise human safety or result in significant damage to human property. Grizzly Bears (less frequently than Black Bears) cause problems such as predation on livestock, raids at park campsites, damage to boats and other equipment, and breaking into dwellings but because they are larger and more powerful, Grizzly Bears can cause more damage and are more likely than Black Bears to be aggressive when confronted. I heard of an incident at a mine in northern BC where a Grizzly Bear came upon two helicopters parked for the night, tearing one apart and moderately damaging the other. As summarized by Ray Demarchi and Carol Hartwig, BC Conservation Officers received 2441 complaints involving Grizzly Bears from 1992 to 1999, and responded by killing 368 bears and translocating 321 others. From government records provided by Matt Austin, Grizzly Bears injured 54 humans and killed 5 in the 19-year period from 1985 to 2003. The average number of incidents with humans increased from 2.5 annually in the years 1985–94 to 3.8 annually in 1995–2003. That increase is likely due in part to increasing use by humans of back-country areas, but some observers also believe that increasing Grizzly Bear populations may result in social pressures that drive dispersing young males into areas occupied by humans. Further, as outlined by University of Calgary professor and bear attack expert Steve Herrero, Grizzly Bear attacks often involve bears that have become habituated to humans in parks, campsites and garbage dumps.

Remarks

Although it is true that Grizzly Bears are poor climbers, particularly in comparison to Black Bears, it is not correct to conclude that they do not climb at all. They can shinny up a straight tree trunk much as a human would. A young male that I had shot with a dart for radio-collaring bit into branches to help pull itself about three metres up into a spruce tree before the drug took effect. An adult female that treed a forest worker near Prince George climbed the tree to about 4.6 metres, breaking off dead branches as she ascended with fearsome speed before being stopped by green branches.

The winter hibernation of Grizzly Bears is a remarkable phenomenon. For the duration of the denning period, six months or more in most cases, they do not eat, drink, urinate or defecate, yet they suffer no deleterious changes in blood chemistry and related life function and do not accumulate toxic waste. While in the winter den they are able to triple their survival time on a given level of energy stores by reducing their metabolic rate. Nevertheless, *survival* for denning bears can be tenuous. In an Alaska study by Grant Hilderbrand and his associates, six adult females that were weighed in fall before denning and then in spring after emergence lost an average of 73 kg. They lost 50 to 60 per cent of their fat and 40 to 50 per cent of lean muscle tissue. Not surprisingly, the two that gave birth and supported cubs in the den lost more muscle mass than did the other four animals, although all used more energy than they had been able to store as fat in the fall.

Selected References: Apps et al. 2004; Craighead 1979; Demarchi and Hartwig 1999; Edge et al. 1990; Farley and Robbins 1995; Fuhr and Demarchi 1990; Garshelis et al. 2005; Hamer et al. 1991; Herrero 1985; Hilderbrand et al. 1999; Hobson et al. 2000; Jonkel 1987; MacHutchon et al. 1993; McLellan 1989; McLellan and Hovey 1995; McLellan et al. 1999; MELP 1995a, 1995b; Miller and Ballard 1982; Mowat and Heard 2006; Pasitschniak-Arts 1993; Pearson 1975; Waits et al. 1998; Wielgus et al. 1994; Wright 1909.

Northern Raccoon   *Procyon lotor*

Other Common Names: Coon, Raccoon, Ringtail; *Raton Laveur.*
Provincial Species Code: M-PRLO.

Description

The Northern Raccoon is a stocky animal about the size of a
Cocker Spaniel. It has a number of distinctive features, including
the characteristic black bandit's mask over the eyes, a conspicuous
black nose-pad, a series of four to seven dark rings on the tail, and
forefeet that are conspicuously hand-like with long, agile finger-
like digits. The facial mask is set off by whitish patches across the
forehead and around the muzzle. The ears are prominent and
rounded, with white around the rims as viewed from the front,
and the snout is pointed. Black and white-tipped guard hairs give
the fur a grizzled appearance that manifests itself as a shade of
grey on most individuals. Some animals have distinctly brownish
tones, especially on the undersides where guard hairs are sparser.
The tail is long and bushy, with the lighter bands between the
dark rings usually the same colour as the animal's back and sides.
We have few measurements and weights for BC specimens, particularly from the interior of the province, but those shown here are
generally within the ranges given for the species in other areas.

Males average 10 to 15 per cent heavier than females, but the biggest local weight differences are related to age, season and feeding opportunities. The largest known specimen, a fat male taken in late autumn in Wisconsin, weighed 28.3 kg.

Measurements:

	All Specimens	Subspecies	
		pacificus	*vancouverensis*
total length (mm):			
male:	855 (750-983) n=20	953 (n=4)	831 (n=16)
female:	802 (650-932) n=12	912 (n=4)	748 (n=8)
tail vertebrae (mm):			
male:	274 (230-319) n=22	287 (n=4)	271 (n=18)
female:	271 (210-315) n=12	313 (n=4)	251 (n=8)
hind foot (mm):			
male:	117 (100-135) n=15	133 (n=4)	112 (n=11)
female:	115 (86-135) n=10	130 (n=4)	106 (n=6)
ear (mm):			
male:	63 (55-67) n=5	63 (n=4)	63 (n=1)
female:	59 (54-64) n=5	60 (n=4)	54 (n=1)
weight (kg):			
male:	7.3 (4.3-16.3) n=22	12.4 (n=5)	5.9 (n=17)
female:	5.3 (3.0-7.6) n=11	6.9 (n=4)	4.4 (n=7)

Dental Formula:
- **incisors:** 3/3
- **canines:** 1/1
- **premolars:** 4/4
- **molars:** 2/2

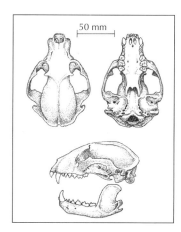

50 mm

Identification
The likelihood of misidentifying a Northern Raccoon (called "Raccoon" in the rest of this account) is small, as no other mammal in British Columbia has the black mask and ringed tail. At a glance, the Raccoon

might be confused with an American Badger because both are medium-sized and stocky, and have greyish, grizzled fur. But Raccoons are most often found in forested areas near water or in urban settings and often climb, while badgers most commonly occur in open grasslands where they burrow but never climb.

Distribution and Habitat

The Raccoon has expanded its North American range westward and northward during the past five or six decades, and that expansion appears to be continuing. It now occurs in Canada in at least the southern extremities of all provinces bordering the United States and in Nova Scotia and Prince Edward Island, in all 48 of the continental United States, and southward through Mexico and Central Mexico to Panama. There are introduced populations in coastal Alaska and it has also been introduced into Europe, deliberately as a fur resource in areas of the former Soviet Union and accidentally in Germany from where it has also spread to France and the Netherlands.

In A.W.F. Banfield's 1974 treatise, *The Mammals of Canada*, British Columbia was shown as the province with the smallest area of Raccoon occurrence, with an occupied range centred on Vancouver Island and the southern mainland coast and extending eastward in a thin band along the border to about Creston. Based on specimen records at that time, northern range limits were at about Namu on the coast and at Okanagan Falls in the interior. Trapper Pete Wise confirms that there were no Raccoons in the Vernon area until the late 1970s, the first he had heard about appearing in 1978. They increased steadily after that, and by the late 1990s his animal control business was receiving over 500 Raccoon complaints per year, resulting in the removal of about 100 animals annually. Farther north, Raccoon populations are now established in the Thompson River and Shuswap watersheds from at least Kamloops across to Sicamous, and the northernmost record in the interior appears to be a road-killed specimen observed by forestry technician Richard Freeman just north of Little Fort (spring 2001, north Thompson River drainage). Trapper and taxidermist Bryan Monroe told me that a Raccoon was treed in a schoolyard in McBride one day in the early 1990s, but he believes that it arrived in that area as a stowaway in a hay truck. The current northern distributional limit on the coast is unclear, but there is a thriving introduced population on the Queen Charlotte Islands.

The Raccoon is an adaptable animal, occurring in a great variety of habitats across its continental range. But it does not cope well with cold temperatures and deep snow, and therefore does not regularly occupy high-elevation or high-latitude areas of Canada. The Raccoon is commonly associated with water, foraging in wetlands, riparian areas along the shores of streams, ponds and lakes, and on ocean beaches and estuaries. It seems to prefer forested areas that have den trees and escape cover from predators, but Raccoons also do well in many areas where forest cover is sparse or absent. Most notably, among North American carnivores the Raccoon is the species most commonly and closely associated with man-made habitats, attaining high densities within cities and in agricultural areas all across the continent. In British Columbia, the species occurs primarily in lowlands of the Coastal Western Hemlock biogeoclimatic zone along the coast and in the valleys of the southern interior, especially in urban and agricultural areas.

Raccoons rest in a variety of sites that provide protection from weather and predators. In forested areas, day-use and maternal dens are usually in hollow trees and logs. For example, in one of the first intensive studies of the species, in Michigan in the early 1940s, all of 34 dens found by biologist Frederick Stuewer were in the cavities of large trees (mean diameter 63.5 cm), at heights of up to 30 metres above the ground. Based on such findings, apparent mid century declines in Raccoon populations in some areas of the central United States were attributed to loss of dens during land clearing operations, and led to local management actions such as purchase of known den trees from land owners and installation of artificial nest boxes.

On the Queen Charlotte Islands, biologist Lisa Hartman found Raccoons denning primarily in tree-trunk cavities and root wads. In areas lacking suitable denning trees, Raccoons readily make use of hollows in debris piles, natural caves and crevices in rock outcrops, and ground burrows excavated by other animals, such as Woodchucks and Red Foxes. In urban and farm areas, Raccoons commonly rest in and under outbuildings, and in the attics and crawl spaces of buildings occupied by humans, and there are reports of well-used dens in other man-made structures such as haystacks, mine shafts, culverts, abandoned vehicles and junk piles.

The attachment to individual den sites varies with habitat, sex, age and season. In one of the few studies using radio telemetry, Alan Rabinowitz and Michael Pelton found female Raccoons using the same den on consecutive days 60 per cent of the time and males doing so 39 per cent of the time. The higher frequency for females was primarily in the spring and summer, during the breeding and

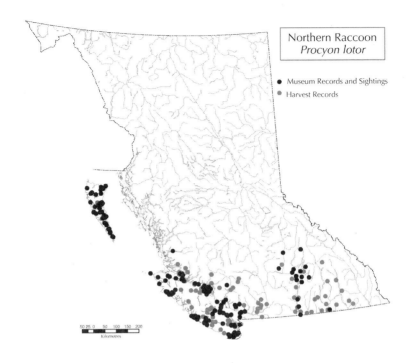

Northern Raccoon
Procyon lotor

● Museum Records and Sightings
● Harvest Records

rearing period. In areas of high density such as in some cities, where individual home ranges can be very small, a Raccoon may use the same den almost continuously, while in natural settings an adult male with a large home range may use different rest sites almost daily. Females with dependent young use dens for longer periods, up to several weeks in the spring, and in northern portions of the continental range, Raccoons of both sexes and all ages may remain in dens for long periods during particularly harsh winter periods. Communal denning may occur during the winter period, and it is assumed that the primary benefit involved is conservation of body heat. Most communal denning involves one or more family groups (females with large young), but other associations have been documented. In one case, reported by biologists David Mech and Frank Turkowski, 23 Raccoons were found together in the cellar of an abandoned house in Minnesota. They included 8 adults (at least 1.5 years of age, one of them male) and 15 young-of-the-year.

Natural History

Feeding Ecology

The Raccoon is opportunistic and omnivorous, eating a wide variety of both plants and animals, and readily scavenging road kills, other carrion and human garbage. Raccoons regularly consume several kinds of nuts, grains, berries and other fruits both wild and cultivated. They eat a lot of corn where it is available; a single cornfield may at times support a large number of feeding Raccoons. They eat many types of animals, but mostly small species such as insects, crustaceans, clams, snails, worms, fish, frogs, toads, salamanders, snakes, turtles, small mammals, and the eggs and young of various birds.

In many areas, the primary food items are crustaceans, especially crayfish in inland waters and various crabs on coastal beaches. On the west coast of Vancouver Island, I examined 330 scats and found that crabs were the dominant food year-round. I found remains of several species, but mostly the small ones that crowd under beach rocks (Purple Shore Crab and Oregon Shore Crab). A single scat contained 288 shore crab legs (representing at least 36 animals), suggesting that the number of shore crabs consumed by a local population of Raccoons in one year would be astounding. The Vancouver Island sample also demonstrated regular and occasionally heavy use of isopods and amphipods, both of which I occasionally observed in large concentrations among beach debris. On the Queen Charlotte Islands, biologist Lisa Hartman also found a high frequency and volume of intertidal invertebrates (especially shore crabs and amphipods) and fish. All of 12 scats collected from a seabird island during nesting season contained bird remains.

Raccoons also eat insects frequently, particularly those that are either generally abundant or concentrated (adults or larvae) in a small area: grasshoppers, crickets, water beetles, ground beetles, bees and wasps; the aquatic larvae of species such as crane flies, caddis flies and dragonflies; and the caterpillars of various moths and butterflies. Mollusks do not appear important in most food studies, but that may be partly because Raccoons do not often eat much of the shell, and the soft parts are not recognizable in stomachs or scats. In my Vancouver Island study, only 37 of 330 scats (11%) contained mollusk shell fragments, mostly small (thin-shelled) mussels. But on a few occasions I observed Raccoons digging and eating clams, mostly Soft-shell Clams, on mudflats near

Tofino. Once I watched a small Raccoon at close range for more than 15 minutes as it ate an 89-mm Butter Clam. The clam was braced, hinge-down, against a large rock, and the Raccoon patiently and labouriously reached into the slightly open shell and pulled small bits of tissue to the edge where they could be gnawed off. When the Raccoon finished, it left the shell intact, but had increased the gap to about 30 mm, and had removed all of the insides except the hinge muscles. During that observation I was tempted to yell, "Break it on the rock, dummy." Subsequently, I found freshly broken shells elsewhere suggesting that some Raccoons do indeed learn to break mollusk shells on rocks.

In inland areas, where much of their foraging activity is in wetlands and along freshwater shorelines, Raccoons are among the major predators of many amphibians (adults and tadpoles), and are notorious predators of the nests of waterfowl, shorebirds and certain seabirds. At a Great Blue Heron rookery in Illinois, researcher Alvin Lopinot watched a Raccoon climb to a heron nest and eat the young, and found several Raccoon scats containing heron feathers in the area. There is also a record of Raccoon predation at a Red-tailed Hawk nest in a tree 15 metres above the ground. Raccoons also prey on turtles, focusing mostly on the nests, where they can eat both eggs and young, but they also attack adult females when they are nesting away from water. In South Carolina, researchers examined 30 female turtles killed and eaten by Raccoons, and they determined that the Raccoons killed the turtles by biting their heads or breaking their necks, and then removed the flesh (head, neck, legs, viscera and eggs) from both ends of the shell.

Mammals do not usually constitute a major component of a Raccoon diet, but a number of species become prey if opportunity permits. In some areas, Raccoons commonly prey on Muskrats, particularly juveniles. In North Carolina, Kenneth Wilson found that Raccoons had damaged 125 (38%) of 329 Muskrat lodges he examined and had clearly removed nestling rats from many of them. Among the other mammals eaten by Raccoons, Meadow Voles and various mice are the most common, but also cottontails (both adults and nestlings), hares, squirrels, chipmunks, Norway Rats, pocket gophers, moles and shrews. Raccoons will also scavenge carcasses of larger animals, such as deer and domestic livestock; and cannibalism has also been reported.

Foraging Raccoons are primarily nocturnal, and are more gatherers than hunters, either casting about here and there for whatever morsels appear, or moving directly to known areas of concentrated food such as a beach at low-tide, corn field, fruit tree, berry patch, bird nesting colony, or sources of garbage (e.g., the dumpster behind the local fast-food restaurant). As with most carnivores, Raccoons use their noses, eyes and ears to help locate food, but they also use their sense of touch, particularly when seeking prey underwater, in small burrows, or holes in trees.

In water they work primarily in the shallows, rarely if ever submerging completely, and reach about in crevices among rocks and debris. As described by Utah zoologist Samuel Zeveloff, the forepaws of a Raccoon have both a high degree of dexterity and a high concentration of sensory receptors, and the portion of its brain associated with tactile stimuli is large in comparison to that of other carnivores. That enables the animal to identify a food object under a rock by feel alone, and to quickly and deftly manipulate it for capture, a clearly useful feature when, as is often the case, the object is a perturbed crab or crayfish.

Raccoons lose weight during the winter months in many areas, especially in the northern portions of their range where staying warm takes more energy and where most food sources become temporarily inaccessible under snow and ice. In Minnesota, David Mech and his colleagues recorded weight losses of 50 per cent or more between mid November and late March, and that occurred to both sexes and all age classes. They weighed one female eight times in two years, with the following results: late fall 1964, 8.6 kg; early spring 1965, 3.3 kg, with litter of at least four young; early summer 1965, 4.8 kg; early fall 1965, 5.3 kg; late fall 1965, 8.6 kg; mid winter 1965–66, 4.8 kg; early spring 1966, 5.0 kg; early fall 1966 (no litter), 9.5 kg. As with bears, Raccoons fatten up and attain the highest weights in the fall.

Home Range and Social Behaviour

As determined in radio-tracking studies, the home ranges of Raccoons vary from less than 0.1 km² in an urban area of Ohio and 0.5 km² on the food-rich Queen Charlotte Islands, to almost 50 km² in a wildlife refuge in North Dakota. Adult males maintain the largest home ranges, averaging four times those of adult females in the Ohio study (0.16 km² versus 0.04 km²), two times those of adult females on the Queen Charlottes (0.6 km² versus 0.3 km²), and three times those of adult females in North Dakota (25.6 km² versus 8.1 km²). During their first year of independence, usually as yearlings, young Raccoons of both sexes use home ranges similar in size to those of adult females. Adult females with young use smaller ranges during the 8–10 weeks in spring when the young are den-bound, but there is little seasonal difference in home-range size and use among other sex, age and reproductive classes.

As is apparent from the high densities found in some areas, Raccoons tolerate each other more than most other carnivores do. In the North Dakota study, where home ranges were the largest recorded, biologist Erik Fritzell found little overlap among adjacent adult males, but considerable overlapping among adult females and yearlings of both sexes, and both groups also overlapped with adult males. An adult male will take over another's territory – a radio-collared animal used a well-defined area consistently for several weeks, but then quickly expanded its home range to include that of a neighbouring adult male shortly after the neighbour's death.

Despite the images evoked by communal denning, extensive home range overlap and observations of numbers of animals foraging in the same general area, Raccoons are not social animals. Individuals remain solitary for most of the year. When two or more animals interact or travel together they are usually male-female pairs during the mating season, females with young after they have left the den or occasionally siblings after they have left their mother. Other apparent groups, most commonly occurring at a location with abundant food, are collections of individuals that try to avoid close contact with each other. As observed by biologist Lloyd Tevis on a California beach, contact in such cases usually involved aggressive responses, although mostly in the form of bluffing and displays (hissing, snarling, raised hackles) rather than actual fighting.

In a behavioural experiment, biologist David Barash live-trapped a number of adult male Raccoons and introduced them, two at a time, into pens where their interactions could be observed. Aggressive posturing and vocalizing occurred in all cases, but in six pairings involving animals captured at sites less than 5 km apart and probably neighbours that had met each other before, only one resulted in physical combat; in the other five pairings, one of the animals became submissive and backed off before a fight could ensue. In contrast, among seven pairings of animals that, by distance between capture sites, were likely strangers, only one did *not* result in fighting. Thus, although behaviour resembling scent marking (neck rubbing, anal dragging) has been observed in captive male Raccoons, it appears that much of the spacing between animals in wild populations may be maintained by physical confrontations and associated experience.

Activity and Movements

Raccoon activity occurs mostly after dark, for a period of eight or nine hours, depending upon season and foraging success. The most typical daily pattern is a fairly direct extended movement from the day bed or den to one or more favoured foraging areas, with reduced and mostly local movement on the foraging areas, followed by another direct movement to the next den or bed site. Recorded straight-line distances between daily resting sites have generally averaged less than 1 km, although actual movements associated with travel and the back-and-forth requirements for foraging are larger than that. During summer in North Dakota, Erik Fritzell documented minimum daily movements of 3.4 km for adult males, 2.4 km for yearling males, 1.6 km for yearling females and 1.5 km for adult females.

The longest movements recorded have been by animals translocated to new areas by humans, and by juveniles dispersing from natal areas in search of vacant areas where they can establish home ranges. Translocated Raccoons (mostly nuisance animals live-trapped in urban areas and moved to rural locations) travel extensively for a few weeks after release, often moving 20–30 km during that period – a Raccoon in Tennessee travelled 295 km after release. Some juvenile Raccoons disperse, from their mothers' home ranges in the late summer or fall of the first year, but most leave in late winter or early spring, when they are 10–12 months old. During intensive radio-tracking studies in North Dakota by Erik Fritzell, only juvenile males dispersed, travelling up to 23.5 km (average 18 km). But based on ear tag returns, at least 5 juvenile females in Frederick Stuewer's Michigan study area dispersed, with a maximum movement of 16 km and an average of 8 km (as compared to a maximum of 27 km and average of 12 km for 9 juvenile males in that area). The two longest natural dispersal distances recorded, both for juvenile males, were 265 km between northern Minnesota and southern Manitoba, and 254 km within Manitoba northward from an initial capture location at Delta Marsh.

Reproduction

In the northern half of their geographic range, Raccoons usually mate in the late winter, from February through early March. I saw mating activity only once on the west coast of Vancouver Island, near Tofino, on February 22, 1969. Females in their first breeding season (i.e., at about 10 months of age) may produce young, but the successful proportion appears to be low in areas or years in which food is limited or winter is severe. The pregnancy rate among yearlings typically ranges from 30 to 50 per cent; among adults (two years old or older) it is usually 90 per cent or more. Most yearling males are capable of breeding, but rarely have the opportunity where adult males are present.

Gestation averages about 63 days, and the young are born in April or May in most areas. In 1969, near Tofino, Roddy Mackenzie and the Dalby boys found dens containing small Raccoons with their eyes still closed on June 14 and 23. Based on their stage of development, the animals were born no earlier than the third week of May, and probably in early June. Family groups (females with small young) didn't appear on local beaches in that area until about August. Litters of up to eight young have been recorded, but the average is three or four. During the course of animal control activities in the Vernon area, Pete Wise observed litters of two to five kits, and estimated the average at three. On the west coast of Vancouver Island in the late 1960s and early 1970s, I recorded observations of up to 21 litters (possibly with some duplication), as follows: 3 with one kit, 14 with two kits, 3 with three kits, and 1 with four kits (average: 2.1). Wild Raccoons do not produce more than one litter each year.

Newborn Raccoons weigh about 60 to 70 grams and are sparsely furred, especially on the undersides. With eyes and ears that remain closed for 3–4 weeks, they are fully dependent upon the mother, which provides all parental care. The young remain den-bound, either in the original natal den or in alternative maternal dens to which the mother may move them, for 10–12 weeks. During that period, they grow at about 15 to 20 grams a day, all on milk, because the mother does not bring solid food back to the den. After leaving the maternal den, usually in June or early July, the young travel and forage with the mother until at least the late fall and often through the first winter.

Health and Mortality

Several aspects of Raccoon life history combine to create an ideal environment for the transmission of parasites and diseases. They can attain a high density, particularly in urban areas, and their social system allows close contact of many animals (such as at concentrated food sources and in communal dens). Their propensity for extreme weight loss and deteriorating physical condition in winter can result in a period of reduced resistance to ailments. Because some of the parasites and diseases involved may also infect humans and domestic pets, the literature relating to Raccoon health issues is voluminous.

Raccoons have been known to harbour a number of species of fleas, ticks and lice, some of which may infect them with bloodborne diseases. A Raccoon can be infested by one species of mite that attacks its ears and another that causes mange. The list of known internal parasites includes numerous species of roundworms, tapeworms, spiny-headed worms and flukes. Two of the roundworms, one that infects the stomach and the other the lungs, are harmful to individuals, but do not affect populations. Of particular interest in relation to human health is the Raccoon Roundworm. The eggs of this intestinal parasite are passed in Raccoon feces, and may be picked up by other hosts (including humans) from accidental inhalation or ingestion of dust containing eggs. The larvae can migrate to the central nervous system, causing serious illness and occasionally death.

Zoonoses associated with Raccoons include leptospirosis, tularemia, listeriosis, tuberculosis, encephalitis and histoplasmosis, but the one of greatest current interest is rabies. In North America, the Raccoon has been diagnosed with rabies more than any other species since 1989, taking over that dubious distinction from the Striped Skunk. Rabies can spread quickly through a population causing significant reductions and it can spread to other species including humans.

Most cases of Raccoon rabies have been in the eastern United States, with a general northward trend resulting in the first Canadian cases in southern Ontario in 1999. As described by Ontario Ministry of Natural Resources scientists Richard Rosatte and his colleagues:

> *Raccoon rabies is considered more of a public health menace than are other strains of rabies virus. That is because Raccoons are concentrated in urban areas and human attitudes are relaxed toward this species. Many people attract Raccoons to their homes by feeding and treating them as pets. This leads to a high potential for Raccoon/human-companion animal contact and exposure to rabies. The threat of Raccoon rabies to Ontario is being viewed as a very serious potential health problem and should be treated accordingly.*

With a predetermined action plan and the help of professional trappers, Ontario contained the spread of the disease in the 1999 cases and, according to an article in *Canadian Geographic* magazine by Scott Gardiner, had allocated an annual budget of about half a million dollars for Raccoon rabies control activities by 2000. As quoted in that article, Richard Rosatte estimated that a full-blown outbreak of the disease would cost the people of Ontario about $20 million per year.

Fortunately, Raccoon rabies has not yet moved west, and there have been no cases reported in British Columbia. In a study of specimens provided by BC trappers to veterinarian Vicky Adams, none of 108 Raccoons examined tested positive for the disease. But we should not be complacent about Raccoon rabies, because Raccoons can move quickly across the country as stowaways in transport trucks or rail cars carrying produce, or as pets or zoo stock.

Raccoons can also be seriously affected by canine distemper, parvovirus and feline distemper. The winter decline in physical condition that occurs in some years and areas has been implicated in malnutrition-related losses of radio-collared animals, either directly as starvation or indirectly through increased susceptibility to diseases and predation.

Many other carnivores prey on Raccoons – Cougars, Bobcats, Grey Wolves, Coyotes, Red Foxes, Fishers – and so do Great-Horned Owls and Alligators; but no study to date has identified predation as a major mortality factor. That may be because most studies have been either in urban areas, where the larger carnivores do not regularly occur, or in areas where human factors (legal harvest, problem wildlife removal, road accidents) are predominant. Little is known of Raccoon mortality patterns in this province, but anecdotal reports indicate that Cougars are a major predator of Raccoons on Vancouver Island, and Coyotes have that role in the lower Fraser River valley.

Large numbers of Raccoons are killed on North American highways. For example, personnel working for the state wildlife management agency in Indiana routinely conducted road kill surveys during their travels from 1966 through 1985. Over that 20-year period, they counted 13,777 road-killed Raccoons. A similar study in Nebraska, but only along a single highway (732 km length), tallied 3,976 dead Raccoons over 7 years (average of 568 per year).

The oldest known Raccoons were two captive females, a 17-year-old in Illinois and a 15-year-old in Manitoba. Among wild specimens, a Michigan female ear-tagged as a yearling was recaptured within 1 km of its original capture site at the age of 12.5 years. As reported by Samuel Zeveloff, juvenile mortality is high in most populations, such that average longevity is just 2 to 3 years, and only about one Raccoon out of a hundred lives longer than 7 years.

Abundance

Over a five-day period in 1948, a team of men captured and translocated 100 Raccoons from a 42-hectare section of the Swan Lake National Wildlife Refuge in Missouri. In another five day effort in a 71-hectare park in Fort Lauderdale, Florida, in 2000, authorities removed 160 Raccoons and observed 9 for a minimum pre-removal density of 238 Raccoons per km^2. Those accounts confirm that Raccoons can be very abundant where they are not exploited and food resources are sufficient. With scientific interests focusing on disease transmission and depredations, most detailed information available on Raccoon population dynamics is from the central and eastern United States, in areas where they have access to artificial food sources such as human garbage or agricultural crops. As reviewed by biologist Seth Riley and associates, density estimates ranged from 1 to 56 animals per km^2 (average 15) for 21 areas considered rural (including at least some without farm-crop availability), and 36 to 125 animals per km^2 (average 75) for 6 urban areas.

There are no comparable data for British Columbia, although a few anecdotal records indicate that certain coastal habitats and urban areas in this province can also support high Raccoon numbers. Commercial crab fisherman Doug Arnet of Tofino told me that once, while doing an early-morning low tide in the mid 1960s, he saw 57 Raccoons at once, all foraging on an exposed mudflat in Tofino Inlet. During my own biological studies in that general area in 1968 and 1969, I observed as many as 14 Raccoons at a time during mid day low tides and saw 20 or more after dark in a relatively small area while doing spotlight surveys for American Minks. The Tofino Inlet Raccoon population crashed in early 1970, probably from a disease outbreak, but appeared to be rebuilding by early 1972.

The only area of the province where a formal Raccoon study has been carried out is the Queen Charlotte Islands, but the study did not include density estimates. As reported by local trappers, Raccoon numbers in the Fraser River valley have decreased in recent years, likely because of predation by the increasing numbers of Coyotes in that area. But the continuing expansion of the species' range in British Columbia in recent decades suggests that the overall population for the province is increasing.

Human Uses

Raccoons were eaten by some aboriginal peoples and by European explorers and pioneers in the eastern United States. Samuel Zeveloff reports that Raccoons were a food staple for Christopher Columbus's sailors on his second voyage to the New World, and that the crew hunted the species to extinction on the island where the expedition's headquarters was established (present-day Dominican Republic and Haiti). In the American south, many slaves and poor families ate Raccoons regularly; night-hunting of the species with hounds is still a cultural tradition in that area. According to Samuel Zeveloff, early pioneers used Raccoon fat as a leather softener, a lubricant and occasionally as a substitute for lard, and certain First Nations in the northeastern United States used the uniquely shaped baculum (penis bone) as a pipe cleaner. But since the 1700s, Raccoons have been hunted primarily for their fur. Racoon fur has been an important trade item to Europe, and has also had periods of high popularity on this continent. Raccoon coats were high-fashion items during the roaring twenties and demand for that symbol of the old frontier, the coonskin hat, heightened in the 1950s and 1960s following Walt Disney's TV series popularizing the life and times of Davy Crockett.

Taxonomy

In his 1950 revision based on skull morphology and pelage traits, Edward Goldman recognized 22 North American subspecies, two occurring in British Columbia, but they have yet to be verified by genetic studies

Procyon lotor pacificus Merriam – restricted to the coastal lowlands and Cascade Mountains of Oregon and Washington, ranging north to and across southern BC. This is a dark form with a small broad skull.

Procyon lotor vancouverensis Nelson and Goldman – restricted to Vancouver Island and adjacent islands, this subspecies is distinguished from *pacificus* by its smaller body, skull and teeth. Cowan and Guiguet assigned Northern Raccoons on the Gulf Islands to *vancouverensis* and that designation would likely apply to populations inhabiting islands off the north and west coasts of Vancouver Island. But the subspecies affinities of Raccoons inhabiting islands closer to the mainland in the Strait of Georgia and Johnstone Strait (e.g., Cortes, Cracroft, Hardwicke, Read, Sonora, Texada and Thurlow) remain unknown. The introduced Raccoons of the Queen Charlotte Islands are derived from *vancouverensis*.

Conservation Status and Management

In British Columbia, the Raccoon is classified and managed both as a furbearer and as small game. Its propensity for trouble and its ability to maintain populations have combined to result in very liberal regulations, with year-round hunting seasons and no bag limits in some regions. The number taken by hunters is not known, but trapper harvests in the province averaged under 200 animals per year between 1993 and 2001 (range: 85–297). For perspective, the total annual North American harvest (hunting and trapping) often exceeds 4,000,000 Raccoons.

Although Raccoons everywhere pose human health risks, especially in urban areas, and are frequently the subject of nuisance wildlife complaints, the primary issue for the species in this province to date has been its depredations on a globally significant bird population. Introduced to a single location on the east central Queen Charlotte Islands in the early 1940s, Raccoons had spread throughout the archipelago by the mid 1980s. At that time they were identified as a serious threat to seabirds, particularly Ancient Murrelets of which 75 per cent of the world population nests on the Queen Charlottes. Such burrow-nesting species have evolved their nesting strategies on remote islands in the absence of mammalian predators. Based on intensive studies of the extent of murrelet predation and the inter-island mobility of Raccoons in the area, Lisa Hartman and Don Eastman estimated that approximately 80 per cent of the burrow-nesting seabirds on the Queen Charlotte Islands are potentially at risk. A Raccoon monitoring and control program has been implemented in response, and is proving successful.

Remarks

Observations of the steady and rapid forepaw manipulation of aquatic food items led to the long-held belief that Raccoons wash their food, which earned the species its Latin name *lotor* meaning washer. As is now well known, Raccoons readily eat a variety of foods where no water is available.

A blind adult male Raccoon in Minnesota, radio-collared and tracked for three months, showed no appreciable differences in home-range size or extent of movements from those of a normal Raccoon, except that its activity usually started earlier in the evening (before darkness) and moved more slowly. The blind Raccoon was familiar enough with its home range to travel 4 km in less than eight hours to a rest site it had used two days earlier.

Both albino and melanic Raccoons have been reported. Two half-grown, pure albino kits were seen in the company of a normally coloured female in Florida, and a strain of black Raccoons released from a fur farm in Indiana had spread over two small islands in southeast Alaska by the late 1940s. Of some interest in BC is the existence of a blonde colour variation among Raccoons on some islands near Nanaimo. BC biologist Dave Fraser and naturalist Bill Merilees have both confirmed that a small proportion of the animals in the area show that colour variation, and this is not a case of a single unusual occurrence.

Selected References: Fritzell 1978, Gardiner 2000, Harfenist et al. 2000, Hartman 1993, Hartman et al. 1997, Hartman and Eastman 1999, Mech et al. 1968, Rivest and Bergeron 1981, Rosatte et al. 1997, Sanderson 1987, Stuewer 1943, Zeveloff 2002.

Sea Otter   *Enhydra lutris*

Other Common Names: *Loutre de Mer, Loutre Marine.* Provincial
Species Code: M-ENLU.

Description
Although a member of the weasel family, the Sea Otter has a large,
chunky body more similar in general aspect to that of a seal than to
any of its weasel relatives. The front legs and feet are so small that
naturalist Elliott Coues described them as having "an appearance
which suggests amputation at the wrist", while the hind feet are
large and flipper-like. The tail is short, making up less than one-
fourth of total length, and varies little in width from base to tip.

The head is rounded, with a prominent, broad nose-pad, a mustache of coarse, usually white whiskers, and small inconspicuous ears. The thick, luxurious pelage is generally brown over most of the body, but with varying amounts of grizzling resulting from lighter tipped guard hairs. On most adult males and some females, the head, neck and shoulders are paler, often appearing white in contrast with the rest of the body. The only tail, hind foot and ear measurements for British Columbia, as listed below, are from a female of unknown age found dead near Kyoquot (northern Vancouver Island) in 1983.

Measurements:

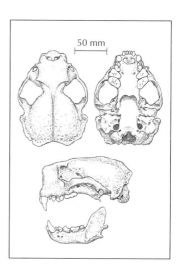

total length (mm):
    male: 1378 (1190-1490) n=16
    female: 1263 (1190-1370) n=10

tail vertebrae (mm):
    male: none
    female: 230 (n=1)

hind foot (mm):
    male: none
    female: 225 (n=1)

ear (mm):
    male: none
    female: 26 (n=1)

weight (kg):
    male: 33.5 (20.3-38.7) n=16
    female: 24.3 (19.0-27.1) n=10

Dental Formula:
    incisors: 3/2
    canines: 1/1
    premolars: 3/3
    molars: 1/2

Identification
The Sea Otter occurs only along marine shores, and might be confused only with one or two other species. The most common case

of misidentification involves the Northern River Otter which, in addition to its occurrence in fresh water systems throughout the province, is common and conspicuous along the BC coast. It is often assumed, incorrectly, that an otter emerging from salt water must be a Sea Otter, but the two species differ considerably, both physically and behaviourally. The Sea Otter is a thick-set animal with a short tail and, in some adults, the head is distinctly paler than the rest of the body. It most often rests, eats and swims while floating on its back, and it is awkward and clumsy on land, going ashore only rarely. In contrast, the River Otter has a more streamlined shape with a thick, tapered tail that is almost as long as the rest of the body, it never has a significantly lighter-coloured head, it usually maintains an upright (belly down) position in the water, and it goes ashore frequently and with moderate agility. As indicated in the description, the Sea Otter is somewhat seal-like in body shape and there may also be some potential for mistakenly identifying it as a seal or sea-lion when it is in the water. But those species are larger, do not float or swim high in the water on their backs and have a smooth and shiny appearance when wet (unlike the Sea Otter's furry look).

Distribution and Habitat

After being hunted to near extinction, the Sea Otter received international protection in 1911. It originally occurred along most of the Pacific coast from Baja California to Kodiak Island, Alaska, along both sides of the Alaska Peninsula and the Aleutian Islands chain eastward to the Commander Islands, and then south to northern Japan. Most of the original distribution in the north, from Prince William Sound westward, has been recovered. In the south, the distribution is fragmented and mostly involves reintroduced populations. The notable exception is a Californian population centred on Monterey Bay, which was maintained from a remnant population, and involves the only members of the southern subspecies. Reintroductions to Oregon, Washington and British Columbia came from Alaska.

One of the world's 13 remnant populations identified in 1911 was on the Queen Charlotte Islands, although its extent at that time and previously is not clear. The only evidence of the species was a skull found in an abandoned cabin by a National Museum of Canada expedition to Graham Island in 1919. Most researchers have concluded that the Queen Charlottes population was extirpated by that time.

The current known BC distribution includes two separate populations. The larger population is on the northwest coast of Vancouver Island, from Nootka Sound to Quatsino Sound; it likely expanded from its centre at Checleset Bay, where it was reintroduced in the 1970s. The other is in the vicinity of the Goose Islands, near Bella Bella on the mainland coast. The Goose Islands population is more than 200 shoreline kilometres from the northern edge of the Vancouver Island distribution, and may represent colonization from some other source population. In recent years, Sea Otters (mostly individuals) have also been seen at a number of other locations in BC's coastal waters and further expansion of the provincial population is likely.

The Sea Otter is almost entirely aquatic. It hunts, eats, rests, fights, breeds, and, at least occasionally, gives birth in the water, and lives only in the near-shore marine waters of the North Pacific Ocean and the Bering Sea. The range limit in the north is believed to be set by sea ice, and that in the south by warm water. Within that broad area, the best local habitats appear to be those with rocky shores and reefs exposed to the open ocean, in relatively shallow water (mostly less than 35 metres), and often in association with extensive beds of kelp.

Sea Otters forage primarily on the ocean bottom, but will also hunt among kelp forests higher in the water column. It was long thought that the maximum dive depth was about 50 metres, but in 1975 an adult male was captured in a crab trap at a depth of 97 metres. While the term "near-shore" generally means within 2 km of shore and is applicable to Sea Otter activity in most areas, the species regularly occurs up to 42 km offshore in the shallow waters of Bristol Bay, Alaska. Animals go ashore during severe winter storms, or when they are ill or injured.

Natural History

Feeding Ecology
Most of the Sea Otter's diet consists of marine invertebrates, particularly sea urchins, crabs and larger shellfish such as abalones and clams, but it also eats mussels, snails, marine worms, sea stars, barnacles and certain slow-moving fishes when more preferred food species are in short supply. Otters succeeded in obtaining food on 77 per cent of more than 10,000 dives observed by Kristin Laidre and Ronald Jameson off the Washington coast; dive times ranged from 3 to 300 seconds, and averaged 55 seconds. Individuals learn

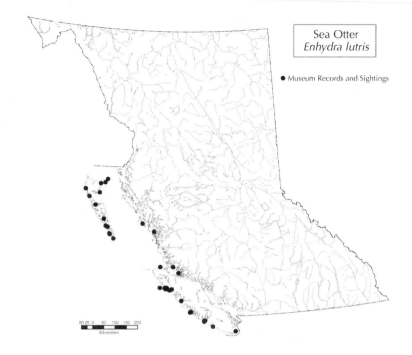

Sea Otter
Enhydra lutris

● Museum Records and Sightings

50 25 0   50   100 150 200
Kilometres

to exploit particular hunting situations. Biologists Kim McCleneghan and Jack Ames observed a young male in California surface with discarded aluminum soft-drink cans on eight dives within about 20 minutes. The otter removed and ate at least six small octopi and several small abalones that had taken refuge (obviously unsuccessfully) in the cans.

Sea Otters are the only mammals other than primates known to regularly use tools when foraging. They have frequently been observed breaking the shells of crabs, clams and abalones by pounding them against a stone or another shell held against the chest, and there is evidence that they occasionally use a stone or other hard object to dislodge prey, such as abalones, from underwater rocks.

To maintain itself in its cold-water environment, the species has high energy requirements, consuming 25 per cent or more of its body weight and foraging up to 12 hours per day. Thus, in areas supporting long occupation or high numbers of Sea Otters, the urchins, abalones, or crabs may be depleted, resulting in reduced carrying capacity for otters and sometimes causing significant, long-term changes to local biological communities.

Home Range and Social Behaviour

Although individual Sea Otters, primarily males, have been documented making regular movements of 80–145 km within their population range, others remain in fairly restricted home-range areas on a seasonal or longer term basis. The smallest home ranges appear to be non-overlapping male breeding territories of 4 to 125 hectares. Male Sea Otters actively and aggressively maintain these territories for periods of a few days up to a full year. The home ranges used by females and non-territorial males are usually from 400 to 1400 hectares, with considerable overlap for all. Sea Otters generally have separate resting and foraging areas, and home-range size is determined largely by the distance between those locations.

Activity and Movements

During much of the year, Sea Otters may be segregated by sex and age, with young (non-territorial) males occurring at high density and sometimes in large groups in peripheral areas, and females and territorial adult males at lower densities in core habitat areas. Individuals, especially adult males and females with young, may be solitary for long periods, while other animals regularly gather into large groups, especially in resting areas. The largest group on record in Bristol Bay, Alaska, in the mid 1970s, was estimated at 2000 animals, by Alaska biologist Karl Schneider.

Reproduction

Breeding may occur throughout the year, but there appear to be peak periods in northern waters, including British Columbia, such that most pups are born in April through July. The total duration of pregnancy, which includes delayed implantation, has not been determined with certainty, but it appears to be at least seven months and perhaps as long as nine months in some cases. Based on the above information, most successful breeding occurs in late fall through early winter (October through January). In Alaska, Karl Schneider documented the highest incidence of estrous females and breeding activity in October and November.

Females reach sexual maturity at three to four years and males at five to six years. In some areas adult males have to travel extensively to find estrous females, while in others they appear to remain within their breeding territories and mate only with passing or locally resident females. A mating pair may be together for three to four days, copulating many times, or have only a single, short contact with no pair bonding. Mating, which occurs in the water, is a rough and tumble affair, and the male maintains control for the 30 minutes or more involved by holding the female's nose or face in his jaws, often causing open wounds in the process. The resulting nose scars, which vary in size, shape and colour, have been used by biologists in some studies as individual recognition features for female Sea Otters.

Although there have been a few records of twin fetuses in necropsied females, I found no report of live litters larger than one. Pups are well developed and fully furred at birth, averaging almost 2 kg in weight. The mother provides all the parental care, starting immediately after birth with an intensive grooming effort (approximately 2.5 hours in one observation) to dry the pup's fur. Once dry, the fluffy natal fur provides insulation from the cold sea water and is so buoyant that pups are unable to submerge until they are about six weeks old. Until that age, at which time they also begin to take solid food, the pups spend more than 80 per cent of each day riding on their mothers' chests, being groomed and nursed frequently. Observations in California by researchers Susan Payne and Ronald Jameson indicated that most pups were self-sufficient by six to seven months of age.

Health and Mortality
The only known ectoparasite of the Sea Otter is a mite that (rarely) occurs in the nasal sinuses. Sea Otters can acquire a number of internal parasites, including roundworms, tapeworms, flukes and spiny-headed worms, but infestations causing illness or death are not known to be common. The most common causes of death are nutrition related: animals unable to obtain adequate food either die of starvation or develop a fatal bacterial infection (Enteritis) in the intestines. Food deprivation may result from local depletion of prey by the Sea Otters themselves, from inexperience of young animals, from infirmities such as broken or worn teeth and forepaw infections from urchin spine punctures, and from prolonged storm conditions (usually in winter) that make foraging difficult. Karl Schneider and James Faro documented mass mortality along the Alaska Peninsula in 1971, when unusually cold conditions caused an extension of sea ice farther south than usual. Large groups of Sea Otters were seen moving long distances across the ice and even inland in search of open water, and hundreds did not find it in time.

Predation on Sea Otters by sharks and Killer Whales has been documented, but was not thought to be significant in relation to population levels in most areas. That conclusion has recently changed, following documentation of a large scale (88%) decline in the Sea Otter population in the Aleutian Archipelago in the 1990s. Based on a variety of evidence, researchers Angela Doroff, James Estes and their co-workers have concluded that Killer Whale predation was the primary cause. The only other predation of note is by Bald Eagles on Sea Otter pups. On Amchitka Island, pup remains were found in 10 to 20 per cent of eagle nests examined between 1969 and 1973, and it was apparent that some eagles had learned to exploit the vulnerability of pups when their mothers were foraging underwater. In that study, researchers Steve Sherrod, James Estes and Clayton White found remains of nine pups in one nest over a four-week period and none in other nests. Although it is probably more common and regular than other predation mortality in most areas, predation by eagles is not considered to be a limiting factor for any of the Sea Otter populations that have been studied.

Except for a traditional aboriginal harvest in Alaska, which removes less than one per cent of the projected state population, the Sea Otter has been protected from hunting over most of its range since 1911. Other sources of human-caused mortality include illegal shooting and collision with boats in southern areas (mostly California), entanglement in fishing nets, and the effects of environmental contaminants, especially oil. The 1989 *Exxon Valdez* oil spill in Prince William Sound, Alaska, killed more than 1000 Sea Otters.

With no significant harvest of the species in almost a century, there has been no mandate and little opportunity to study age structure in Sea Otter populations, and little is known about natural longevity. Of 143 Sea Otters found dead on Amchitka Island, Alaska, over a three-year period in the early 1990s, 62 (43%) were older than 9 years, 6 were older than 15 years, and 2 animals were 20 years old.

Abundance

From 1751 to 1911, before the advent of modern resource management, the Sea Otter was hunted relentlessly for its fur and was reduced to as few as 2000 animals in 13 remnant populations. Subsequent growth of those populations and a number of successful translocations have resulted in a current world population estimate of 150,000 animals.

The historical abundance of Sea Otters in British Columbia has not been documented, but they were virtually gone from our shores by the early 1900s and did not recover naturally in more than half a century following. The species was reintroduced to BC with transplants of 89 animals from Alaska between 1969 and 1972 and, after a period of adjustment, the population has grown at a rate of almost 20 per cent per year since 1977. As reported by biologist Jane Watson and her associates, the most recent assessment (1995) put the total BC population at "a minimum of 1522 Sea Otters", most within 100 km of the release site on the northwest coast of Vancouver Island.

Human Uses
Skeletal remains of Sea Otters are occasionally found at archeological sites in BC, suggesting some historical uses by aboriginal peoples, but I was unable to determine the nature of that use (i.e., whether for food or fur). The record is clear that the Sea Otter's soft, luxurious fur was the reason for the intensive human pursuit of the animal for more than 150 years, from the mid 18th century to the early 20th century. Sea Otter pelts obtained from Vancouver Island by Captain Cook's expedition in the 1770s brought very high prices on Oriental markets, and helped stimulate the rush for fur that followed. In the early 1900s, when the demand was still high and the supply dangerously low, a single pelt could be sold for as much as $2000, far exceeding the average annual income for most people at that time.

Taxonomy
Based on skull morphology, Don Wilson and his colleagues recognized three subspecies. Both the original historical populations that inhabited British Columbia and the current populations derived from Alaskan transplants are members of the subspecies *kenyoni*.

Enhydra lutris kenyoni Wilson et al. – occurs from the Aleutian Islands and Alaska Peninsula to Oregon. Although strongly differentiated from the California subspecies *nereis*, this subspecies is very similar in morphology and DNA to *E.l. lutris*, the western Pacific subspecies which ranges from Russia to Japan.

Conservation Status and Management
After exploitation to near extinction, the Sea Otter received full international protection in 1911, with the signing of a treaty between the United States, Great Britain (for Canada), the Soviet Union and Japan. That protection has been maintained in British Columbia since that time under both the Federal Fisheries Act and provincial regulations, and the species was officially listed as endangered by COSEWIC in 1978 and by the province in 1980. Due

to the subsequent increase and expansion of the main (Vancouver Island) population, the species was downlisted by COSEWIC to special concern in 2007, but is still on the provincial Red List (considered imperilled).

The primary threat in BC continues to be the risk of a major oil spill that, depending on location, volume, wind and water currents, could affect most members of the Vancouver Island population. The only other issue relates to the potential effects of an expanding Sea Otter population on marine organisms of commercial or recreational value to humans, a matter that has been receiving some attention in recent studies by Jane Watson and her colleagues from the Pacific Biological Station in Nanaimo and the Vancouver Aquarium. It has been a hotly contested issue in California, highlighted by the existence in the 1980s of two opposing advocacy groups known as Friends of the Sea Otter and Friends of the Abalone.

The history and ecology of Sea Otter population recovery is a rich field for conservation biology, with the papers by James Estes and his associates, in particular, providing some thought-provoking scenarios, conclusions and hypotheses. For example, Estes reports that the costs to capture and rehabilitate oil-soaked Sea Otters after the *Exxon Valdez* disaster exceeded $80,000 per animal, and that the animals known or believed to have been saved by those efforts were insignificant to the population. In addition, the recent dramatic decline of Sea Otters in the Aleutians is one of several examples demonstrating that protection from human exploitation does not automatically ensure that populations will thrive over the long term.

Remarks

Unlike other sea mammals, the Sea Otter has no significant layer of fat (blubber), and relies completely on its fur for insulation and protection against the cold waters in which it resides. Sea Otter fur, which is twice as dense as that of any other mammal at more than 100,000 hairs per square centimetre, retains its insulating qualities and prevents penetration of water to the skin only when it is clean, thus grooming is a regular and continuing feature of a Sea Otter's daily activity. Adult females observed by researcher Finn Sandegren and his associates in California spent 20 per cent of their daylight hours grooming their young and 10 per cent grooming themselves. Contaminants such as oil cause the Sea Otter's fur to lose its protective qualities, resulting in death by exposure.

As might be expected, based on their infrequent occurrence on land, Sea Otters can apparently drink and process sea water.

Selected References: Bigg and MacAskie 1978; Coues 1877; Cronin et al. 1997; Doroff et al. 2003; Estes 1980, 1991; Estes and Duggans 1995; Estes et al. 1998; Garshelis 1987; Kenyon 1969; Laidre and Jameson 2006; MacAskie 1987; Monson et al. 2000; Morris et al. 1981; Munro 1985; Raum-Suryan et al. 2004; Watson and Smith 1996; Watson et al. 1997; Wilson et al. 1991.

Wolverine   *Gulo gulo*

Other Common Names: Devil Bear, Devil Beast, Glutton, Indian Devil, Skunk Bear, Quickhatche; *Carcajou, Gluton, Goulu*. Provincial Species Code: M-GUGU.

Description

The Wolverine is the largest terrestrial member of the weasel family in North America. It is a medium-sized, stocky, bear-like animal with short, thick-set legs and large paws. Its tail is conspicuous and bushy, though not long (less than one-third of the body length), and its head is large and broad to accommodate the heavy jaw musculature required for the consumption of bone and frozen flesh. The ears are short, round in profile and well-furred, and the eyes are small and wide set. The pelage is long, thick, glossy and generally dark brown to black, although paler variants do occur. Most specimens have light-coloured lateral stripes extending from the shoulders along both sides and meeting at the base of the tail. The stripes range from cream to yellow to shades of brown, their conspicuousness depending upon the degree of contrast with the basic body colour. Most animals also have a pale forehead and irregular-shaped white to bright orange patches at one or more locations on the chin, throat and chest, and some have one or more

white toes or feet. The sexes are similar in appearance, although adult males average 30 to 40 per cent heavier and about 10 per cent larger in linear measurements than females. The few sets of measurements available for BC specimens are consistent with those obtained elsewhere. In a sample from the Yukon Territory, mean total lengths and weights were 1006 mm and 14.1 kg for 52 adult males and 912 mm and 9.4 kg for 21 adult females. The heaviest animals in that sample were 16.4 kg for males and 10.8 kg for females, but male weights up to 27.5 kg have been reported.

Measurements:

total length (mm):
male:	964 (903-997) n=5
female:	912 (892-954) n=4

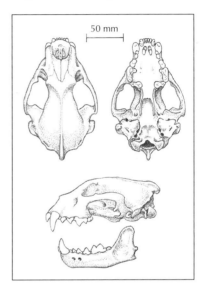

tail vertebrae (mm):
male:	203 (166-241) n=5
female:	212 (202-230) n=5

hind foot (mm):
male:	174 (161-191) n=5
female:	170 (163-180) n=5

ear (mm):
male:	56 (55-57) n=2
female:	56 (55-57) n=3

weight (kg):
male:	11.5 (10.4-12.7) n=3
female:	9.1 (6.7-10.4) n=4

Dental Formula:
incisors:	3/3
canines:	1/1
premolars:	4/4
molars:	1/2

Identification
Most Wolverines are recognizable by their contrasting (light-coloured) lateral stripes, which are not present on any other species. The bands on the back of a skunk, a much smaller animal, are occasionally wide enough to show from the sides, but they are white

rather than cream to brown and they contrast with a jet-black rather than brown background colour. Young Grizzly Bears may show light markings on the chest or the appearance of collars around the upper body, but not along the sides. Wolverines with faint or no lateral stripes can be distinguished from Fishers by their chunky rather than elongated body, and by a tail that appears short and thick rather than long and flowing when the animal is moving. The conspicuous tail also separates Wolverines from both bear species.

Distribution and Habitat

The Wolverine is a circumpolar species, occurring in the boreal zone of the northern hemisphere across the whole of Eurasia from Scandinavia eastward and in North America from coastal Alaska to Hudson Bay. The present North American range of the Wolverine is essentially the northern forest and tundra, with southward extensions into the western mountains. The species is believed to have occurred historically in most continental (non-island) Canadian provinces, although probably less abundantly in the east. It was reportedly extirpated in New Brunswick by 1850, and is encountered only rarely in the southern portions of Quebec,

Ontario and the Prairie provinces, although there is reason to question whether it ever maintained viable populations in those areas. There has been no known change from the original distribution in British Columbia or the Territories, and BC is commonly regarded as the primary source for natural recolonization of Wolverines in the western United States.

Within BC, Wolverine distribution is still imperfectly known. Based on harvest records, the species occurs widely and is relatively abundant in much of the northern two-thirds of the province, and there is also a centre of abundance in the Columbia Mountains of southeastern BC. The only areas of the province where Wolverines do not regularly occur are the Lower Mainland, dry sections of the Fraser River and Okanagan valleys in the southern interior, and the Queen Charlotte Islands. The known distribution includes Vancouver Island, although the most recent specimen record is from 1978 and the last documented sighting was in 1981. There are few records for the coast north of Howe Sound but, given the rich feeding opportunities in that area (e.g., spawned salmon, berries, marine mammals and invertebrates), that probably reflects an absence of people to make reports rather than an absence of animals.

The Wolverine is generally associated with remote wilderness habitats, particularly the boreal forests and arctic tundra of the north, but it also occurs in montane forests and alpine tundra in the southern portions of its range. Field studies have confirmed its affinity for high latitude and high elevation (and associated forests of spruce and fir), and have documented a tendency for seasonal vertical migrations in mountain areas (up in summer and down in winter). They have also recorded the use of a broad range of other habitats. Overall, it appears that a Wolverine's habitat selection may be keyed more to the distribution and availability of food than to particular topographic or plant associations.

It is also apparent that one of the most important features of the Wolverine's environment is snow, and the species' occurrence in northern and upland habitats is probably at least partly reflective of that. It is physically well-adapted for winter conditions, with pelage renowned for its frost and cold repelling qualities, and with relatively large feet that give it better mobility in deep or soft snow than many other species. Further, the negative effects of deep snow and cold on other species, especially ungulates, constitute a major benefit to Wolverines with their scavenging lifestyle. Wolverines are also behaviourally adapted to snow, both able and inclined to use its protective, insulating characteristics when necessary. In an arctic study area, where a wind chill lower than minus 31°C occurred more than 50 per cent of the time from December to February, Alaskan biologist Audrey Magoun found Wolverines creating systems of caves and tunnels under the snow 30 metres or more in length. It is in such subnivean caves that the young are born, mostly in late winter when the snowpack is thickest.

There is little information on the daily bed-sites of Wolverines, but they often use suitable locations along the daily travel route rather than returning to any particular site. During snow-tracking studies in Montana, Maurice Hornocker and Howard Hash found beds "in snow, on open outcrops in timber types which afforded cover". Construction of snow caves and tunnels, especially near ungulate kills and other large supplies of carrion, has also been reported. Biologist Dave Day of Calgary reported that a Wolverine resided under a staircase at a ski lodge in Banff National Park during several winters.

Wolverines usually tunnel into an area of deep snow to make a natal den, often taking advantage of structural features such as caves, rock overhangs, stumps or fallen trees on the ground below. One den in Alaska was in an abandoned Beaver lodge. All of 12 natal dens documented in recent BC studies were at high elevation, between 1550 and 1775 metres in Eric Lofroth's study area in the Williston Reservoir area, and between 1500 and 1800 metres in John Krebs' Columbia Mountains study area in the southeast.

Natural History

With more aliases than the average Mafia hitman (see Other Common Names, page 182), the Wolverine has a reputation for toughness and fearlessness. Its propensity to lay claim to whatever it finds, including supplies in wilderness cabins and animals taken by humans for food or fur, put it in frequent conflict with North American pioneers. That behaviour is now understood as normal for the species, but in the Wolverine's extensive mythology, it was long interpreted as malevolence. As described by Elliot Coues in 1877:

> *Probably no youth's early conceptions of the Glutton were uncoloured with romance; the general picture impressed upon the susceptible mind of that period being that of a ravenous monster of insatiable voracity, matchless strength, and supernatural cunning, a terror to all other beasts, the bloodthirsty master of the forest.*

Feeding Ecology

The Wolverine has been referred to as the "hyena of the north", because it makes a significant portion of its living as a scavenger. That is particularly the case in winter, when the primary fare is large carrion, mostly ungulates killed by other carnivores, severe climatic conditions, starvation or accidents (e.g., avalanches), but also including beached marine mammals and spawned-out salmon in coastal areas. The Wolverine is the ultimate generalist and opportunist when it comes to feeding, and the list of species it consumes is long. Summer foods include rodents, birds and eggs, and berries. In a large winter sample of digestive tracts from the Yukon, Vivian Banci identified remains of all 6 available species of wild ungulates (Moose, Caribou, Elk, Mule Deer, Mountain Goat and Dall's Sheep), at least 14 species of small mammals (including ground squirrels, marmots, tree squirrels, voles, chipmunks, Beavers, Porcupines, Snowshoe Hares, pikas and shrews), 7 species of carnivores (including other Wolverines), birds (including grouse, gulls and an owl), fish and insects. Analyses of stomachs and scats collected in Eric Lofroth's BC study area near Mackenzie produced a somewhat similar list, but with fewer bird remains.

Despite their general proclivity for carrion, Wolverines do systematically seek and capture some animals, particularly certain rodents. A small sample of summer scats from northern BC primarily contained voles and lemmings, and accounts of Wolverines digging for ground squirrels and marmots in habitats ranging from arctic tundra in Alaska to the mountains of California suggest that along with the hyena image, "tundra badger" might also apply. The extent to which Wolverines hunt Porcupines is unclear, but quills are common in Wolverine hides and carcasses. A Wolverine that I trailed for about 2 km in Glacier National Park, BC, deposited a scat that consisted entirely of Porcupine hair and quills, mostly the latter.

In a recent study in northern British Columbia, researcher David Gustine and his co-workers found that Wolverines were the primary predators of Caribou calves up to two weeks old. Wolverines eat adult ungulates when they find them as carrion, but they also occasionally succeed in preying on live adults. In 1951, BC biologist Charles Guiguet reported a Wolverine attacking a group of Mountain Goats twice in four hours. On the first occasion it was easily outrun, but on the second it managed to secure a grip on the hind leg of an adult female. It was subsequently thrown by the animal's horns and unable to make a kill. It is likely that most successes involve ungulates in poor health or disadvantaged by poor conditions, such as deep, soft snow. In Scandinavia, Wolverines kill domestic reindeer and sheep, and there are also records of successful predation on Moose. Comparable records from North America include accounts of Wolverines killing or seriously injuring Moose, Elk, Caribou and Mountain Sheep.

In his north-central BC study, Eric Lofroth recorded six incidents of Wolverines killing Caribou, and in four of those the Wolverines were adult females. In one case, involving a male Caribou, both predator and prey wore radio-collars. During a radio-tracking flight in early April 1996, Caribou researchers Mari Wood and Fraser Corbould observed an adult bull Caribou being attacked from the rear by a female Wolverine. The following day, the Wolverine remained at the site (determined by her radio-signal), and the Caribou was dead. A site investigation revealed the Wolverine had attacked several times and made the kill after the wounded and bleeding Caribou had bedded down.

Biologists who have followed tracks in snow have reported the impression that foraging Wolverines are simply looking for something to eat rather than hunting in the usual sense. If they find something dead they eat it or cache it, and if it is alive they may try to kill it. They commonly cache food by burying remains in snow or soil. Cached food is usually marked with musk, urine or both, reducing its attractiveness to other species and probably helping the Wolverine to find it again. Much of the winter foraging may involve locating and digging up caches made by themselves, other Wolverines or other carnivores. Indeed, evidence from tracking indicates that Wolverines commonly follow the tracks of Lynxes, Red Foxes and Grey Wolves, presumably in hopes of finding their caches.

Wolverines have an excellent sense of smell, and use it to great advantage when scavenging. On an aerial radio-tracking survey in the Spatsizi Wilderness Park area in June 1986, I was attempting to locate a collared Caribou that had died, and had tracked its signal to a large, permanent ice-field still covered with some of the previous winter's snow. Spotting a Wolverine track going straight up the slope, I confidently followed it to a deep crevasse, where scattered bloodstains and an increased signal volume confirmed the Caribou's location and fate.

Wolverines must often endure extreme conditions during the scavenging season. In northern Alaska, Audrey Magoun found Wolverines gnawing cached Arctic Ground Squirrels from frozen ground, ingesting as much soil as squirrel in the process, and in one winter subsisting on little more than the bones and hide scraps from long-dead Caribou. In February 1989, I found evidence on two Beaver lodges that Wolverines had expended great effort to dig, claw and chew their way in. In one lodge, blood stains indicated that the Wolverine had been successful, but the other one apparently came up empty. Overall, it is clear that occasional and sometimes intense hunger is accessory to the Wolverine's lifestyle, particularly in winter, since it relies largely on extrinsic forces (mostly climate) and other carnivores to provide the next meal.

Home Range and Social Behaviour

Although the movements of some Wolverines suggest a generally nomadic existence, telemetry studies since 1980 confirm that individuals remain in established home ranges for months or even years. Home-range sizes in those studies have varied considerably, but the general pattern is for males to have larger ranges than females, and females without kits to have larger ranges than those with kits. In five studies outside BC, the average home ranges were about 500 km² (21 males and females), 300 km² (21 solitary females), and 100 km² (9 females with kits); the largest was 963 km² for a Montana female without young. In British Columbia, Eric Lofroth documented home ranges in the Williston Reservoir area averaging 1366 km² for adult males and 405 km² for adult females, while John Krebs in the Columbia Mountains reported averages of 1005 km² for males and 311 km² for females. Consistent with their more itinerant nature, subadult males in both areas were tracked over ranges exceeding 2500 km², to a maximum of almost 4600 km² in the Williston area. Many Wolverines maintained different areas of activity in different seasons, and range boundaries were not necessarily the same for each individual year after year, likely reflecting differences in local food availability over time.

Based on observations of marked animals, the normal home-range spacing pattern is intrasexual, meaning considerable overlap between the sexes, but very little within. Thus, biologists speculate that female ranges are determined primarily by the distribution of food and those of males are determined by the distribution of females. The maintenance of separate territories appears to be accomplished mostly by indirect means rather than confrontation. In that regard, Wolverines produce a variety of odorous substances that may be important in communication. They use musk from the ventral abdominal gland and urine most often in ritualized scent-marking behaviour and like the skunk, spray a powerful-smelling fluid from the prominent anal glands, primarily when afraid or defending themselves. In addition, Wyoming biologist Stephen Buskirk and associates have described structures on the hind feet (plantar glands) that they believe broadcast the animal's presence whenever the hind feet touch an object (which is often).

Observations during snow-tracking confirm that scent-marking is a frequent and conspicuous activity among free-ranging Wolverines, although the chemical signals do not completely prevent trespass. Indeed, several researchers have identified the practical difficulties and logical inconsistencies of a strict territorial system for this species. Given the large range a Wolverine requires for its scavenging lifestyle, territorial maintenance would be very difficult and, given the unpredictability of food availability in some seasons, possibly unrewarding. Maurice Hornocker and associates concluded that the Wolverine has a flexible social system that "may be regarded as a positive adaptation to different and changing environments and is of value to the species' survival". The primary function of scent marking appears to be the maintenance of temporal rather than areal spacing, by indicating that an area is presently occupied and thereby facilitating avoidance of potentially dangerous encounters between individuals.

Activity and Movements
Except for brief periods when mating or caring for young, Wolverines are solitary. They are among the most diurnal (day-active) of the carnivores, which may benefit them by giving them access to cues, such as Common Raven activity, to help locate carrion, and may reduce the chances of dangerous encounters with other large carnivores that are more nocturnal in habits.

Based on snow-tracking, Wolverines have been characterized as extraordinarily active and vigorous. Biologist Victor Scheffer reported a case in which an observer followed fresh tracks for two days, from a location in Washington northward to the BC border, without overtaking the animal (which, by then, had become a landed immigrant). Audrey Magoun calculated rates of travel for animals in Alaska at 8.6 km/hr for males and 4.6 km/hr for females. Thus, it is not surprising that daily movements of 30 km or more have been recorded, some with no obvious signs of rests or pauses. That accords with several of my own winter aerial observations, where sets of Wolverine tracks proceeded in nearly straight lines across valleys, and up and down mountains over considerable distances, with no signs of the animals changing pace or seeking easier routes (figure 51). Wolverines travel in a characteristic loping gait that is almost mechanical, the action reminiscent of a coiled spring. Once, in Alaska, I became aware of a peculiar series of puffs moving across a recently burned landscape. As I watched, it materialized into a Wolverine, raising a small plume of ash each time it hit the ground, and not obviously varying in either cadence or distance between jumps for more than 10 minutes and 2 km.

Wolverines can cover considerable ground in a relatively short period but telemetry studies show that such movements are not a daily feature of every individual's existence. Some animals have been documented making long distance excursions away from established home ranges, returning some days or weeks later. Those excursions probably reflected temporarily reduced feeding opportunities in the home areas. The converse also appears to be true: wolverines with adequate food supplies can be quite sedentary for a time. For example, a young male in Alaska appeared to remain near a moose carcass for 27 days.

Telemetry studies have also documented dispersal movements to find new home ranges, mostly by young males. The longest involved a two-year old male that remained in a south-central Alaska study area for at least 19 days after being radio-collared, then disappeared and was caught by a trapper 20 months later in Yukon Territory, 378 km to the east. Dispersing juveniles that fail to establish permanent home ranges probably adopt a transient lifestyle, moving regularly in response to food availability or social pressures and showing the largest apparent home ranges and movements in the process. Transients are most likely to augment populations in the southern portions of the range in the continental United

Figure 51. Wolverine tracks, Spatsizi Wilderness Provincial Park, northern BC.

States. They are also the ones that appear in unusual places, including a female killed by a farmer in an Iowa cornfield in 1960, a road-killed male in a lowland, agricultural area of south-central Washington in 1991, and an unclassified adult I observed crossing a wheat field south of Lethbridge, Alberta in 1993.

Reproduction

Wolverines mate in summer, from late May through early August, probably mostly in June, and the young are born in winter, mostly in February and March. The long gestation period, up to 270 days, results from delayed implantation. The actual length of pregnancy after implantation is 30–40 days. Snow cover is important as insulation for the young in the natal den. But the primary reason for winter birthing may be nutritional, relating to the availability of carrion from ungulates killed by other predators or winter weather.

Females can breed in their second summer, when they are about 18 months old (subadults) and produce their first young at about two years old. Most males become sexually mature at 14–15 months, but are rarely in a position to compete with resident males for mating privileges. Reproductive success appears to be related mostly to nutritional factors. Vivian Banci speculated that the energetic demands of raising young may "reduce the probability that a female ... will successfully reproduce the next year" and that "females may not implant unless they are in sufficiently good condition to maintain pregnancy". Based on embryo and fetus counts, litter sizes as high as six are possible, but I found no records of live litters larger than four and the average in all areas has been less than three.

Although males have occasionally been seen near active natal dens, there is no evidence that they participate in parental care. The young are pale to white at birth and weigh 80 to 90 grams. They grow quickly, are weaned at about 7–8 weeks, and may begin to accompany the mother on foraging trips by 10–12 weeks. They grow to adult size by 6–7 months and some, particularly males, may disperse at that time in early winter.

Health and Mortality

The known parasites of Wolverines include at least one species of fluke, six roundworms (including Trichina larvae), three tapeworms, and various ticks, fleas and mites. One or more gastrointestinal worms were found in 38 of 39 Wolverines (97%) from the Northwest Territories and in 69 of 80 (86%) specimens from Alaska. The most frequent and widespread of those identified to date is the tapeworm, *Taenia twitchelli,* whose life cycle is believed to involve Wolverine as the primary host and Porcupine as the essential intermediate host, suggesting a more regular predator-prey relationship between those two species than is known. While some parasite infestations can harm an individual Wolverine, there are no records of any diseases or parasite disorders affecting populations.

John Krebs and a number of other researchers compiled records for 239 Wolverines that were radio-collared and monitored in 12 North American study areas between 1972 and 2001. Of 54 deaths for which cause could be determined, 29 (54%) were from natural causes and 25 (49%) were human-caused, including 22 (41%) harvested by trappers and hunters and 3 (5%) killed on roads and railways. The most frequent cause of natural mortality among the collared animals was starvation (18 of 29), and the remaining 11 deaths were due to predation by Grey Wolves, Cougars and other Wolverines. There are several accounts of wolf predation on Wolverines and others of Wolverines being injured by wolves or by climbing trees to escape from them. Most of those incidents were at or near wolf-killed ungulate carcasses being scavenged by a Wolverine, thus illustrating another foraging hazard; the killed Wolverines were rarely consumed, suggesting that those fights were over food or that Wolverines are relatively unpalatable. There is still too little information to speculate on the population implications of Wolverine interactions with larger carnivores, but it appears that a dynamic tension exists, with the Wolverines both partially dependent upon them for food and in danger from them as predators and competitors.

Wolverines in captivity have often survived to 15 years of age, and the oldest was 17 years. In contrast, wild Wolverines rarely live more than 8 years as determined from harvested samples. Of 576 animals harvested in Alaska and 142 in Yukon Territory, 95 per cent were under 8 years old, and only 1 per cent were 9 or older (to a maximum of 13 years); but a male Wolverine killed by a Grey Wolf in the Northwest Territories was aged at 14 to 15 years.

Abundance

British Columbia may be the continental centre of abundance for Wolverines, with an official minimum estimate of 5000 animals in 1987. But the species normally occurs at low density, and an inventory of populations over large areas would be impractical if not impossible. The first attempt at local census in North America appears to be that of Horace Quick in the early 1950s. Based on snow track counts, he estimated that there were 250 Wolverines in his 52,000 km^2 northeastern BC study area (one per 207 km^2). Subsequently, using a variety of methods, researchers have calculated densities ranging from one per 65 km^2 (Montana) to one per 177 km^2 (Yukon Territory).

It is generally believed that Wolverine populations in eastern North America declined in historic times and have not recovered, although evidence is lacking that there were ever viable populations in that area. But there is evidence of population increases and range expansions in the mountains of the western United States over the last few decades, particularly in Montana, Washington and Oregon, and dispersal from viable populations in British Columbia appears to have contributed to those increases.

Human Uses

Aboriginal people in the north have long used Wolverine fur as trim, especially around parka hoods, because it is easily rid of the frost build-up that plagues other materials used for that purpose. Wolverines are still used for garment trim, but the primary demand is now for taxidermy. The harvest in British Columbia has averaged about 165 animals per year since the mid 1980s, with a high of 246 in 1986. That is considerably less than recorded historic harvests (annual average of 313 in 1960–80, with a high of 634 in 1973–74). The lower recent numbers probably reflect both reduced trapper effort and the more conservative management program now in place.

Taxonomy

Taxonomists have been unable to agree on the number of North American Wolverine subspecies. Some authorities classify all populations as a single subspecies and others recognize as many as four. Recent DNA studies appear to be most supportive of the one-subspecies position, but without a formal revision, two subspecies have traditionally been recognized in British Columbia.

Gulo gulo luscus Linnaeus – broadly distributed across Alaska, Canada and the northwestern United States; this subspecies occupies the entire mainland of the province.

Gulo gulo vancouverensis Goldman – restricted to Vancouver Island, this type was described as a distinct subspecies based on minor skull and pelage colour features. But Vivian Banci compared the eight known museum specimens from Vancouver Island with specimens from the BC mainland and found little distinction in their morphology. Given that there have been no confirmed sightings or specimen records in over 25 years, resolving the taxonomic status of the Vancouver Island Wolverine will likely require a study of DNA extracted from historical specimens housed in museum collections.

Conservation Status and Management
The Wolverine is Blue-Listed in British Columbia, meaning that it is considered vulnerable and therefore a priority for management consideration. There is some concern that the increase in backcountry recreation in BC mountains is encroaching on areas that, in the past, Wolverines pretty well had to themselves. The implications of that are not clear, but biologists are beginning to consider the questions involved. The northern wolf control programs of the 1950s, which involved widespread distribution of large poison baits, probably had significant effects on Wolverine populations. Fortunately, those programs will not be repeated. Habitat management for Wolverines requires a broadly integrated approach that promotes and protects biological diversity and richness.

Remarks
BC biologist Tom Sullivan has demonstrated that odours from Wolverine urine will suppress local feeding by Black-tailed Deer and Snowshoe Hares, and therefore has potential for use in reducing damage to seedlings in conifer plantations.

For trivia buffs, it may be of interest that despite the well-documented mobility of Wolverines, there has never been a valid record of a Wolverine from Michigan, a jurisdiction known as "the Wolverine State".

Selected References: Banci 1982, 1994; Banci and Harestad 1988, 1990; Gustine et al. 2006; Hatler 1989; Hatler and Beal 2003a; Hornocker and Hash 1981; Krebs and Lewis 2000; Krebs et al. 2004; Lofroth 2001; Magoun 1987; Pasitschniak-Arts and Lariviere 1995; Tomasik and Cook 2005; Wilson 1982.

Northern River Otter   *Lontra canadensis*

Other Common Names: Canadian Otter, Fish Otter, Land Otter, North American Otter, Otter, River Otter, Waterdog; *Loutre, Loutre de Riviere, Loutre du Canada*. Provincial Species Code: M-LOCA.

Description
The Northern River Otter, a semiaquatic member of the weasel family, is a medium-sized carnivore about the size of a Basset Hound. It has a long, thick-set body, short legs, and a thick, powerful tail that constitutes approximately one-third of the total length. The head is broad and flat, with small inconspicuous ears, a wide nose-pad, and long, prominent vibrissae (whiskers). Adults continue to grow for a few years after attaining maturity and vary in size, but males average about 15 per cent heavier than females of the same age. The pelage is usually some shade of brown, appearing almost black when wet, although the undersides are often greyish in general tone.

Measurements:

	All Specimens	Subspecies *mira*	*pacifica*	*periclyzomae*
total length (mm):				
male:	1251 (1093-1380) n=52	1263 (n=36)	1221 (n=13)	1242 (n=3)
female:	1166 (1047-1350) n=45	1174 (n=30)	1140 (n=7)	1162 (n=8)
tail vertebrae (mm):				
male:	444 (359-503) n=16	397 (n=2)	443 (n=12)	497 (n=2)
female:	429 (361-495) n=14	452 (n=4)	424 (n=7)	411 (n=3)
hind foot (mm):				
male:	139 (128-157) n=16	132 (n=2)	137 (n=12)	155 (n=2)
female:	127 (110-136) n=14	124 (n=4)	128 (n=7)	130 (n=3)
ear (mm):				
male:	27 (23-29) n=11	23 (n=1)	27 (n=10)	none
female:	25 (19-27) n=7	none	25 (n=7)	none
weight (kg):				
male:	11.3 (9.5-14.2) n=18	12.6 (n=3)	11.0 (n=15)	none
female:	9.1 (7.3-10.9) n=10	10.2 (n=2)	8.6 (n=7)	10.6 (n=1)

Dental Formula:
- **incisors:** 3/3
- **canines:** 1/1
- **premolars:** 4/3
- **molars:** 1/2

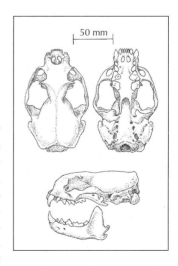

50 mm

Identification

Because the Northern River Otter (called "River Otter" in the rest of this account) is in or near water most of the time, it is most likely to be confused with two other semiaquatic mustelids in BC, the Sea Otter and the American Mink. It is often assumed, incorrectly, that an otter observed emerging from salt water must be a Sea Otter but the two species differ considerably, both physically and behaviourally. The River Otter has a streamlined shape with a conspicuous long, tapered tail, its head is generally the same colour as the rest of its body, it usually maintains an upright (belly down) position in the water, and it goes ashore frequently and with moderate agility. In contrast, the Sea Otter is a thick-set animal with a short, thin tail, and some adults have a distinctly pale head compared to the rest of the body. Sea Otters spend most of their lives on sea water floating on their backs, and are awkward and clumsy on land. Like the River Otter, the American Mink occupies both marine and fresh water habitats in British Columbia, and it also has the same general body shape and predominantly brown pelage, but the River Otter is up to ten times larger. A very young otter pup might be similar in size to a large American Mink, but has a broader nose-pad, more conspicuous whiskers, less prominent ears and, unlike most minks, never has white patches on the chin, throat or chest.

Distribution and Habitat

The River Otter is believed to have occurred historically in suitable habitats over all of North America except the Arctic and the arid southwest, but populations (possibly small to begin with) were depleted in much of the Prairies and Plains and in the eastern United States by the end of the 19th century. Many of those populations increased naturally or were restored by reintroductions in the 20th century. The current distribution in Canada is coast-to-coast, still with areas of no or low occurrence in the southern Prairies. The species occurs throughout British Columbia including all offshore islands, and the distribution is not known to have changed from pristine times.

River Otters subsist in many kinds of aquatic habitats, including near-shore marine waters, coastal and freshwater marshes and estuaries, and inland streams, rivers, lakes and ponds. The basic requisites for all habitats are the presence of suitable prey, primarily fish, and upland structures that serve as denning cover. In freshwater stream habitats, most otter foraging is in areas where fish tend to concentrate such as beaver ponds, eddies, backwater sloughs, the mouths of tributaries, and the pools above or below rapids and falls. River Otters cover most of the shoreline within their home ranges while travelling between hunting spots, and do not avoid rough water, rugged shores or human developments while doing so. They are commonly seen within the municipal boundaries of Vancouver and Victoria, and L.D. Mech gives an account of a River Otter living in a highly urbanized area in St Paul, Minnesota.

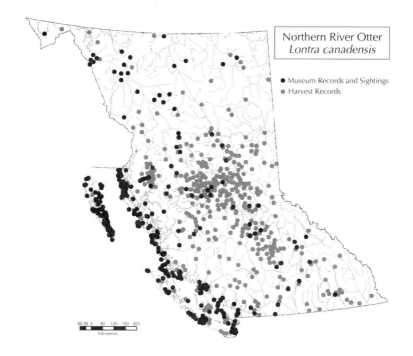

Northern River Otter
Lontra canadensis

● Museum Records and Sightings
● Harvest Records

In the Prince William Sound area, Alaska, biologist R.T. Bowyer and associates found River Otter sign (trails, scats) to be more frequent along shores adjacent to old growth forest than in logged areas or areas with non-forest vegetative cover. That presumably related in part to the presence of large structures suitable for denning cover. But on the west coast of Vancouver Island, I found evidence of River Otters resting on and hunting from most of the unforested offshore islands and islets in and around Barkley Sound and Clayoquot Sound, including islets that were primarily rock, supporting little or no vegetation.

In the mountainous terrain of interior British Columbia, River Otters occur most commonly in valley bottoms and the lower sections of freshwater drainages, especially in winter, but individuals will include fish-bearing streams and lakes at higher elevations in their travel routes during the open-water season. Wetland systems, particularly those created by Beavers, are also regular River Otter haunts. Biologist Don Reid and his colleagues identified a close relationship between those two species during winter in boreal Alberta, with the River Otters frequenting Beaver ponds for hunting habitat and using Beaver lodges and burrows for denning. During winter in such northern areas, River Otters may conduct much of their activity beneath the ice, resting in dens accessible through underwater entrances and breathing at air spaces and pockets wherever they may be found. When on the surface in winter, River Otters use their streamlined shape to travel in a distinctive push-slide motion over ice, the slide component often extending as far as five metres (figure 52).

River Otters often den in abandoned Beaver lodges, females for birthing and early rearing of young and others for daily resting. They also den in burrows made by other animals (such as Woodchucks and Red Foxes), in natural cavities (such as those under the roots of large trees), among large boulder rubble and in hollow logs.

Feeding Ecology

Fish has been the predominant River Otter food in most areas studied in western North America, both in freshwater habitats (Idaho, 97%; Montana, 93%; California, 89%; northern Alberta, 92% in one study, 85% in another) and on the coast (southeastern Alaska, 99%; British Columbia, 98%). The list of species taken is very large, attesting to the opportunistic nature of River Otter hunting, but the bulk of the diet is comprised of relatively slow-moving or schooling species: suckers, sculpins, sticklebacks, Northern Pikeminnows, chubs, sunfish and perch in fresh water; and sculpins, rockfish, surfperch, flounders, greenlings, pricklebacks and gunnels in marine water. In July 1976, Mary Morris and Brian Eccles saw a River Otter with a Pacific Lamprey on the Kitimat River. Whitefish, Kokanee and salmon are seasonally important in some areas, when they are concentrated for spawning in shallow near-shore waters. River Otters I have observed caught mostly small fish, less

Figure 52. River Otter tracks along the Nahlin River, northern BC.

than 20 cm long, but I also recorded several occasions along the west coast of Vancouver Island when they caught much larger fish: a Red Irish Lord about 40 cm long, two Cabezons at 76 and 78 cm, a Lingcod of about 90 cm, and a Wolf Eel that was 167 cm long. River Otters also eat birds, mostly species associated with water (ducks, grebes, gulls, alcids, cormorants). They usually take birds in summer, when flightless waterfowl, the young of most species and concentrations of colonial nesting birds are available. On islands off the west side of the Olympic Peninsula, Washington, Steven Speich and Robert Pitman observed two different locations where River Otters had killed more than 100 adult storm petrels. In July 1972, I found the remains of several Leach's Storm Petrels and five Rhinoceros Auklets in a River Otter resting area on Seabird Rocks, near Bamfield. When preying on birds in nesting colonies, River Otters sometimes kill more than they can eat, leaving carcasses consumed only partially or not at all. None of the otter-killed auklets I observed at Seabird Rocks had been fully eaten and one was intact.

Nicholaas Verbeek and Joan Morgan monitored River Otter predation on gulls from July 7 to August 7, 1977 at Mandarte Island, near Victoria, and found partially consumed remains of 19 adults, 153 chicks, and one adult Pigeon Guillemot. They believed that all were killed by a single River Otter, and calculated that approximately six per cent of that year's gull recruitment was lost. It seems apparent that, while River Otters have both the opportunity and predilection for bird predation, most do not have a strong taste for birds and do not exploit them as much as they could. Considering the mobility of the species, that is probably fortunate.

Mammal remains were found in seven per cent (35 of 498) of River Otter scats from two areas of northeastern Alberta, but were in only one to four per cent of the samples from other freshwater areas and were entirely absent in study samples from the coast. The most commonly occurring species were Muskrat and Beaver, the latter only rarely, but the list also included hares, voles, mice, shrews and a Red Squirrel. Amphibians, mostly frogs but including salamanders and newts, were found in 115 (46%) of 252 scats in one area of Montana and 135 (12%) of 1122 from another, and in 12 per cent of stomachs examined in Oregon; but they were rare or entirely absent in food-study samples from inland California, Idaho and northern Alberta. Reptile remains were found in 15 (16%) of 94 scats examined in the California study, but were in less than 1 per cent in all other areas; most of the reptiles were garter snakes, but the California study also found a few occurrences of Western Pond Turtle.

In terms of frequency, invertebrates are second only to fish in most areas, although the overall dietary importance of some is low because of their small size (e.g., insects such as stoneflies, dragonfly nymphs and aquatic beetles). Though abundant in marine waters, crustaceans do not appear to be heavily exploited by River Otters. Douglas Larsen recorded several species in 41 (15%) of 272 scats from southeastern Alaska, but Gordon Stenson and his coworkers found crab remains in less than 2 per cent of almost 600 study samples from British Columbia's shores. However, others in some freshwater systems appear to eat significant quantities of crayfish; they were found in 25 per cent of the study samples from Oregon and 16 per cent of those from California, but were not present in the materials from Idaho, Montana or Alberta.

River Otters usually hunt in the water, diving and pursuing their prey mostly at depths of 5 metres or less, although dives to 20 metres have been documented. In my experience, River Otters usually go to shore to eat large prey, but remain afloat on the water while eating small items. In June 1969, I watched a River Otter hunting for 35 minutes in marine waters near Tofino. In 24 dives it caught and ate 14 fish – 9, too small to identify, it dispensed with in a few quick chews; the other 5 were pricklebacks or gunnels 20–30 cm in length. The dives ranged in duration from 11 to 40 seconds, and averaged 25.5 seconds. Another River Otter, observed in Barkley Sound in June 1971, was successful in catching small fish in 50 of the 57 dives for which I was able to make a determination. Its dives ranged from 7 to 43 seconds and averaged 20 seconds.

In February 2000, I observed a River Otter hunting at the frozen surface of Griffin Lake, near Revelstoke, diving through a man-made ice-fishing hole. It dove three times while I watched, returning to the ice surface in each case with a small fish that it quickly consumed.

Home Range and Social Behaviour

The annual home ranges of adult male River Otters are generally larger than those of females, but the difference varies with the nature and extent of shoreline habitat available, and probably also with population density. In an Alberta study, the home ranges of six adult males averaged 231 km², which was more than three times that of three females (70 km²). Adult home ranges in a coastal Alaska study were smaller (9–25 km²), doubtless because of the richer, more concentrated food supplies in that habitat. Other studies have described home ranges in terms of shoreline or waterway length, with individual ranges varying from 8 to 78 km along freshwater systems and

1 to 23 km along coastal shores. Winter home ranges are smaller, particularly in interior areas where freeze-up occurs.

River Otters appear to have a somewhat flexible social system. In common with other mustelids, they are believed to mark locations within their home ranges with excrement scented with secretions from their well-developed anal glands, thus there is a basis for assuming that a territorial system exists. But while adults maintain distinct home ranges over at least part of the year, overlapping habitat use by both sexes and all ages is fairly common, probably more so when populations are high. Since River Otters occupy mostly aquatic portions of the landscape, particularly for foraging, they do not have as many options for avoiding trespass that the more widely-ranging terrestrial species have. Thus, scent-marking may simply help individuals avoid direct confrontations.

Females with dependent young in the first month or two following birth maintain the most exclusive territories and smallest home ranges. They remain separate from each other, although their territories often overlap with those of local adult males. Adult male River Otters are usually solitary, but unlike most other mustelids, they sometimes congregate in small bachelor groups. The largest male group documented was 7 animals, but groups of up to 30 (of unknown sex and age) have been observed in coastal waters, and I am aware of one similar sighting in a freshwater habitat in central BC: one October day in the mid 1970s, guide outfitter Allen Ray observed 26 River Otters together on a rock outcrop on the shore of Tahultzu Lake, in the Nechako River drainage about 30 km south of Fraser Lake. In light of recent observations by Thomas Gorman and his research associates in Minnesota, those large groups may have been composed of several family groups.

Activity and Movements

River Otters are most active between dusk and dawn, but they sometimes travel and forage during daylight hours, especially in winter. Daily movements can be extensive, up to 42 km for a dispersing animal in one study, but they usually average 4 to 5 km in summer and 2 to 3 km in winter. During the breeding season, males may travel longer distances; radio-tagged males in Don Reid's northern Alberta study were difficult to track during that period.

The dispersal of juveniles in Wayne Melquist's study area in Idaho involved both sexes and occurred in the spring when the animals were 12 to 13 months old. They travelled at about 4 km per day and moved up to 192 km in total. On March 2, 1980, in a helicopter, I followed a River Otter track along a tributary of the Tahltan River near Telegraph Creek (Stikine River drainage). It crossed the extensive plateau on the upland of Level Mountain (elevation 1600 metres) and then descended to the Nahlin River (Taku River drainage) – a straight-line distance of 57 km. The otter, probably a dispersing juvenile or a male searching for a mate, had no access to water for at least 40 km of the journey. As further indication of the mobility of this species, reintroduced River Otters in the United States have commonly moved 80 km or more from their release sites; the apparent record is 965 km by an otter that was released in Nebraska and recovered in Missouri.

Reproduction

The breeding season for River Otters in British Columbia has not been clearly determined, but it appears to be in late winter, from February through April, in most areas. I am aware of only two BC records: Steve Johnson observed a copulating pair on the marine waterfront at Sidney on April 6, 2000, and trapper Colin Brookes watched an extended mating episode at Mabel Lake, near Enderby, in late May 2006. Although these records are not supportive, coastal River Otters may be less constrained by seasons than those in inland freshwater systems. Pregnancy, which includes a delay in implantation (see page 18), lasts up to 12 months. Once implantation occurs, gestation is about two months.

Although there are records of yearling pregnancies, females usually mate for the first time in the late winter or spring of their second year, producing their first litter at about three years of age. Males are also sexually mature at two years of age, but many do not mate until they are at least five years old. The participation and success of young males in breeding may depend, in part, upon how much they have to compete with mature males. As appears to be typical of mustelids, mating is strenuous and vigorous to induce ovulation, the male subduing rather than courting the female. In Steve Johnson's Sidney observation, the male caught the female in the water with a bite to the nape of her neck and held her in that grip, staying in the water and copulating intermittently over a period of 77 minutes. She broke free briefly after that but was caught in the water again and held in another copulatory episode lasting 15 minutes, after which she escaped. During the copulatory embrace, the female was frequently held completely underwater and had to exert considerable effort to surface and breathe. Colin Brookes's observation at Mabel Lake was similar in all respects.

River Otters can bear litters of up to five young, but the average is three or less in most areas. On the west coast of Vancouver Island, during the late 1960s and early 1970s, I observed 20 apparent family groups with litter sizes (and frequencies) as follows: one (1), two (3), three (10), four (5), five (1) and six (1). Most of those consisted of one larger animal (presumably an adult female) with the number of smaller animals indicated, although I was unable to confirm that there was only one adult in the largest group, or that it was a single litter. The smallest young I observed, in a group of three, were about the size of an American Mink, near Tofino on July 9, 1971. It is possible that I observed their first trip to the water, as they were relatively unskilled swimmers. That being the case, they were probably born in mid to late May.

River Otter pups are fully furred at birth, weigh about 130 grams, and are blind, toothless and helpless. Their eyes open at about three weeks of age, but they remain den-bound and on their mother's milk until about six weeks old. They take solid food at about two months, are weaned at about three months, and then travel and hunt in the family group until they are at least twelve months old, into the following spring. The mother provides most of the parental care but there are records of coastal family groups with an additional female apparently helping the mother. It is likely that such females are related to the mother, probably daughters from previous litters. Juvenile River Otters may remain together in sibling groups for some time after they disperse from the natal range.

Health and Mortality

River Otters are host to a number of roundworms, tapeworms, flukes and spiny-headed worms, and to external parasites such as ticks, lice and at least one species of flea; but none of those are known to cause chronic health problems or affect populations. Likewise, a number of diseases including canine distemper, hepatitis, jaundice, pneumonia, and various respiratory and urinary ailments have been diagnosed in individual River Otters, but are not known to occur commonly anywhere. I am aware of one case of rabies, found in a Pennsylvania animal that had been captured for a reintroduction project.

In the water, North American River Otters are subject to predation in marine systems by Killer Whales and in southern areas frequented by alligators, but only rarely. On land, Grey Wolves, Coyotes, Bobcats, Cougars and domestic dogs prey on River Otters, probably most often juveniles. In February 1990, during a helicopter survey, I found evidence of a Grey Wolf preying on a River Otter. An otter track we had followed for about 10 km on the ice of Meziadin Lake came to an abrupt end, with only a slight scuffle and a small blood spot, at its intersection with a wolf track. Back-tracking showed that the wolf had broken into a run from a distance of about 800 metres, quartering in from the rear, and had likely overtaken the otter before it knew it was being pursued.

The extent of natural mortality to predation, and to accidents such as flooding and scouring associated with ice jams is largely unknown, but is believed to be low. River Otters live relatively long, attaining ages of up to 25 years in captivity: the oldest recorded wild specimens, including one from coastal British Columbia, were 14 years old. In most areas where River Otters have been studied, their main causes of death are human-related, such as incidents involving highways traffic, boat propellors and fishing gear. Three different commercial crab fishermen told me that they had caught otters in crab traps, in one case, an entire family group (female and three young). River Otters are also harvested legally by trappers, but the total in BC has been low, averaging about 500 annually over the past two decades.

In Alaska, R.T. Bowyer and associates conducted extensive studies on River Otters following the 1989 *Exxon Valdez* oil spill. They found that surviving otters in the oiled area showed definite physiological, behavioural and demographic effects in the first two to three years following the spill, but had recovered by 1999.

Abundance
The number of River Otters in a particular area is related to the number and quality of water bodies and wetlands present, and to the abundance and accessibility of aquatic food. While there have been few applicable studies, and none in BC, it is evident that marine environments generally support higher otter numbers than freshwater systems. Densities determined in good habitats in western North America to date have ranged from about one River Otter per 1.3 km of marine shoreline in southern Alaska to an average of one per 3.9 km along a river system in Idaho.

Human Uses
With its dense underfur and short, silky guard hairs, River Otter pelage has long been valued for its beauty and durability. Of its flesh, the 19th-century naturalist Elliott Coues warned: "the habits of the

otter, and its rank fishy taste, have procured for it the distinction of being permitted by the Church of Rome to be eaten on maigre days."

Taxonomy
Based on skull morphology, Constantinus van Zyll de Jong recognized seven subspecies, of which three occur in British Columbia, but there have been no genetic studies to validate them.

Lontra canadensis mira Goldman – southeastern Alaska, including the Alexander Archipelago, and most of coastal BC (Vancouver Island and the coastal mainland, including all islands except for the Queen Charlottes). This is the largest of the subspecies, distinguished by a large skull and robust teeth.

Lontra canadensis pacifica Rhoads – an immense range across the western United States, Alaska, and western and northern Canada, including the entire mainland of BC east of the coastal mountain ranges.

Lontra canadensis periclyzomae Elliot – restricted to the Queen Charlotte Islands. This subspecies differs from *mira* by its smaller and narrower skull and less robust teeth.

Conservation Status and Management
River Otters are ecologically important in aquatic systems and, because of their position at the top of the food chain in those habitats, are commonly used as bio-indicators for water quality and pollution monitoring. Because they are widespread and common throughout British Columbia, they are of no special conservation concern. In this province, the River Otter is classified and managed as a furbearer, and therefore may be legally taken only by licensed trappers within prescribed areas and seasons. It is harvested in each of the province's eight administrative regions, but most commonly in the Cariboo, Skeena and Omineca-Peace regions.

Historically, the River Otter has not been regularly involved in conflict situations with humans. In their natural habitat, otters most often prey on non-game or coarse fishes and do not impact human recreational or commercial fisheries. There have been occasional complaints involving the mess of River Otter feces in boats or on docks, and otters occasionally prey on young birds in island seabird colonies. More recently, with the growth of the aquaculture industry, a new form of conflict has arisen in the form of predation by River Otters on captive fish in farm pens.

Remarks

River Otters have several adaptations for their aquatic lifestyle. They have a large lung capacity and are able to automatically shut down the blood supply to parts of their body when diving, allowing dive times of up to four minutes. They have more subcutaneous fat than most other mustelids, providing an insulation layer; their dense fur also provides excellent insulation, and with frequent grooming, repels water. All the feet are webbed; but the rear legs and tail provide the primary propulsion and steering when the animal is swimming. The River Otter's eyes and ears are high on its fairly flat head, enabling it to see and hear above the water with only a small portion of the head protruding. The ears close when the animal submerges and the long sensitive whiskers help it navigate and locate prey in dark or murky water.

An account of River Otters is not complete without reference to what we generally regard as play. The most common form of play is when the animals repeatedly slide down banks made slippery by wet vegetation, mud, ice or snow. The naturalist John James Audubon observed two otters clamber up a steep bank and descend the same slide 22 times in succession over a short time period. Another apparently playful behaviour is "porpoising", in which the animal swims rapidly near the water's surface in a series of shallow, arching dives, undulating in and out of the water. I once observed an adult female and her four young porpoising in single file, in what appeared to be a follow-the-leader type of activity. The result, on that foggy morning, was an apparition as close to the mythical sea serpent as any I have observed.

My favourite story of River Otter playfulness took place on a small islet off the west coast of Vancouver Island in winter 1968. One side of the islet had once been used as an aboriginal burial ground. I had seen skeletal remains there previously, while conducting my studies of American Minks in the area, so I was not surprised to find a human skull on an otter trail near the islet's centre. My personal policy in such matters is to show respect and avoid desecration, so I left things as I found them. The following day I saw the same skull on a different otter trail, and on the next day it had been moved again. Clearly the otters were playing with that skull, and I fancied that its former owner would have approved of being so engaged with wildlife.

River Otters are fairly vocal, emitting a variety of growls, barks, hisses, snorts, whistles, and a low-frequency "chuckling" that may be akin to purring. At Vargas Island near Tofino, an adult female that I startled rose high in the water to observe me, then let out a sharp, whistling chirp. That was an apparent alarm call to her four half-grown young, onshore nearby, which all immediately dove into the water and did not surface again within my view.

Selected References: Bowyer et al. 2003; Elliott et al. 1999; Footit and Butler 1977; Gilbert and Nancekivell 1982; Gorman et al. 2006; Lariviere and Walton 1998; Larsen 1984; Melquist and Dronkert 1987; Melquist and Hornocker 1983; Reid et al. 1988, 1994a, 1994b; Stenson et al. 1984; van Zyll de Jong 1972; Verbeek and Morgan 1978.

American Marten   *Martes americana*

Other Common Names: American Sable, Canadian Sable, Hudson Bay Sable, Marten, Pine Marten, Marten Cat; *Fouine, Martre, Zibeline.* Provincial Species Code: M-MAAM.

Description

The American Marten is small, the largest individuals about half the bulk of an average-sized house Domestic Cat. In common with its weasel relatives, it has a long, thin body and short legs. The neck is long and the head is triangular in shape when viewed from above, with a pointed muzzle, prominent, fox-like ears, and bright dark eyes. The pelage is thick and luxurious, with guard hairs up to 4 cm in length along the sides, making the animal appear larger than it actually is. The tail is conspicuous, making up about one-third of the total length, and is well-furred, thick and bushy. The basic body colour on most individuals in all seasons is brown, but there is considerable variation, from pale cream, through shades of tan, reddish and chocolate browns and greys, to nearly black. The most dramatic specimen I have seen was brilliant orange. Most American Martens are grey on top of the head, and the legs and tail are darker than the rest of the body. Most also have a bib-like patch of contrasting colour, usually

yellow to orange but sometimes white, on the throat and occasionally extending down onto the chest (Plate 4). In a sample of 967 British Columbia animals from my records, 918 (94.9%) had a throat or chest patch, and it was white (with no yellow or orange tones) on only 9 (less than 1%). As with most small mustelids, there is strong sexual dimorphism in size. In a large sample examined by David Nagorsen, males from the interior averaged 50 per cent heavier than females and those from the coast were about 60 per cent heavier.

Measurements (see next page)

Dental Formula:
   incisors: 3/3
   canines: 1/1
   premolars: 4/4
   molars: 1/2

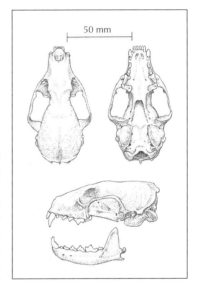

Identification
The two BC carnivore species most likely to be confused with the American Marten (called "Marten" in the rest of this account) are the Fisher and the American Mink. The Fisher is a close relative and is superficially similar to the Marten in body shape and features of the head, but it is larger and usually appears stockier. Adult female Fishers are 50 to 100 per cent larger than the largest (male) Martens, approaching or exceeding the size of a Domestic Cat, and adult male Fishers are larger – up to twice the size of the females. Unlike Martens, most Fishers have a grizzled appearance from head to shoulders, from light-tipped guard hairs, and usually do not have the lighter coloured bib commonly found on Martens. Many Fishers have irregular-shaped patches on the chest and abdomen, but only in white to cream and never yellow or orange. The American Mink is similar in size to the Marten, but has shorter fur, a less bushy tail, smaller ears that are rounded rather than triangular in profile and the bib patch, when present, is only in shades of white. American Minks are usually seen near or in water (marshes, streams, ponds, lakes), where they swim and dive with alacrity, while Martens enter water only rarely and reluctantly, and never dive beneath the surface.

Measurements:

	All Specimens	abietinoides	Subspecies actuosa	caurina	nesophila
total length (mm):					
male:	617 (577-685) n=511	629 (n=198)	621 (n=4)	604 (n=253)	637 (n=56)
female:	554 (515-599) n=256	562 (n=70)	554 (n=3)	546 (n=156)	578 (n=27)
tail vertebrae (mm):					
male:	196 (162-230) n=511	192 (n=199)	176 (n=4)	199 (n=253)	202 (n=55)
female:	178 (144-202) n=255	172 (n=71)	164 (n=3)	180 (n=155)	183 (n=26)
hind foot (mm):					
male:	93 (83-104) n=497	96 (n=199)	96 (n=4)	91 (n=233)	94 (n=61)
female:	81 (70-89) n=256	83 (n=70)	85 (n=2)	80 (n=156)	83 (n=28)
ear (mm)					
male:	48 (44-54) n=201	48 (n=197)	49 (n=4)	none	none
female:	44 (40-47) n=72	44 (n=70)	43 (n=2)	none	none
weight (g):					
male:	1045 (750-1625) n=517	1031 (n=198)	1050 (n=4)	985 (n=250)	1317 (n=65)
female:	664 (514-986) n=264	674 (n=69)	709 (n=2)	631 (n=165)	829 (n=28)

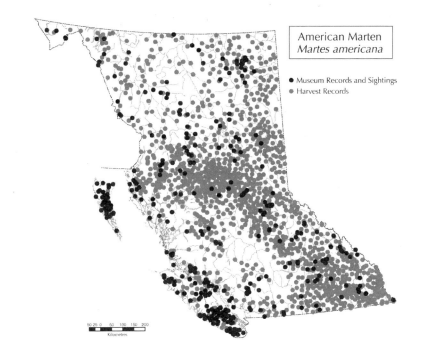

American Marten
Martes americana

● Museum Records and Sightings
● Harvest Records

50 25 0 50 100 150 200
Kilometres

Distribution and Habitat

The Marten occurs throughout most of forested Canada and Alaska, north to the treeline. South of the Canadian border its distribution is limited to forested areas of the mountainous west (in the Rocky Mountains south to northern New Mexico, and discontinuously in the Cascades and Sierra Nevadas), portions of the Great Lakes region in Michigan and Minnesota, and northern New England (Maine, Vermont, New York). In British Columbia, the Marten can be found everywhere except the open interior grasslands and desert country of the central and southern interior. It is common in coastal forests, including on Vancouver Island and the Queen Charlotte Islands, but has not been found on most of the smaller near-shore islands along the coast.

The Marten has long been described as an old-growth-dependent species, suggesting that it only occurs in deep, dark coniferous forests. But observations by biologist Judy Baker and biologist/trapper Jim Hatter on Vancouver Island indicate regular occurrence in young coastal forests, I have seen abundant sign in more open, mixed-forest and shrub areas in northern BC, biologists

Kim Poole and C. Maundrell have found apparently viable populations in deciduous-dominated forests in the Peace River region (near Chetwynd), and many others have reported high Marten densities in expanses of extensively burned areas within a few years after a fire.

Although Martens thrive in old-growth forests, they are not completely dependent upon them. Recent research has indicated that it is the structural make-up of the forest, not its age, that is of greatest importance. Forest habitats other than pure old growth can support the necessary structure, particularly the coarse woody debris that provides cover for Martens and their prey. In Alaska, biologist Thomas Paragi and associates found higher Marten densities in a 10-year-old burn than in a 30-year-old burn; they related the higher density to a higher amount of coarse woody debris and a greater diversity and abundance of small mammals.

In British Columbia, Martens are most commonly found in moist forests (spruce/fir in the central and northern interior and cedar/hemlock on the coast), occurring at all elevations from valley bottom to high subalpine in those habitats. "Pine Marten", a former name, is a misnomer, because large expanses of pure (dry) pine usually do not constitute good habitat, particularly where coarse woody debris and understory vegetation is sparse or lacking, although wet basins and riparian areas within pine stands are commonly used, especially in the north. Martens do not regularly occur in open parkland, shrub or grassland habitats, and in the southern interior, most occurrence is at high elevation, particularly in or near the Engelmann-Spruce-Subalpine Fir Biogeoclimatic Zone. Large logged areas in which woody debris is removed mechanically or burned become poor Marten habitat for decades, particularly in the interior where regrowth is relatively slow.

Marten habitat has been the focus of a number of radio-telemetry studies in recent years, and the daily resting sites used by collared animals have frequently been documented. Among almost 1200 such locations observed by biologists Evelyn Bull and Thad Heater in a northern-Oregon study, two-thirds were in trees at an average of about 12 metres above the ground. Occurring mostly during the snow-free period (May through October), the tree sites were either on platforms (43%) or in cavities (23%). Platform resting sites were mostly in or on branch tangles called witch's broom (caused by fungal or Dwarf Mistletoe infections), but were occasionally on large branches or in branch forks. Most platform sites had thick overhead cover, presumably offering protection both from precipitation and from aerial predators. Other sites used included hollow logs, underground dens and logging-slash piles.

A Marten's long, thin body makes it particularly susceptible to heat loss, and the presence of well-insulated resting sites seems to be critical in winter. Most documented winter resting sites have been beneath the snow, usually in association with stumps, hollow logs, or other woody debris, but also in rock cavities and in burrows dug by Martens or other animals. Martens also commonly use cabins and other buildings in wilderness areas, much to the consternation of their owners, since Martens can make quite a mess. Steve Johnson and I recorded a memorable example of that in spring 1993, at an outfitter's lodge near the confluence of the Spatsizi and Stikine rivers. A Marten had clearly had the run of the place over winter. It had knocked pots, pans, cans and other items from kitchen shelves to the floor, chewed holes in several foam mattresses, opened a flour bin and tracked flour everywhere, and deposited about 300 scats throughout the dwelling (including a large pile on one of the beds); many of the scats had a flour base, and resembled hardened plaster.

Most documented natal and maternal den sites have been in tree cavities, burrows or rock crevices, but some have been found in stumps, hollow logs and slash piles. A female may move her young several times to different sites in the 10–12 weeks they are denbound. Trees and logs used by denning females are usually large, those in the Oregon study averaging 75 cm in diameter and more than 20 metres in height or length. In Wyoming, Leonard Ruggiero and his team of biologists found 10 of 18 natal dens and 12 of 97 maternal dens in structures associated with Red Squirrel feeding middens (conifer cone piles), and they suggested that those locations provided more than protection for the Marten families involved. The only BC records are from Vancouver Island, where Judy Baker found two dens occupied by females with young, both under large stumps.

Although they are rather poorly adapted to cold, Martens are well-adapted to snow, travelling easily over the surface with their large foot-surface-to-weight ratio, and readily travelling, hunting and denning beneath the surface where suitable subnivean structures exist (mostly coarse woody debris, but also the air spaces under spreading shrubs and around tree trunks).

Natural History

Feeding Ecology

The food habits of the Marten have been documented at many locations in North America and, in most areas, small mammals predominate, particularly the Red-backed Vole. But the species is a generalist, with a long list of known prey, including shrews, bats, pikas, Snowshoe Hares, Northern Flying Squirrels, Red Squirrels, Yellow-bellied Marmots, ground squirrels, chipmunks, pocket gophers, voles and lemmings, deer mice, woodrats, jumping mice, Old World mice and rats, and weasels. On the Queen Charlotte Islands, where the Marten subspecies is large and many of the introduced Black-tailed deer are small, two observers have reported Martens attempting to prey on fawns. Both saw a frantic, bleating fawn running through the forest with a Marten on its back, but were unable to determine the outcome.

Martens commonly prey on birds in most areas, and in two coastal British Columbia locations studied by David Nagorsen birds were important winter foods. On Vancouver Island, he examined 701 Marten digestive tracts and found that birds were the most frequent prey at just under 30 per cent. He identified these species: Red-breasted Sapsucker, Hairy Woodpecker, Downy Woodpecker, Northern Flicker, Winter Wren, Golden-crowned Kinglet, Ruby-crowned Kinglet, Varied Thrush, American Robin, Dark-eyed Junco, Steller's Jay, Blue Grouse and Ruffed Grouse. Mammals ranked a close second in that sample at 29 per cent, including Deer Mouse, Red Squirrel, Townsend's Vole, House Mouse, Black Rat, Norway Rat, Eastern Cottontail and a Townsend's Big-eared Bat. Fishes (mostly spawned-out salmon) and ungulate carrion (mostly Black-tailed Deer) were also important dietary items, found in 22 per cent and 20 per cent of study samples respectively. Other items consumed by Vancouver Island Martens were garter snakes, Pacific Treefrogs, Red-legged Frogs, and invertebrates, such as beetles and wasps.

On the Queen Charlotte Islands, where there are no voles and few other small mammals, Nagorsen found birds in 53 (55%) of the digestive tracts of 97 Martens caught in winter. Included were three species of woodpeckers, Winter Wren, Golden-crowned Kinglet, Varied Thrush, Dark-eyed Junco, Song Sparrow, Pine Siskin, Steller's Jay, Cassin's Auklet and American Coot. Other winter foods he documented for that area were deer carrion, some local rodents (Red Squirrel, Deer Mouse, Muskrat and Black Rat), and various invertebrates including some from the marine intertidal zone.

In Oregon, Evelyn Bull found evidence of radio-collared male Martens killing Common Ravens on seven occasions, all in winter. She considered it likely that the ravens had been killed at their night-roost sites, as is probably the case for most of the birds consumed by Martens. Based on tracks in snow, I have documented Marten kills of Spruce Grouse, Ruffed Grouse and Willow Ptarmigan and all appeared to be at night. But Alaskan biologists Thomas Paragi and George Wholecheese reported that a Marten stalked and attacked an adult female Northern Goshawk in daylight; this was a bit of "turn-about-is-fair-play" since Goshawks occasionally prey upon Martens.

In keeping with its generalized diet, the Marten eats a variety of fruits when they are available and scavenges proficiently. In late summer and fall, many Marten scats consist almost entirely of the seeds and skins of berries (Blueberry, Huckleberry, Crowberry, Rose hip, Raspberry, Currant, Mountain Ash, Saskatoon, Soopolallie and others), fruits are likely a particularly important food source for newly independent and dispersing juveniles. Martens are quick to identify feeding opportunities at remote human camps, raiding supplies left unattended and garbage when it is available. In December 1980, a Marten appeared and attempted to carry off a frozen package of hamburger from my food box less than five minutes after my arrival at Turnagain Lake in northern BC. Hours later, I watched the same Marten snarling furiously as it attempted to pull a complete moose hide (at least 50 kg) through a 6 x 20 cm crack between the logs of a storage building. Martens also scavenge kills by human hunters and wild predators; more than once I have seen them scurrying from the body cavity of wolf-killed Moose.

There are relatively few accounts of Marten hunting methods and behaviour. As reported by biologists Wayne Spencer and William Zielinski from observations in northern California, Martens hunted by following tracks of prey in snow, by waiting in ambush near burrows, by locating and robbing bird nests (eggs and nestlings), by excavating burrows and enlarging openings in tree-cavity nest sites, and by scanning from high perches such as tree branches and elevated logs. They also employ the typical mustelid search activity I call "bird-dogging", where the animal zig-zags across the landscape, investigating every nook and cranny. Martens cache larger food items for later use, especially in winter, either burying them in snow or moving them to a resting site.

Martens have a high surface-to-volume ratio resulting from their long, thin shape, their fur does not insulate them very well and they maintain little body fat – thus they are not well-adapted to retain body heat. They compensate for that with a fairly high metabolic rate, requiring them to eat frequently (20–25% of their own body weight or the equivalent of 10 to 12 Red-backed Voles per day for a large male), and by minimizing their exposure to the coldest conditions. Warm resting sites are particularly important. Martens also reduce their activity in winter, often limiting it to daylight hours, when ambient temperatures are warmest.

Home Range and Social Behaviour

There are many publications reporting on characteristics of Marten home range and social organization. In the earliest detailed live-trapping studies (1950s), in Glacier National Park, Montana, biologists Fletcher Newby, Vernon Hawley and Richard Weckwerth found that a Marten population is made up of a nucleus of relatively secure animals (residents, mostly adults) that remain in an area for months or even years, and a more mobile component (mostly younger animals referred to as either "temporary residents" or "transients", depending upon the number of times they were observed) that do not demonstrate a clear, long-term site attachment. Those more mobile Martens constituted the larger portion (65%) of 85 animals captured and marked. The mobile component is the source of replacements for residents that die or emigrate.

Males have larger home ranges than females, averaging in various locations, years and seasons from about 1 km^2 to about 16 km^2. Average home-range sizes for resident males and females, respectively, in study areas in northern and western North America were 2.4 km^2 and 0.7 km^2 (northern Montana), 2.7 km^2 and 1.4 km^2 (northern Oregon), 6.2 km^2 and 4.7 km^2 (southern Yukon), and 14.2 km^2 and 6.8 km^2 (Northwest Territories); there are comparable data from two university graduate studies in British Columbia, both in the early 1990s. On southern Vancouver Island, Judy Baker recorded home ranges of 2.5 to 7.8 km^2 (mean 4.8 km^2) for males, and 1.1 to 4.5 km^2 (mean 2.3 km^2) for females; and in the Bulkley Valley near Smithers, Eric Lofroth documented home ranges of 4.1 to 6.3 km^2 (mean 5.3 km^2) for males, and 1.3 to 4.4 km^2 (mean 3.2 km^2) for females.

In most studies, resident animals that were radio-tracked for long periods maintained about the same home ranges year-round (i.e., without seasonal changes) and in consecutive years. However, boundary adjustments may occur when neighbours disappear or new animals appear in an area, and in some cases apparently stable home ranges may be vacated either temporarily or permanently. In a study in Maine, David Phillips and associates documented home-range abandonments by five adult females, and speculated that those were due to social stresses associated with high density in the untrapped population involved. In the Yukon, Ralph Archibald and Harvey Jessup also recorded home range changes by resident adults in winter. Reports from trappers in BC suggest that home-range changes may be fairly common in late winter, probably related to changes in food abundance or availability.

Observations throughout the species' range confirm that Martens have the typical social organization of mustelids: individuals remain solitary for most of the year and maintain a territorial system in which home ranges of resident animals overlap little with those of their neighbours of the same sex. For a small sample of radio-collared animals in Oregon, biologists Evelyn Bull and Thad Heater calculated the overlap between adjacent resident animals at 2 to 4 per cent for females (average 3%) and 1 to 53 per cent for males (average 13%), while overlap between the sexes was 1 to 100 per cent (average 64%). It appears that females maintain their territories more rigidly, and researchers believed that relates to their resource and privacy needs while producing and rearing young. The mechanism for territory maintenance is not fully understood, but seems to involve response to scent marks and fecal deposits more than physical confrontation. In addition to the anal scent glands characteristic of mustelids, Martens have a mid-ventral scent gland, and individuals have been observed rubbing it on tree branches and ground structures.

Activity and Movements
Martens are active primarily at night, but individuals may continue hunting into daylight hours when having difficulty obtaining sufficient prey, or during particularly cold winter conditions. The naturalist E.T. Seton characterized the Marten's active behaviour:

> *One cannot long watch this creature, even in a cage, without getting an impression of absolutely tireless energy. For hours he will race up and down, leaping from perch to wall, to ground, to perch, to wall, to ground, to perch, over and over again, doing endless*

gymnastic feats, giving countless surprising proofs of strength, with bewildering quickness, all day long, without a quickening of his breath.

Martens are capable of extensive, short-term movements. For example, a recently transplanted animal in Wisconsin moved at least 23 km in 30 hours. Daily movements within home ranges vary considerably, and are difficult to measure, but straight-line distances of 1 to 3 km between resting sites are common. Prior to the advent of radio-tracking, biologists often followed animal tracks in snow and in one such effort, in Idaho, William Marshall documented daily movements of up to 14 km. As with most carnivores, the largest Marten movements involve juveniles dispersing from their natal ranges and adults that have abandoned or have been displaced from previously stable home ranges. Some dispersal movements recorded outside British Columbia are 40 km (juvenile male, Montana), 28–43 km (three animals, Oregon), 50 km (juvenile female, NWT) and 61 km (juvenile male, Manitoba). In north-central BC, three animals radio-collared by Eric Lofroth moved out of his study area and were taken by trappers 51, 82 and 82 km away.

Reproduction

The Marten has a relatively low reproductive potential, due to slow maturation and small litters. Martens breed in July and August, and with delayed implantation, females give birth to young eight to nine months later, in the following spring (March or April). Because mating occurs before the young-of-the-year mature, young females do not mate any earlier than their second summer (about 14 months of age) and, with the long gestation to follow, do not produce young until they are at least two years old. Young males are sexually mature as yearlings, but are not structurally or behaviourally ready for mating until at least the following year.

Mating is vigorous, involving much chasing and squealing, and the male grips the female firmly by the neck with its teeth while mounting. In captivity, pairs copulated as many as 15 times in a few days, in episodes lasting up to an hour or more. Unable to escape for even a short period, several captive females died from injuries inflicted by the males while mating. Steven Henry and his colleagues observed several courtship and mating episodes involving free-ranging radio-collared Martens in California. Those animals also engaged in sustained intensive activity and multiple matings, but the females suffered no adverse effects. In a few cases, mating took place in trees up to 8 metres off the ground, and sometimes the female's kits from the previous year were present while the mating activity was going on. Martens in captivity gave birth to one to five kits per litter (136 litters) averaging about three. Information on wild Marten litters is sparse, and comes mostly from counts of ovarian structures from the carcasses of trapped animals: each *corpus luteum* (CL) represents one ovulation (release of an egg cell), the potential for at least one kit. In Ontario, Marge Strickland and Carmen Douglas compiled data from almost 900 female Martens from the mid 1970s through mid 1980s, obtaining an average of about 3.46 CLs per female for that period. In the Yukon, Ralph Archibald and Harvey Jessup found CLs at a rate of 3.3 to 4.1 in three years of study (average 3.8), while in the Northwest Territories, biologists Kim Poole and Ron Graf obtained CL averages exceeding 4.5 in some years. Most reported sightings of live litters have ranged from one to three kits, although I found an anecdotal report of four, in Colorado in 1950. There is evidence that reproductive performance relates to nutritional condition, with pregnancy rates and CL counts for both yearling and adult females increasing when food is abundant and decreasing when it is scarce. In the Northwest Territories study, Poole and Graf suspected that the primary food variable was Snowshoe Hare abundance.

At birth, Martens are small (about 30 grams), naked and helpless, but like most mustelids, they grow and develop quickly. From observations in captivity, they develop a fine grey pelage by about three weeks, start weaning at about six weeks, shortly after their eyes open, and are near full size (but not weight) by three months. The only applicable observations of wild Martens appear to be those of biologists Steven Henry and Leonard Ruggiero in Wyoming. They saw young Martens first venture outside the den at 10 weeks and follow their mothers and climb trees at about 12 weeks; they observed one animal exploring by itself, away from the den, by 17 weeks. The mother provides all parental care; she begins bringing solid food to the den when the young are 6 to 8 weeks old.

Health and Mortality
Various ticks, lice and fleas have been found on Martens, the variety reflecting the long list of prey species from which they often acquire them. As most trappers know, flea infestations can be severe in years of high population and on individuals in poor condition, and some of those fleas readily dine on humans. Martens

carry a few tapeworms and roundworms in their gastrointestinal tracts, and other roundworms (including Trichina Worms) in kidneys, bladder and muscle tissue, but infection rates and parasite loads are usually low. Given the numbers of animals that have been handled in labs across the continent, the reported incidence of diseases is remarkably low, and none significantly affecting populations has been identified.

With their high metabolic rate and normally low fat reserves, Martens are susceptible to negative effects from food deprivation, especially in winter. An injured or unlucky individual that is unable to obtain food can lose weight and deteriorate into poor condition in just a few days; in an effort to recover it must increase its activity and exposure to the elements. Such animals, often the temporary residents or transients, are particularly vulnerable to baited traps, but even in the absence of trapping it is likely that many of them do not survive. An animal that makes its living by pursuing prey while avoiding being preyed upon is at a considerable disadvantage when it becomes weak.

Among the known predators of Martens are large raptors such as Great Horned Owls, Northern Goshawks and Golden Eagles, and other carnivores such as Grey Wolves, Coyotes, Canada Lynxes, Bobcats, Red Foxes, Fishers and, under certain circumstances, other Martens. Most radio-tracking studies have documented just one or two incidences of predation, but one study in an untrapped area of northeastern Oregon reported significant deaths from predators. There while monitoring 35 collared Martens, Evelyn Bull and Thad Heater found that 18 of 22 animals that died were killed by Bobcats (8), raptors (4), other Martens (4), and Coyotes (2); 4 were females, taken in early summer before their kits were independent, so that predation likely resulted in the deaths of the kits as well as the adult involved.

Interestingly, four of eight Martens killed by Bobcats were buried and uneaten, and the other four were only partially consumed, and one of the two taken by Coyotes was about one-quarter eaten. In north-central Alaska, I found the desiccated carcass of a male Marten, scarcely touched, in a Golden Eagle nest; it appeared that the two large nestlings preferred the ground squirrels and ptarmigan that their parents provided in abundance. These observations indicate that predation on Martens does not always directly benefit the perpetrator.

Among human-caused Marten mortality, the most obvious and significant is fur trapping, although regulations and guidelines are in place to keep Marten harvests sustainable in BC. Unlike the case for many of the other carnivores, Marten habitat is largely separate from most urban and agricultural settings and conflicts with humans are rare. Road kills are fairly common in some areas, and destructive individuals in wilderness camps are occasionally shot.

Martens live relatively long for their size, up to 15 years in captivity, although few reach that age in the wild. In a sample of almost 6500 trapped animals examined by Marjorie Strickland and Carmen Douglas in Ontario over a 12-year period, one female was aged at 14.5 years and one male was 13.5 years old. Just 184 of 4372 males (4.1%) and 107 of 2076 females (5.2%) were 5 years old or older. Based on the survival of radio-collared Martens in their untrapped Oregon study area, Bull and Heater projected that no more than 15 of 100 animals would attain an age of 5 years.

Abundance

The Marten is an important species ecologically and economically, and has been the subject of population studies at many locations throughout its North American range. Most density estimates are based on exhaustive live-trapping and marking, but they are minimum figures, because it is impossible to capture all the animals in a large study area. Further, the population is typically higher in the fall and early winter, when the young are present, than in late winter or spring after many have died or dispersed. With those caveats, most of the estimates that have been reported continent-wide fall within the range obtained by biologists Ian Thompson and Patrick Colgan at one intensively studied area in Ontario: 0.4 to 2.4 animals per km². In that case, the lower figure was obtained in the spring during a year of prey scarcity and the higher in the fall during a year of relative prey abundance.

It is apparent that food supply is a major determinant of population levels, with production of young and numbers of Martens generally increasing when small mammal prey are abundant and decreasing when they are not. Among the factors involved in prey population changes are natural cycles, local weather effects on prey productivity and survival, and changes in habitat quality. Population cycles of about 4 years between peaks for various voles and 10 years for Snowshoe Hares are most apparent in northern latitudes, including the northern half of BC, and the combined effect of predation by Martens and other predators is thought to influence the recovery rate of the prey populations involved. A period of extended snow cover during a late spring is a negative weather factor, usually affecting the prey population for only a year or less. The most important agent of habitat change in much of the Marten's range is logging, particularly when it results in large cutover areas with little or no coarse woody debris or other cover left behind.

In BC, particularly in the north, trappers are well aware that Marten populations may fluctuate dramatically, with several years of relative abundance followed by a period of one or two years when there are almost no juveniles and overall sign is scarce. The last year of abundance, which is sometimes synchronous over large areas, is often characterized by mass dispersal resulting in increased observations of Martens in inappropriate habitats such as expanses of deciduous forest and even lowland agricultural areas. With a population conservatively estimated by the provincial management agency at about 160,000 animals in the mid 1980s, British Columbia probably supports more Martens than any other North American jurisdiction, with the possible exception of Alaska. Still, the long-term trend is almost certainly downward, particularly in the southern half of the province, because of the cumulative effects of logging on habitat quality.

Human Uses

Marten bones have been identified at archeological sites, suggesting that the species was occasionally used for food by some early societies, and it is likely that the people would also have used the fur for winter clothing or ceremonial purposes. Throughout written history and up to the present, humans have used Martens primarily for their fur, a true luxury item.

Taxonomy

American Martens fall into two distinct groups (*americana* and *caurina* forms) that differ in both their skull structure and DNA. The *americana* forms range from eastern North America to Alaska and British Columbia; the *caurina* forms inhabit the Pacific coast from California to Alaska, and parts of the southern Rocky Mountains. Although it has been suggested that the two groups may be distinct species, there is morphological and genetic evidence that they interbreed in a few regions and most taxonomists continue to classify them as one. Of thirteen subspecies currently recognized, four occur in BC, including two from the *caurina* group (*caurina*, *nesophila*) and two from the *americana* group (*abietinoides*, *actuosa*).

Martes americana abietinoides Gray – western Alberta, northern Idaho and Montana, and south-central BC east of the Coast Mountains.

Martes americana actuosa Osgood – Alaska, the Yukon Territory, the Northwest Territories, northern Alberta and northern BC. It differs from *abietinoides* primarily in its larger size.

Martes americana caurina Merriam – from Oregon to northern BC and southeastern Alaska. In BC, this subspecies inhabits Vancouver Island, the coastal mainland and associated islands, and parts of the southern interior. The precise boundaries separating this subspecies from the *americana* forms, *abietinoides* and *actuosa*, in the interior are unknown.

Martes americana nesophila Osgood – restricted to the Queen Charlotte Islands. The largest of the subspecies in the *caurina* group, this form is strongly differentiated from other coastal populations by its large, broad skull and robust teeth. Maureen Small and her co-workers found that it also shows minor differences in DNA from the *caurina* forms on islands of the Alexander Archipelago in southeastern Alaska.

Conservation Status and Management

In British Columbia, the Marten is classified and managed as a furbearer, which can be harvested only by licensed trappers in prescribed areas and seasons. There are no management or sustainability issues in that regard, and none of the four subspecies are considered to be at risk. The most important issue throughout the species' range in North America is the maintenance of suitable habitat during timber harvesting. Referred to as the "Spotted Owl" of eastern Canada by biologist Ian Thompson, the Marten is endangered in Newfoundland despite the fact that it has been protected from hunting and trapping there since the early 1930s. Determination of its needs in relation to those of the forest industry is the subject of considerable study and debate in that province. The Marten is listed as an indicator (focal) species for forest management in areas of Nova Scotia, New Brunswick and Ontario, and in several areas in the continental United States. Over the past two decades in BC, habitat biologists have consistently listed the Marten as a species of special interest and concern in forest management. With trappers leading the way at public planning tables, there is some evidence that government managers and foresters are beginning to take notice.

Remarks

Biologists Ingrid Belan, Philip Leiner and Tim Clark identified seven different vocalizations of adult Martens: "huff", "growl", three kinds of "chuckle", "scream" and "whine". All were recorded while the animals were in livetraps, and probably related mostly to fear or aggression.

There is little published information on the swimming ability of Martens, but they will take to the water occasionally. In Minnesota, biologists David Mech and Lynn Rogers reported that a male crossed a small lake (64 metres width) on at least two occasions in May 1973. I have seen Marten tracks near open water on many occasions in winter, and have documented stream crossings on log bridges or by jumping, but I have never found evidence of swimming in winter.

In a recent study in Alaska, Jena Hickey and her associates found that berry seeds ingested by Martens readily germinate and, given the wide-ranging lifestyle of the species, the Marten plays a potentially important role in seed dispersal and "shaping landscape patterns of berry-producing plants".

Selected References: Archibald and Jessup 1984; Baker 1992; Bull 2000; Bull and Heater 2000, 2001a, 2001b; Buskirk and Ruggiero 1994; Clark et al. 1987; Cowan and MacKay 1950; Giannico and Nagorsen 1989; Gyug 2000; Hagmeier 1961; Hatler et al. 2003; Hawley and Newby 1957; Huggard 1999; Lofroth 1993; Marshall 1951; Mowat et al. 2000; Nagorsen 1994; Nagorsen et al. 1989, 1991; Paragi et al. 1996; Proulx 2006; Quick 1955; Small et al. 2002; Stordeur 1986; Strickland and Douglas 1987; Therrien 2002; Thompson 1991; Thompson and Colgan 1987; Thompson and Harestad 1994; Weckwerth and Hawley 1962.

Fisher   *Martes pennanti*

Other Common Names: Black Cat, Fisher Cat, Pekan, Pennant's Marten, Wejack; *Martre de Pennant, Pékan*. Provincial Species Code: M-MAPE.

Description

Adult female Fishers are about the size of a house cat, and adult males are almost twice that size. Like other members of the weasel family, the Fisher has an elongated body – although it is somewhat stockier than most of the others – with short legs and large feet. Its head is triangular in shape when viewed from above, narrowing at the muzzle, and its rounded ears are fairly large. The Fisher's fur is long and silky, except on the head and neck where it is short, and its conspicuous tail is long and thick, tapering to a pointed tip. It is primarily brown, but from a distance it may appear almost black because of dark guard hairs that cover the legs, belly and most of the rear half of the animal, including the tail. Most Fishers look grizzled

in front, from above the eyes to the shoulders, due to pale silver-tipped guard hairs there. They also have irregular white or cream-coloured patches on any or all of the chest, abdomen and genital area. Measurements for British Columbia specimens, listed below, show the strong sexual dimorphism in size in this species, with the males averaging almost twice the weights of females. The largest known individual, a male from Maine, weighed just over 9 kg.

Measurements:

total length (mm):
    male: 1021 (940-1135) n=53
    female: 912 (845-1043) n=82

tail vertebrae (mm):
    male: 377 (340-410) n=30
    female: 350 (305-391) n=54

hind foot (mm):
    male: 132 (130-134) n=5
    female: 119 (107-126) n=9

ear (mm):
    male: 54 (52-57) n=4
    female: 51 (46-53) n=9

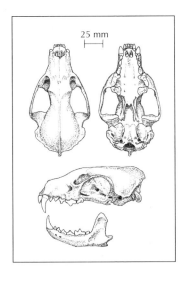

25 mm

weight (g):
    male: 4969 (3600-6835) n=30
    female: 2930 (2050-3750) n=55

Dental Formula:
    incisors: 3/3
    canines: 1/1
    premolars: 4/4
    molars: 1/2

Identification
In the field, there is some possibility of confusing Fishers with American Martens and American Minks (both less than half the size) and Wolverines (about twice the size). The resemblance is greatest between Fishers and American Martens, which are closely related. But in addition to being the larger of the two, the Fisher can usually be distinguished by its round rather than pointed ears,

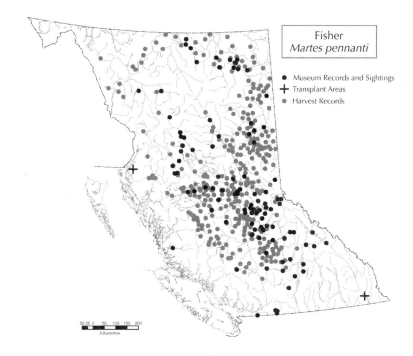

its heavier build, its lighter coloured head, neck and shoulders relative to the rest of the body and the absence of a bib-like, yellow or orange throat patch. Compared to the American Mink, a Fisher has longer fur, a thicker, bushier tail, more prominent ears, a more pointed snout, and the grizzled head and shoulders that Minks do not have. Fishers and Minks also exhibit major behavioural differences – American Minks often swim and dive in streams, lakes and ponds, while Fishers, despite their name, rarely venture into water. Compared to the Wolverine, a Fisher is more slender and streamlined, with a longer, more conspicuous tail, and it does not have the light-coloured lateral lines (one on each side) that show conspicuously on most Wolverines.

Distribution and Habitat

The Fisher occurs only in North America, and the largest portion of its continental range is in Canada. It appears to be most common in the east, with a centre of distribution covering most of Ontario and extending from there southward around the Great Lakes in northern Wisconsin and northern Minnesota, and eastward through southern Quebec, New Brunswick and the northern New England states from Maine to New York. It was believed to have been extirpated in Nova Scotia prior to the 1920s, but has since been reintroduced in that province. West of Ontario, the Fisher occurs in the forested central and northern portions of the Prairie Provinces, but is found in only the southern extremities of the Northwest and Yukon Territories. In the western United States the species still occurs, apparently in low numbers, in northwestern Montana, northern Idaho, and forested portions of Washington, Oregon and northern California.

The Fisher occurs throughout British Columbia, except for the coastal islands. It is not possible to distinguish residents from dispersers among the sighting or specimen records, so it is difficult to delineate areas of regular occurrence in the past. Overall, there is little evidence that Fishers have ever been well-established anywhere west of the Coast Mountains, particularly in the heavy snowbelt to the north, or in the open country and dry forests of the southern interior. Naturalist Allan Brooks reported that the Fisher had previously been fairly abundant in the Fraser River valley area around Chilliwack, but by 1902 it was "very rare" and that is still true. Based primarily on trapping records since the mid 1980s, the Fisher occurs primarily in the central interior (Cariboo-Chilcotin and Omineca regions) and the Peace Region in the northeast. But dispersing animals may appear anywhere, and I have reliable records of recent Fisher sightings in the Okanagan Valley area, outside of their regularly occupied range. Trapper and wildlife control specialist Pete Wise saw a Fisher near Pinaus Lake (Falkland area) in winter 1996–97, and he saw fresh tracks in the same general area in late 1999. Trapper Robert Hooker observed a Fisher just northeast of Enderby in January 2000, and Alison Beal saw one at close range on the Pennask Plateau west of Kelowna in October 2004. The current distribution may now include a local population in the East Kootenays, as a result of a reintroduction program in the 1990s, although biologist Irene Teske advises that the success of that program is still unclear.

In parts of eastern North America, the Fisher subsists in a variety of habitats including lowland, mostly second-growth forests and abandoned farmlands, but it is generally associated with older coniferous forests in the west. It is not considered to be old growth dependent and regularly forages in riparian areas, forest edges, and the thick conifer and shrub patches that often regenerate in areas opened up by forest fire, extensive windthrow, and some logging practices. Biologist Mike Badry and associates found that

Fishers transplanted to aspen parkland in Alberta established home ranges and apparently did well in that rather uncharacteristic habitat. The current view among most biologists who have studied the species is that, although the Fisher is clearly a forest animal (not regularly occurring in open habitats such as grasslands, alpine tundra and extensive new clearcuts), it is fairly catholic in its forest-habitat predilections. In most cases, it appears that the structural make-up of a forest is more important than its age or tree-species composition. Among the structural features involved are coarse woody debris, snags and multiple layers of overhead vegetation (shrubs, saplings, trees), all of which provide cover for both Fishers and their prey.

Most studies of Fisher habitat use have documented an apparent preference for high canopy closure (thick overhead cover), and it has been speculated that this is a response to the potential for predation by raptors. A few incidences of such predation have been reported, but it seems more likely that the Fishers' canopy preferences are more reflective of the security needs of the prey they are seeking rather than their own. In addition, in winter the snow interception and shedding features of a thick forest canopy may be very important to Fishers. A number of studies have shown that they are poorly adapted for travel in deep soft snow, to the extent that University of Maine professor William Krohn and his associates have suggested that both the continental and local distribution of the species may relate primarily to snow-cover characteristics.

In British Columbia, Fishers are most commonly associated with sub-boreal forests in the central interior (Sub-boreal Spruce, Sub-boreal Pine-Spruce and Engelmann Spruce-Subalpine Fir biogeoclimatic zones) and in the mostly lowland boreal forests to the north (Boreal White and Black Spruce and Spruce-Willow Birch zones). They do not regularly occur in coastal forests or mountainous areas with heavy snowfall.

Although they commonly hunt for prey in more open habitats, Fishers usually rest or den in areas with high structural diversity. The day-resting sites of radio-collared animals in various North American studies have most often been high in live trees, either on platforms of branches or witch's broom (abnormal twig growth caused by fungal or Dwarf Mistletoe infections) or in cavities, but some have been in snags, hollow logs, stumps, squirrel and raptor nests, brush piles, rockfalls, burrows, and even abandoned Beaver lodges (accessible from land). In a study by Stephen Arthur and his associates in Maine, the resting sites used most often by Fishers in winter were Woodchuck burrows. In the Williston Reservoir area of north-central BC, Rich Weir found that the selection of resting sites in winter varied, with the animals moving beneath the snow and using spaces in or under large coarse woody debris when temperatures were below freezing. Resting sites are generally used only once, although an individual Fisher may use a particular site for several days in succession if it is near a large food source such as a deer carcass, and there is some evidence that resident animals remember and periodically return to certain sites within their home ranges.

Natal (birth) and maternal (rearing) den sites are almost impossible to find without radio-tracking, but in all areas most have been in cavities in large trees. In British Columbia, Rich Weir located 19 such dens in two studies (13 in the Williston area and 6 in the Quesnel area), and all were in large Black Cottonwood or Balsam Poplar trees (both primarily riparian species). Based on detailed assessments in the Williston area, den trees averaged 108 cm dbh (diameter at breast height), the den cavities averaged 15 metres (maximum of 26 metres) above the ground, and most cavities were associated with heart rot in the tree. Trees of suitable size and decay characteristics were not common in the area and Weir suggested that their availability may be a limiting factor for some local populations. In an ongoing study in the Chilcotin region, where cottonwoods are rare, Larry Davis has located Fisher natal dens in cavities in Trembling Aspens, Lodgepole Pines and Douglas-fir.

Natural History

Feeding Ecology

Although Snowshoe Hares are common and important in its diet in most areas, the Fisher is a versatile hunter and an opportunistic scavenger, and it consumes a large variety of other foods. As listed in studies from several areas across North America, its other prey include many small mammals (shrews, moles, mice, rats, voles and chipmunks) as well as many larger ones including Grey Squirrel, Flying Squirrel, Red Squirrel, ground squirrels, Yellow-bellied Marmot, Woodchuck, Muskrat, Porcupine, Opossum, and several of its fellow carnivores, including American Mink, American Marten, Striped Skunk and Northern Raccoon. The Fisher eats large mammal carrion (usually deer, Elk or Moose) whenever available. Mammals commonly constitute 70 per cent or more of its diet, but the Fisher also eats birds and eggs (grouse, jays, crows, small passerines), snakes, lizards, frogs, insects and a long list of fruits. In the Sierra Nevada of California, at the southwestern margin of the Fisher's range, biologist William Zielinski and his associates found evidence that at least some of the Fishers in their study area were feeding on fungi (truffles).

The ability of Fishers to kill Porcupines is legendary, but the extent of the relationship between the two species is sometimes exaggerated. In 17 major food-habits studies, Porcupine was found in more than 10 per cent of the scats or gastrointestinal tracts examined in only 6 studies, and in 20 per cent or more in only 2. The highest frequency of Porcupine occurrence was 24 per cent, in an Ontario study by Anton de Vos in the early 1950s, and was second only to that of the Snowshoe Hare (30%). Porcupine was the predominant food item only once, also in Ontario but 25 years later, occurring in 22 per cent of the food samples studied. In that study, by biology graduate student Mark Clem, Snowshoe Hare was second at 15 per cent. In the only BC study to date, in the Cariboo Region near Quesnel, Rich Weir found Porcupine in 16.1 per cent of the stomachs examined, putting it fourth in frequency of food items behind the Snowshoe Hare (31.4%), Red Squirrel (26.9%) and Red-backed Vole (18.5%).

Although Fishers do not prey on them frequently, Porcupines are relatively large animals and a kill provides a Fisher with an important amount of food. The species' predatory relationship with Porcupines has been studied by Malcolm Coulter, the first serious student of Fisher biology (1960s), and biologist Roger Powell, a currently recognized authority on Fishers. Both found that the Fisher's usual approach is to repeatedly attack and bite the Porcupine's face, which is free of quills, gradually causing it to weaken until it is unable to defend itself. Powell noted that the Fisher is probably the best adapted carnivore for that strategy, built low enough with its short legs so that it is able to attack straight ahead (not reaching up or down) when on the ground, quick enough to avoid swings of the Porcupine's studded tail, and able to descend trees head-first so that the facial attack can still be undertaken even if the Porcupine tries to escape by climbing. At the age of two years, a captive female Fisher raised by Powell killed a Porcupine by the facial attack method instinctively, that is, without previous experience or parental instruction. The killing process took a long time, ending with the Fisher "pulling the top off the Porcupine's head with her final, killing bite".

Fishers are active mostly around dawn and dusk and at night, but may also hunt during the day when hunger dictates. They hunt by zig-zagging through patches of cover where prey is abundant, and by darting in straight lines from one patch to another. Winter hunting and scavenging success in Roger Powell's Michigan study area was low, with only 14 food encounters recorded in 123 km of snow-tracking (believed to represent about 21 Fisher-days of activity). The foods involved were one Snowshoe Hare, one Porcupine, two squirrels, two mice, some deer carrion, and seven occurrences of bits of hide and hair at old kill sites. Powell calculated that, to maintain weight in winter, a large Fisher needs to eat the equivalent of about 1/14 of a Porcupine, 1/3 of a Snowshoe Hare, 1 Red Squirrel or 12–16 mice per day.

Home Range and Social Behaviour

The home ranges of Fishers have been determined in several areas, although the numbers of animals studied have usually been small (seven or fewer). Resident males usually maintain ranges that are two to three times larger than the ranges of females, with a combined average of 31 km^2 and 12 km^2, respectively, for studies in Maine, New Hampshire, Michigan, Wisconsin and California. The largest average home-range size for samples of five or more animals was 33 km^2 (males, Maine) and the smallest was 8 km^2 (females, Wisconsin). In British Columbia, Rich Weir and associates reported much larger home ranges, averaging 169 km^2 for males and 35 km^2 for females in two study areas (Williston and Quesnel). They speculated that the larger ranges were due to a lower density of resources for Fishers in those areas.

Studies have generally confirmed that, as with most of the other mustelids, Fisher social organization takes the form of intrasexual territoriality. Established residents of both sexes remain solitary for most of the year, maintaining their home ranges as distinct territories that overlap very little with those of their neighbours of the same sex. During the breeding season, males in search of mates may travel widely outside their own territorial boundaries and, therefore, within those of others; transient animals (mostly juveniles) in search of vacant habitat also travel extensively and, unavoidably, through established resident territories. Biologists still do not know exactly how Fishers maintain their territories. Some aggressive encounters have been documented, but scent marking with urine and possibly glandular secretions has also been observed and probably plays the major role. In Ontario, naturalist Ronald Pittaway observed a male Fisher apparently laying claim to a White-tailed Deer carcass by depositing and rubbing urine on it.

Activity and Movements

As determined by radio-tracking, Maine Fishers studied by Stephen Arthur and William Krohn moved an average of 2 to 3 km every 16 hours. Seasonally, adult males and females moved similar distances in summer (about 2 km), but males moved more than females in other seasons, especially in spring during mating. Typically, the largest recorded movements made by Fishers have been those of young transient animals dispersing from their mothers' home ranges in fall and early winter and by animals transplanted into unfamiliar surroundings. Among the longest juvenile dispersal movements documented outside of British Columbia are 30 km (male, Maine), 42 km (male, Idaho) and 60 km (sex not specified, Manitoba). In Rich Weir's Williston study area, a juvenile male successfully established a home range after a dispersal of 20 km, while a juvenile female that was unsuccessful in its dispersal attempt moved at least 132 km, including a straight-line distance of 74 km in 8 days, before dying of apparent starvation at a distance of 77 km from where it had been released after capture. Males probably disperse farther than females, but there are as yet too few juvenile dispersal records to confirm that. Dispersal distance relates in part to the availability of vacant habitat, so in the heavily trapped area studied by Stephen Arthur and associates in Maine the distance was relatively short (average of 10.8 km for males and 11.3 km for females). The researchers noted that "dispersal of nearly all juveniles allowed them to quickly replace adults removed by fur trapping".

Fishers have been the focus of reintroduction programs in several areas, and radio-collared animals in those efforts have also provided perspective on the extent of movements by transient animals. Translocated Fishers have commonly travelled distances of 90 to 100 km between release sites and newly established home ranges, and one animal in Montana moved 163 km. In BC, Fishers translocated from the Chilcotin River drainage to Rich Weir's study area southeast of Quesnel wandered extensively, moving up to 276 km (females) and 1055 km (males), crossing large rivers and rough topography in the process. One male moved 53.2 km in less than 68 hours.

Reproduction

Late winter and early spring must be a difficult period for most adult female Fishers. Having survived the winter, they have to contend with pregnancy, birth and early rearing of young, and then the rigours of mating. Most young are born in March and April, and females come into heat and may breed again within a few weeks, sometimes within days, of giving birth. A nearly year-long pregnancy, featuring delayed implantation, follows. In Fishers, the delay is 10–11 months, and the actual period of pregnancy after embryos implant and resume development is 35–40 days. In British Columbia, the average birthing date for 10 litters of radio-collared female Fishers monitored by Rich Weir in the Williston Reservoir area (1997–2001) was April 6 (range: March 30 to April 19); another 10 litters produced by wild-caught females in a holding facility for a reintroduction project in the East Kootenay region (1997 and 1998) were born between March 17 and April 4.

Female Fishers breed for the first time at the age of one year (at the earliest), and therefore may produce their first litters at about two years. Males reach sexual maturity as yearlings, but are not known to take part in mating until they are two years old. Although most direct observations are of captive animals, it is evident that mating is a vigorous and energy draining experience for both sexes. Males travel extensively in search of receptive females, often outside the boundaries of their familiar range and probably forage less than usual and with less success. There are few descriptions of actual mating, and none from the wild, but copulations as long as five hours in duration have been reported. Assuming that the male may bite the female during the process of getting her into position, as is the case for American Martens and some other mustelids, there is potential for serious injury. A radio-collared female monitored by Rich Weir in the Williston Reservoir area was found dead in spring 1997 of injuries believed to have been caused by a male.

The pregnancy rate averaged 97 per cent for almost 1200 adult females examined by Carmen Douglas and Marjorie Strickland in Ontario over a 12-year period. But rates as low as 60 per cent have been detected in smaller samples elsewhere, and both nutritional deficiencies and inadequate numbers of adult males have been speculated as causes. In BC, adult females studied by Rich Weir in the Williston area were known or suspected to have produced young in 13 of 19 cases (68%), although some of those that were unsuccessful may have lost young in the womb or soon after birth (i.e., the actual pregnancy rate may have been higher). In an East Kootenay Fisher reintroduction project undertaken by Anna Fontana and associates, 13 of 23 adult females (56%) were pregnant when captured (all from the Cariboo Region) and 10 of those (43%) produced young in captivity.

Although females in captivity have given birth to litters as large as six, many of those in the wild bear one to four young in a litter and the average is usually two or three. Relevant information from British Columbia includes examination of carcasses provided by some northern trappers in the late 1970s, in which biologist Karen Liskop and associates found evidence of two or three kits per litter; and of the ten litters born in captivity at the East Kootenay holding facility, one had one kit, three had two, five had three, and one had four (average 2.6).

Newborns are blind and naked except for a light grey fuzz along the upper back; they are helpless, but they grow and develop quickly. Kits born in the East Kootenay holding facility were about 50 grams at birth. Generally, newborns open their eyes and begin eating solid food at about 7 weeks, at which time they have increased in weight by eight to ten times, they are fully mobile (walking and climbing) by 10 weeks, and are able to kill prey and are becoming independent (though not yet fully grown) at 18–20 weeks.

The mother provides all parental care, which is not an easy chore. In the East Kootenay reintroduction project, "kits were very aggressive nursers and the mother's abdomen and teats were often scratched and reddened. The kits' tiny claws were caked with blood." Further, biologists Roger Powell and Richard Leonard calculated that a female Fisher with a litter of seven-week old kits has an energy requirement 2.3 times that of a non-reproductive female, spending up to 21 hours away from the den (presumably hunting) and travelling up to 13.4 km in that time period. In BC, Rich Weir monitored a natal den in the Williston Reservoir area continuously from April 8 to May 1, 1999 – the mother was away from the den for 8 to 20 hours per day (average 10).

Health and Mortality

Fishers appear to be fairly free of ectoparasites, with only one species of tick, one flea and one mite reported throughout the species' North American range. In one case of infestation involving the mite, an adult male in Maine had a serious case of mange, with "large areas of encrusted skin on the ventral surface of the body and on ... the legs"; the animal was thin and in apparent poor condition. Six species of tapeworms, at least one fluke, and nineteen roundworms (including Trichina Worms) have been found in Fisher specimens, but none seriously impact populations. Carmen Douglas and Marjorie Strickland examined Fishers in Ontario and found the following incidence of diseases and parasites: Aleutian Disease, 2.5 per cent; leptospirosis, 5.5 per cent; toxoplasmosis, 41.0

per cent; Guinea Worm, 2.5 per cent; Kidney Worm, 0.2 per cent; trichinosis, 5.2 per cent. The high incidence of toxoplasmosis, a potentially debilitating disease caused by a protozoan in the blood is of potential concern, but no clear effects have been documented. Because Fishers normally occur at low density and are mostly solitary, they are not particularly disposed to disease transmission.

Even in areas of relative abundance, Fishers are rarely seen in the wild, alive or dead and there were few records of natural mortality prior to the use of radio-tracking. We now know that carnivores and large raptors occasionally kill Fishers, although usually dependent young, animals that are injured or ill, and unsettled animals such as dispersing juveniles and those recently translocated to new areas. The known predators of Fishers include Cougars, Canada Lynxes, Bobcats, Wolverines, Coyotes, other Fishers, and Golden Eagles. Eight radio-collared Fishers monitored by Rich Weir in the Williston area died from Lynx predation (one adult female), over aggressive mating by a male Fisher (one adult female – see page 236), apparent starvation (two juvenile females), injuries during capture (one juvenile male), and fur trapping (one adult and two juvenile females). In the East Kootenay reintroduction project, 3 of 6 kits and 8 of 37 adults released with radio transmitters were known to have died. The 3 kits all appeared to have been killed by predators (Bobcat, raptor and unknown), and the adults died from predation (3 – raptor, Cougar or Bobcat), fur trapping (2) and unknown causes (3).

Despite their proficiency at handling Porcupines, Fishers occasionally take on quills both directly and by ingestion. The presence of one or more quills were found in 5 (29%) of 17 carcasses or pelts examined in northern British Columbia, 120 (33%) of 365 in Maine, 13 (15%) of 89 in New Hampshire and 312 (9%) of 3464 in Ontario. During necropsies of some Ontario specimens in the 1950s, M.J. Daniel of the University of British Columbia found a quill embedded in an adrenal gland, another in lung tissue, and several protruding into the body cavity from stomachs and intestines. Most such occurrences do not appear to have any lasting effect on the Fishers, but there are a few reports of Fishers being blinded by quills in the face. One of Roger Powell's captive animals suffered a serious infection from a quill in its chest, and Powell believed it might have died if it had not received veterinary treatment.

The most common human-caused Fisher deaths are from fur trapping and road accidents – in BC mostly trapping. Fishers do not appear to live long, although that view may be biased because most data are from areas that have been heavily trapped. In a sample of more than 6000 specimens from Ontario, only 1.2 per cent of males and 5.4 per cent of females were older than 5 years; 4 of 3262 females were 10 years old, but none of 2747 males had attained that age. As reported by Rich Weir, the oldest known Fisher was a 12-year-old female, apparently still reproductively active, from BC.

Abundance

Fishers naturally occur in low numbers over large areas, and are difficult to count. In the mid 1940s, based on tracks encountered in the snow along some trapline trails in the Fort Nelson area of northern British Columbia, biologist Horace Quick produced what is probably the first density estimate for the species, about one animal per 206 km^2. Since then, more sophisticated study methods have found densities ranging from one per 2.6 km^2 to one per 20 km^2, a 10 to 100 times higher density than that obtained by Quick. Those figures are from the northeastern United States where most of the intensive Fisher field research has been undertaken.

Little is known about Fisher populations in the west. One of the primary food species in most areas is the Snowshoe Hare and evidence, presented independently by Ian McTaggart Cowan for the 1820s through 1850s and by Oxford biomathematician M.G. Bulmer for the century following, suggested that Fisher populations were affected by the well-known ten-year cycle in hare abundance. Specifically, a decline in Fisher numbers appeared to regularly follow the crash in hare populations by about three years. Anecdotal evidence, mostly from trappers, suggests that Fisher populations have declined in some areas of the province over the past 30–40 years. Given habitat changes associated with human developments and resource use (detailed in Conservation Status) in BC Fisher range, it is indeed likely that current numbers are lower than they were historically. Government estimates for the total BC Fisher population have ranged from as high as 15,000 animals in the mid 1970s (apparently based mostly on conservative extrapolation from densities calculated in eastern populations) to just 1100 to 2700 animals in 2003 (based on conservative extrapolation from one study area). The actual number no doubt varies over time as affected by climate and prey abundance, and more extensive study is required to determine the range of that variation.

Human Uses

Evidence at archaeological sites indicates some early aboriginal use of Fishers for food and ceremonial purposes, but the primary use since European contact has been in the fur trade. In the Great Depression years, when a 16-hour day of hard labour payed a man about $1, Fisher pelts commonly fetched $200 or more. But due to changes in fashion, the market value of Fisher pelts has fluctuated greatly over the years, and has averaged only about $40 since the mid 1990s.

Taxonomy

Early taxonomists recognized three subspecies and two of those were listed for British Columbia by Cowan and Guiguet. But University of Victoria professor Edwin Hagmeier found that the minor differences in skull structure among Fisher populations did not warrant separation to subspecies, and recent DNA studies support that view.

Conservation Status and Management

The history of Fisher occurrence in North America is an important success story in wildlife conservation. From all accounts, Fisher populations over much of the continent had declined to very low levels by about the 1940s. Several factors were reportedly involved, including large scale habitat loss from timber harvesting and agricultural clearing, several decades of intense, mostly unregulated trapping and hunting stimulated by high pelt prices, and incidental losses during predator poisoning campaigns. Wildlife authorities responded with a variety of management actions, such as restricted trapping seasons, the establishment of quotas, local closures, and local transplant and reintroduction projects. Those measures, in concert with habitat improvements due mostly to natural regrowth of cleared areas, resulted in recovery and even increased populations in most areas.

The Fisher's history in BC is not entirely clear, but there is no evidence that declines occurred at the same scale or in the same time period as those in the east. Indeed, some of the transplant programs in other areas in the 1950s and 1960s (Montana, Idaho, Oregon) used Fishers captured in this province. But there is anecdotal evidence of more recent local declines or changes in distribution in some BC areas. For example, Robert Hooker, who grew up on a trapline in the Horsefly area (northeast of Williams Lake), told me that Fishers were common in that area through the 1950s and

American Martens were seldom seen, while the reverse was true 40–50 years later. In that and other such cases, both habitat changes and trapping have been hypothesized, but no clear cause-and-effect relationships have been demonstrated. Some of the habitat changes relate to wildfire suppression and intensive silviculture (both of which prevent or reduce establishment of forest structural stages best suited to Snowshoe Hare production), habitat fragmentation by timber harvesting, and extensive flooding behind hydroelectric dams. Valley flooding may be of particular importance where it affects the availability of natal and maternal den trees, which Rich Weir's research has indicated may be mostly large riparian species, such as Cottonwoods in some areas.

In British Columbia, the Fisher is classified as a furbearer, and may be taken only by licensed trappers within prescribed areas and seasons. There have been no open seasons in four of the province's eight administrative regions (those in the south) since the late 1980s. The difficulty of monitoring Fisher populations over large, remote areas such as the vast wilderness of BC and the uneasiness associated with the knowledge of its historical decline patterns elsewhere have contributed to this species' assignment to the provincial Blue List, meaning that it is considered "vulnerable" and of high management priority. A two-year province-wide trapping closure in 1991–93, later shown to be a "false alarm" in terms of actual versus interpreted population trends, was initiated by the government and accepted by the BC Trappers Association on short notice in favour of erring on the side of caution. In relation to potential habitat concerns, the Fisher is the only BC furbearer designated on the Forest Practices Code list of "Identified Wildlife" (i.e., those requiring special consideration in forest planning).

There have been three Fisher translocation projects in BC. The first was undertaken by the BC Forest Service in the Khutzeymateen area north of Prince Rupert in the mid to late 1980s, in an attempt to control Porcupines thriving in tender young Western Hemlock plantations in the area. The suitability of the area as Fisher habitat was not fully considered and the transplant failed, but did produce some interesting stories. In one, a large male Fisher being transported by a trapper escaped from a cage inside a vehicle and, after an exciting few minutes, leaped out an open window and transplanted itself approximately midway between its Burns Lake area origin and the intended destination near Prince Rupert.

In the early 1990s, Rich Weir moved 15 Fishers from the central to eastern Cariboo, to provide a sufficient base population for his graduate study program in the latter area. Several of the translocated animals eventually settled into home ranges, but the implications of that for the local population are not known. The most recent transplant program, some aspects of which are still in progress, is that undertaken in the East Kootenays by Wildlife Branch biologists Anna Fontana, Irene Teske and their associates. With an objective of reintroducing the species to an area believed to have supported it historically, they brought 61 Fishers from the Cariboo between 1996 and 1999. Recent monitoring has obtained evidence of some post-release reproduction in the area, although the overall success of the program is yet to be determined.

Remarks

Some authorities believe that the name "Fisher" came from "fitch", a European mustelid that to early North American settlers may have resembled our Fisher. In any case, the name has long been problematic to biologists, as Elliot Coues explains in his 1877 treatise on mustelids:

> The name "Fisher", very generally applied to this species ... is of uncertain origin, but probably arose from some misconceptions of its habits, or from confounding them with those of the mink. The name is entirely inapplicable, as the animal is not aquatic, does not fish, nor habitually live upon fish, and [the name] should be discarded, as likely to perpetuate the confusion and misunderstanding of which it has always to a greater or less extent been the cause. "Pekan" is a word of unknown, or at least of no obvious application, but is less objectionable, inasmuch as it does not mislead.

Selected References: Badry et al. 1997; Banci 1989; Cowan 1938b; Douglas and Strickland 1987; Fontana et al. 1999; Hagmeier 1959; Powell 1982; Quick 1953; Weir 1995, 2003; Weir and Harestad 1997, 2003; Weir et al. 2004, 2005.

Ermine   *Mustela erminea*

Other Common Names: Short-tailed Weasel, Stoat; *Hermine*.
Provincial Species Code: M-MUER.

Description

The Ermine is a small animal, usually noticeably smaller than a
Red Squirrel. It has the characteristic long, thin body of a weasel,
with short legs, a long neck, and short, dense fur. The tail, which
has a conspicuous black brush at the tip in both winter and summer pelage, averages less than half (about 40%) of the head-and-
body length. The head is flat in profile and roughly triangular
when viewed from above, and the ears are rounded and short. In
summer, the Ermine is distinctly bicoloured, uniformly pale to
chocolate brown on the head, back and upper sides, sharply separated from a white to cream colour on the chin, throat, chest and
belly. In most areas it becomes completely white in winter, except
for its dark eyes and the black tip of the tail; but in areas where
snow does not usually accumulate such as the lower Fraser River
valley and along the coast, the change to white may not occur or
may be incomplete. The subspecies vary considerably in size. The
northern subspecies (*richardsoni*) is distinctly larger than the others,
and the southern interior subspecies (*invicta*) is larger than those on
the coast. Among the subspecies in the interior, males average

nearly twice as large as females. Sexual dimorphism in size appears to be less extreme in coastal populations, but there are as yet too few specimens from those areas to confirm that.

Dental Formula:
 incisors: 3/3
 canines: 1/1
 premolars: 3/3
 molars: 1/2

Identification

The Ermine is by far the most common and widely distributed of the three weasel species in BC. It is most likely to be confused with the Long-tailed Weasel which, like the Ermine, is bicoloured in summer (and year-round in snow-free areas) and white in winter, and has a black-tipped tail in both seasons.

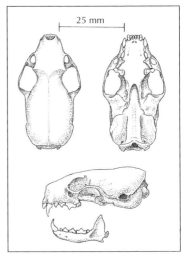

But in summer the Ermine is usually white rather than yellowish on the undersides, and its tail is less than half the length of its head and body, as compared to the Long-tailed Weasel's tail which is more than half its head-and-body length. In addition, in summer pelage the feet of the Ermine are usually white on top, while those of the Long-tailed Weasel are brown. Size differences between the

Measurements:

	All Specimens	anguinae	fallenda	Subspecies haidarum	invicta	richardsonii
total length (mm):						
male:	331 (257-384) n=329	270 (n=25)	294 (n=31)	282 (n=8)	316 (n=62)	350 (n=203)
female:	260 (225-289) n=136	251 (n=9)	238 (n=10)	245 (n=2)	253 (n=28)	267 (n=87)
tail vertebrae (mm):						
male:	96 (60-119) n=340	78 (n=25)	86 (n=33)	71 (n=7)	90 (n=66)	102 (n=209)
female:	72 (60-89) n=140	73 (n=9)	67 (n=10)	61 (n=2)	70 (n=30)	73 (n=89)
hind foot (mm):						
male:	44 (32-52) n=335	35 (n=22)	39 (n=32)	39 (n=7)	42 (n=64)	47 (n=210)
female:	33 (28-39) n=138	33 (n=8)	31 (n=9)	34 (n=2)	32 (n=30)	34 (n=89)
ear (mm):						
male:	22 (15-25) n=247	16 (n=5)	none	19 (n=3)	21 (n=54)	22 (n=185)
female:	17 (14-20) n=89	15 (n=1)	none	none	17 (n=17)	17 (n=71)
weight (g):						
male:	165 (68-225) n=230	75 (n=6)	none	112 (n=4)	144 (n=53)	177 (n=167)
female:	67 (50-100) n=87	54 (n=1)	none	69 (n=2)	79 (n=18)	64 (n=66)

two species are not reliable as a distinguishing feature, since a large male Ermine may be as large as a female Long-tailed Weasel. The Ermine is more easily distinguished from the Least Weasel, which, despite its similar colour pattern in summer and winter, has a more vole-like shape, with a very short tail that does not have a black brush at the tip. The Ermine is also easily distinguished from the American Mink, which is larger (about half the size of a small domestic cat) and mostly brown (rather than bicoloured) in all areas and seasons.

Distribution and Habitat

The Ermine occurs widely in Europe, Asia and northern North America, and was introduced to New Zealand in an unsuccessful attempt to control introduced rabbits in the 1880s. The North American distribution includes Alaska, most of Canada, the northwestern United States west of the Rockies, and the tier of northern states from about Minnesota to New York. It is the most widely distributed of the three weasel species in Canada. In British Columbia it occurs over the entire mainland, on Vancouver Island and the Queen Charlotte Islands, and on some of the other large coastal islands. Two museum specimens reportedly came from Saltspring Island, one in 1935 and one in 1974.

Occurrence and distribution on the islands and in the lower mainland is of special interest, since the species appears to be uncommon in those areas. As shown on the distribution map, there are records from both coastal and inland habitats over most of Vancouver Island, but there have been few in the past two decades. Recent records include six specimens provided by trappers William Martin and Jim Hatter in 1989–91 (four from the vicinity of Duncan and two from Jordan River), and tracks observed at 7 of 155 study sites on northern Vancouver Island by biologist Garth Mowat and his associates.

Recent studies on the Queen Charlottes Islands, led by Don Reid, have added substantially to distributional knowledge in that area, showing historical occurrence over most of Graham and Moresby islands, and also on two smaller islands (Louise and Burnaby) in that archipelago. Most of that information was gathered by interviewing knowledgeable residents, and more than half of the 162 records obtained were recent, involving sightings and specimen records from the 1980s and 1990s.

In contrast, recent contacts with the subspecies *fallenda*, which ranges from the Fraser River valley near Chilliwack to the coast, have been sparse. The only records since 1980 are a sighting on

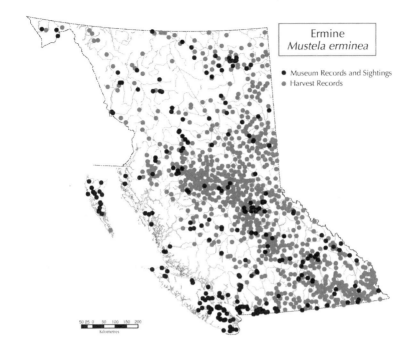

Ermine
Mustela erminea

● Museum Records and Sightings
● Harvest Records

Cortez Island by trapper Wilfred Freeman in 1984 and two animals taken by trapper Doug Williams, one each in 1985 and 1986, on Nelson Island off the Sechelt Peninsula. The last known record from the mainland was a specimen from Haney, taken in November 1972.

The Ermine occurs and thrives in a wide variety of tree-and shrub-dominated and forest-edge habitats, from valley-bottom forests, marshes and riparian areas to high subalpine parkland. It is less common in open forest, grassland and desert habitats at mid to low elevation, such as those in BC's central and southern interior, but will often be found in open areas above timberline. Provided that prey is available, it occupies the same habitats in winter as in summer, using the blanket of snow for heat conservation and protection from predators, and readily hunting in the subnivean spaces and tunnels around and under objects such as tree and shrub stems, boulders, fallen logs and branches (coarse woody debris), and mats of herbaceous vegetation.

In BC, Ermines reach their greatest abundance in the boreal forests of the north and in upland forests in the south, but also occur in openings and along edges of wet forests along the coast.

There is evidence that they prefer young forests. Thus, unlike the case for some of the other mustelids, clearcut logging may improve habitat conditions for Ermines, providing those earlier forest stages with thicker ground cover, less overhead cover and fewer perching opportunities for competitors and predators. In Washington, biologists T. Wilson and Andrew Carey found Ermines using commercially thinned Douglas-fir plantations much more frequently than did Long-tailed Weasels, but in Ontario, snow tracking studies by Ian Thompson and his colleagues found no obvious habitat selection by Ermines regarding the age of forest stands. The only applicable studies in BC have been on the Queen Charlotte Islands, where Don Reid and his associates found roughly equal use of logged and unlogged habitats, and on northern Vancouver Island where Garth Mowat and his co-workers concluded that forest edges and openings were the primary habitats used. Riparian areas were specified as important in both cases and, in fact, on the Queen Charlottes most documented occurrences (77%) were within 100 metres of water, including the ocean.

Day and natal dens used by Ermines are similar, often in burrow systems made by their rodent prey, but also cavities in rock piles, hollow trees, logs, and natural or man-made debris piles. Around human habitations, denning in and under buildings is also common. The actual resting site is usually a nest of grasses, often lined with rodent fur, and it may be used as a home base for periods of up to several weeks, as long as prey are locally available. A female Ermine that appeared a few times inside my house in the Bulkley Valley, near Telkwa, lived in the crawl space beneath for about a month, until she exhausted the local vole and deer mouse populations. In a later maintenance operation, I discovered that she had entered the house through a clothes dryer exhaust hose, in which I found three large steak bones that had become lodged there during transport from my garbage can to her nest.

Natural History

Feeding Ecology

> This little Weasel is fierce and bloodthirsty, possessing an intuitive propensity to destroy every animal and bird within its reach, some of which, such as the American rabbit, the ruffed grouse, and the domestic fowl, are ten times its own size. It is a notorious and hated depredator of the poultry house, and we have known forty well-grown fowls

to have been killed in one night by a single Ermine.
Satiated with the blood of probably a single fowl, the
rest, like the flock slaughtered by the wolf in the sheep-
fold, were destroyed in obedience to a law of nature, an
instinctive propensity to kill.

— John James Audubon

The primary prey of Ermines in most areas are voles (several species) and lemmings. A number of studies have demonstrated that Ermines prey so heavily on those species, particularly in winter, that they negatively affect populations. For example, in a California study based on spring examination of Montane Vole nests, researcher B.M. Fitzgerald calculated that Ermines had eaten 54 per cent of the voles in the study area during one of the winters involved. Female Ermines, better able and more inclined with their small size to hunt in the tunnels and burrows of small rodents and less able to subdue larger prey, are more specialized in their foraging than are males. Thus, most of the Ermines killing animals such as hares, grouse and domestic chickens are the larger males.

Ermines have been known to prey on Jumping Mice, Deer Mice, chipmunks, squirrels, barn rats, Pikas, shrews, Least Weasels, Rock Ptarmigans, small birds (including the eggs and nestlings of Snow Buntings and Lapland Longspurs), frogs, fish, insects and earthworms. A high winter incidence of Meadow Jumping Mice in Ermine stomachs from northern Alberta (30%) and southern Northwest Territories (18%) likely meant that the Ermines had located the rodents' underground hibernation chambers and preyed upon them there. In a study on a Utah wetland, Ermines were the second most important predator (behind Striped Skunks) on waterfowl nests, and in a California study were identified as the primary threat to the nesting success of three species of songbirds.

A number of observers have described Ermines trailing prey with nose to the ground, like bird dogs, and there is one record of an Ermine using its ears to follow and locate grasshoppers. It appears that they use their sight only in close range pursuit. Attacks on vertebrates, especially rodents, are usually directed to the back of the head ending quickly with a bite to the brain or through the neck and spine. As with other weasels, Ermines often wrap themselves around the prey, holding it tightly with body pressure and all four feet.

Ermines have a high metabolism and require an abundant and steady supply of food, requiring each day 19 to 32 per cent of body weight for males and 23 to 27 per cent for females. Without food, individuals in good condition can starve to death within 48 hours. Radio-telemetry studies have shown that Ermines are active (mostly hunting) six hours a day. They hunt mostly at night, but also in daytime, especially if unsuccessful the night before. Excess killing and caching of prey is common, and important for continued sustenance on days when hunting is unsuccessful.

Home Range and Social Behaviour

In Pennsylvania in the 1940s, biologist F. Glover followed tracks of a female Ermine in snow over four consecutive days, and reported that she produced a trail pattern that "resembled a clover leaf", about 250 metres in radius with the home den at the centre. She apparently investigated every nook, cranny and slash pile in that area. More sophisticated studies since, mostly in Europe and New Zealand, have identified home ranges averaging 25–40 hectares for males and 5–15 hectares for females, but with considerable variation depending upon terrain features and the density and distribution of prey.

Studies by Sam Erlinge in Sweden, Carolyn King in Britain and New Zealand, and David Simms in Ontario all indicate that Ermines are solitary for most of the year, and are territorial outside of the breeding season, with resident adults maintaining ranges separate from those of others of the same sex. The range of a male may enclose all or parts of the ranges of several females. Home ranges for this species are difficult to study, because an Ermine population may often comprise 75 to 85 per cent juveniles, most of which have not established home ranges and, as transients, are less secure and more likely to appear in the researchers' baited traps. Further, Ermine life spans and periods of residency are, on average, fairly short, giving little time for the researcher to document and confirm individual patterns. There is some evidence that female home ranges are more stable than those of males, particularly since females do not leave those areas during the mating season.

With large numbers of transients present, it is not possible for an animal to monopolize its territory, particularly if it is dozens of hectares in size. But it is possible to protect high-use areas such as dens and productive hunting spots, mostly by scent-marking and occasionally by physical confrontation. Urine and feces serve as scent markers, but in detailed studies in Sweden, Sam Erlinge identified two behavioural applications of glandular secretions. The "anal drag", which deposits musk from the anal glands on objects such as rocks, tree branches and patches of bare ground, was more common among dominant animals and clearly associated with marking boundaries of important locations in the home area. Chemical analyses showed individual differences in anal gland scent, and subsequent observations in enclosures demonstrated that Ermines detected those differences, reacting in dominant or submissive fashion depending upon their previous experience with the source animal. The other scent-marking behaviour, "body rubbing", involved rubbing the shoulders or belly against objects or along the ground to deposit skin gland secretions. It was used primarily in aggressive interactions, or when an animal was introduced to a nest box used previously by another animal, and was believed to constitute threat behaviour.

Activity and Movements
Normal movements between dens and hunting areas vary, but generally reflect overall home range sizes and are larger for males. In David Simms' Ontario study, males moved up to 1.7 km and females up to 0.9 km between capture locations over a one-week period. But that method does not take animal movements between captures into account, and actual distances moved are much larger – up to 8 km in a single hunting excursion as measured in other studies by following tracks in the snow. During the mating season, resident male movements increase dramatically and may take them several kilometres outside of their established ranges. In a Swedish study, the average range of males during the spring/summer mating season was over 40 times larger than their ranges in the rest of the year, up to 26.4 km^2 in one case.

As with most species, the largest movements appear to involve young males during dispersal from their natal ranges. One tagged Ermine in Alaska was recovered 35 km from its capture location seven months later, and Carolyn King documented several juvenile male movements in excess of 20 km in New Zealand. In contrast, juvenile females usually do not leave their natal range.

Reproduction
Ermines breed in the summer, about June to August, when adult females are still essentially den-bound raising the current year's litter. Adult females are bred while still lactating and, even more unusual, juvenile females may be bred before they leave the natal den. Thus males potentially obtain a "harem" whenever they locate a maternal den, and at the conclusion of the breeding period every female in a population may be pregnant. A road-killed male

I examined near Smithers on June 16, 1985, was in full breeding condition at that time; I am aware of no other BC records. Delayed implantation arrests embryo development for approximately eight months, making the total gestation period about nine months. Most young are born in April or May. Litters may be large, with as many as 13 young, but the average in North America is about 6. The largest BC litter I have heard about was 11, observed by Eckard Mendel near Dease Lake in late July 1978. Aspects of reproductive success, including the proportion of females that give birth, litter size and the survival of young, are all likely to be high when prey is abundant and low to nil when food is scarce. The mother provides all parental care and, given adequate food, the young develop quickly. Small (about 1.7 grams), blind and naked at birth, young Ermines increase in size about 10-fold and are fully haired in three weeks, and take solid food at four weeks. The eyes and ears open and the size difference between the sexes begins to appear at about five weeks, and by six weeks the young males are nearly as large as their mother. Both sexes are 85 per cent grown by twelve to sixteen weeks, and females are sexually mature by that time. Sexual maturity in males does not occur until the next breeding season, at the age of about 10 months.

Health and Mortality

There is surprisingly little information on parasites of Ermines, particularly for North America. Infestation by external parasites is common, and I have seen winter Ermines with so many fleas they appeared to be mottled brown and white rather than pure white. Among internal parasites of Ermines, the one that has attracted the most scientific attention is the Sinus Worm, which can cause severe damage to the skull and may impact the brain in some cases (see American Mink account, page 286). Considering infection rates that have reached 100 per cent in some North American samples, biologists have speculated that Sinus Worms could impact Ermine populations, but that has not yet been demonstrated. Other endoparasites found in Ermines include other roundworms, a few tapeworms and at least one species of fluke. No regular disease condition is known, but Ermines have been diagnosed with canine distemper and have been known to carry tularemia.

Although their flesh does not appear to be preferred, Ermines are killed by a number of larger predators, including the Long-tailed Weasel, American Marten, Fisher, Red Fox, Canada Lynx, Coyote, Domestic Cat, and several species of hawks and owls. Such predation is probably more common when other, more palatable, prey is scarce or when the populations of the predators include high numbers of inexperienced juveniles. Among other causes of natural mortality that have been documented, starvation is probably one of the most important. Given the species' high metabolic demands, an individual that finds itself in circumstances that prevent it from obtaining prey, such as from injuries, illness, or bad judgement, will not survive for long. As an example of bad judgement, biologist William Pruitt reported finding the carcass of a female Ermine high on an icy ridge of North America's highest mountain, Denali, in Alaska. He believed that the animal had arrived there, more than 3 km above the nearest vegetation, "under her own power before dying, presumably of starvation."

Ermines also die from human activities, particularly fur trapping, but also including road accidents and occasional shooting for predator control purposes. None of those are known to affect populations significantly. In New Zealand, Carolyn King studied intensive control and extermination measures taken to protect native bird populations, and concluded that trapping was not effective. It is therefore highly unlikely that regulated sustainable harvests have a negative effect on populations.

The lives of Ermines are fast, hard and typically short. In Sweden, Sam Erlinge calculated that at the age of independence from their mother (about 3–4 months), the average remaining life expectancy was 1.4 years for males and 1.1 years for females. Researchers have noted that it is uncommon to monitor an animal for more than one year, although wild Ermines as old as 4.5 years (male) and 3.5 years (female) have been documented.

Abundance

There have been no Ermine population studies in British Columbia, therefore little is known about densities or trends in the province. An intensive inventory effort on the Queen Charlotte Islands from 1992 through 1998 turned up only two records of occurrence during systematic live-trapping and tracking studies, and in 1997 and 1998 on northern Vancouver Island Ermines were detected at only 7 of 155 sample sites. Those studies appear to confirm that the species occurs in low numbers on the coastal islands.

Studies elsewhere show that Ermine populations fluctuate quickly in response to changes in prey abundance. When prey populations are high, Ermine density can increase to 10 animals per km^2, and when prey is scarce they shrink to almost nil or even temporary local extirpation. Population declines when food is scarce result from reduced reproductive success and survival of young Ermines, as well as increased starvation-mediated predation by larger carnivores, especially American Martens in BC. Researchers studying historical fur-harvest data from eastern North America concluded that Ermine populations in that area were somewhat cyclic, possibly related to the approximately four-year cycles of vole populations. Their conclusion accords with observations of trappers, who report that noticeable Ermine declines occur in BC every few years, but that the populations always "come back strong" a year or so later. In that sense, within the framework of normal ups and downs, the provincial mainland Ermine population can be characterized as stable and may actually be increasing if habitat changes due to logging are indeed favourable for the species.

Human Uses
Many First Nations in North American use the white winter skins of Ermines to decorate headdresses and ceremonial clothing, and stuffed skins are among the ceremonial relics from those societies. That tradition apparently carried over into European culture; Ermine skins were once reserved for use only by royalty and the clergy, and are still used on official ceremonial robes worn by high officials in Canada and other countries.

Taxonomy
In his classic taxonomic revision of the North American weasels (1951), E. Raymond Hall recognized 20 subspecies, with 5 occurring in British Columbia. Although Ermines show considerable variation in colour and size across North America, results from recent morphological and genetic studies suggest that the number of subspecies named by Hall is probably excessive. Nevertheless, until a formal revision is undertaken, the current list for BC retains the original five he proposed.

Mustela erminea anguinae Hall – restricted to Vancouver Island and adjacent larger islands such as Saltspring. Subspecies affinities of Ermines inhabiting islands that lie closer to the mainland (e.g., Cortes Island in northern Johnstone Strait) are unknown. A small form, this subspecies is characterized by minor pelage differences and less pronounced size differences between the sexes.

Mustela erminea fallenda Hall – extreme northwestern Washington and the coastal lowlands and mountains of mainland

BC as far north as Powell River. This is a small form that reportedly remains brown in winter.

Mustela erminea haidarum Preble – restricted to the Queen Charlotte Islands, this is a distinctive form with a broad skull, robust teeth and an extensive black tip on its tail. A recent DNA study by Melissa Fleming and Joseph Cook revealed that Ermines on the Queen Charlotte Islands and islands of the Alexander Archipelago in southeast Alaska are indeed genetically distinct from mainland populations.

Mustela erminea invicta Hall – Alberta, northern Washington, Idaho, Montana and the southern interior of BC as far north as Yellowhead Pass.

Mustela erminea richardsonii Bonaparte – across northern Canada from Quebec to the Yukon Territory; occupying the entire northern and central mainland of the province. Distinguished from other sub-species in the province by minor colour markings and its larger size.

Conservation Status and Management

In most of British Columbia, the Ermine is managed as a furbearer, which can be harvested only by licensed trappers within pre-scribed seasons, and its populations appear to be thriving. However, there is official concern for the two island subspecies, with *haidarum* on the provincial Red List and rated "threatened" nationally by the Committee on the Status of Endangered Species of Wildlife in Canada (COSEWIC), and *anguinae* on the provincial Blue List. There is no open trapping season for either of those sub-species. In the case of *haidarum*, Don Reid and his associates con-cluded that a naturally low food base for Ermines and increasing competition from and possible predation by other carnivores on the Queen Charlottes leaves little hope that the situation for that subspecies can be improved. In the most recent assessment for COSEWIC, biologist Allan Eadie presents a compelling theoretical analysis showing that American Martens are likely the primary limiting factor for *haidarum*, both by competition and predation, and that could easily be the case for *anguinae* as well.

Remarks

The change to white winter pelage appears to occur on the Queen Charlotte Islands, even though snow accumulation is uncommon in most of the lowland areas where Ermines have been detected. Trapper Herb Hughan, now resident in the Nass River valley, has probably seen more Queen Charlotte Ermines than any other human, and he reported that all were white in winter. On Vancouver Island the situation is less clear. A specimen from Quatsino Sound in February 1936 was white "with a little brown around the head", while two taken in mid December 1950 were both "pale brown with no white patches".

In an interesting theoretical account, biologist Roger Powell concludes that the black tail tip is important in winter in confusing potential avian predators. He believes that, against a snowy background, the conspicuous black tip becomes the focus of the predator's attack, and at least occasionally causes it to miss the main target.

Selected References: Eadie 2001; Fleming and Cook 2002; Hall 1951; Hatler et al. 2003b; King 1983, 1989; Lisgo 1999; Mowat et al. 2000; Northcott 1971; Reid et al. 2000; Simms 1979; Svendsen 1982; Wilson and Carey 1996.

Long-Tailed Weasel   *Mustela frenata*

Other Common Names: Prairie Long-tailed Weasel, Prairie
Weasel; *Belette à Longue Queue.* Provincial Species Code: M-MUFR.

Description
The Long-tailed Weasel is the largest of British Columbia's three
weasel species, about the size of our native Red Squirrel. It has the
typical long, thin weasel shape, with short legs, a long neck and
short, dense fur. The feature suggested by its name, the long tail,
averages about three-fifths the length of the head and body, and
has a conspicuous black brush at the tip. The head is flat in profile,
and the ears are rounded and prominent. In summer, the Long-
tailed Weasel is distinctly bicoloured, with dark shades of brown
above, sharply separated from pale cream to yellow below. The
lower cheeks and chin are usually white. In most areas of the
province, this weasel becomes completely white in winter except
for its dark eyes and the black tip on the tail. In areas where snow

does not usually accumulate, such as the lower Fraser River valley and along the coast, the change to white may not occur. Males are up to 40 per cent larger than females in body measurements and are about twice as heavy.

Measurements:

	All Specimens	Subspecies		
		altifrontalis	*nevadensis*	*oribasus*
total length (mm):				
male:	425 (380-486) n=68	401 (n=4)	422 (n=55)	458 (n=9)
female:	359 (316-427) n=22	336 (n=1)	341 (n=15)	407 (n=6)
tail vertebrae (mm):				
male:	158 (140-186) n=77	150 (n=4)	157 (n=64)	168 (n=9)
female:	131 (110-162) n=28	140 (n=1)	125 (n=21)	149 (n=6)
hind foot (mm):				
male:	49 (44-58) n=78	48 (n=4)	48 (n=65)	52 (n=9)
female:	40 (34-53) n=27	40 (n=1)	38 (n=21)	48 (n=5)
ear (mm):				
male:	26 (23-28) n=46	none	26 (n=45)	23 (n=1)
female:	22 (19-25) n=10	none	21 (n=9)	25 (n=1)
weight (g):				
male:	246 (200-350) n=47	none	246 (n=47)	none
female:	142 (120-175) n=9	none	142 (n=9)	none

Dental Formula:
  incisors: 3/3
  canines: 1/1
  premolars: 3/3
  molars: 1/2

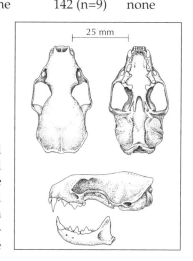

25 mm

Identification
The species that the Long-tailed Weasel is most likely to be confused with is the Ermine. Both are bicoloured in summer (year-round in snow-free areas) and white in winter, and both have conspicuously black-tipped tails. But the

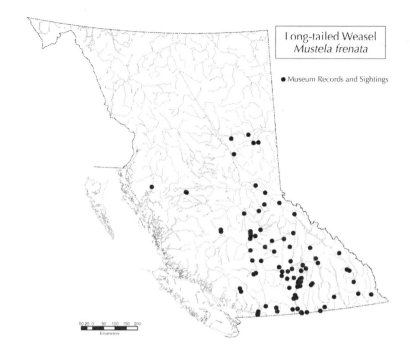

Long-tailed Weasel
Mustela frenata

● Museum Records and Sightings

50 25 0   50   100   150   200
Kilometres

Ermine is white rather than yellowish on the undersides in summer pelage, and its tail is less than half the length of its head and body, as compared to more than half for the Long-tailed Weasel. In addition, in summer pelage the feet of the Long-tailed Weasel are usually brown on top, while those of the Ermine are white. The size difference between the two species is not reliable as a distinguishing feature, since a large male Ermine may be as large as a female Long-tailed Weasel. Long-tailed Weasels are easily distinguished from Least Weasels, which are more vole-like than weasel-like, with very short tails lacking a conspicuous black brush at the tip. American Minks are larger than Long-tailed Weasels, about half the size of a small domestic cat, and appear solid brown (rather than bicoloured) in all areas and seasons.

Distribution and Habitat

The Long-tailed Weasel is the most widely distributed mustelid in the western hemisphere, occurring throughout southern Canada, the United States, Mexico, Central America, Venezuela, and western South America to southern Peru and northern Bolivia. In British Columbia, it is not present on any of the coastal islands, including Vancouver Island and the Queen Charlottes, and there are no records for areas in or west of the Coast Mountains. The areas where most occur are the Fraser River drainage, possibly from the Lower Mainland north discontinuously to at least the Prince George area, and the dry interior of southern BC including the Okanagan and Kootenay regions.

In winter 2003–4, trapper Louis LaRose caught a large male Long-tailed Weasel on the outskirts of Kitimat. His local inquiries indicated that none had ever been seen in that area before, and he speculated that it may have arrived as a stowaway on a railcar. Prior to that, the most northwestern record was at Wistaria, at Ootsa Lake, from three specimens provided by trapper John Shelford in the late 1930s. While that is possibly the northern edge of the species' continuous distribution in the west, the next nearest record is in the Chilcotin River drainage, 250 km southeast. Similarly, records from the Peace River country, near Fort St John, do not necessarily indicate continuous distribution from there south to the Fraser River drainage, a distance of at least 200 km. Rather, they probably represent an extension of a separate distribution along the Peace River system from Alberta. Researchers Gilbert Proulx and Randal Drescher correlated Long-tailed Weasel distribution in Alberta to that of the Northern Pocket Gopher, but that does not appear to be the case in BC.

Over its broad geographic range, the Long-tailed Weasel occupies most habitats available, from valley bottom forests, marshes and grasslands to alpine tundra. It does not occur in the driest expanses of western deserts, but is found in the more vegetated sites associated with watercourses in desert country. In BC, it occurs primarily in the southern half of the province, suggesting that some component or combination of winter conditions may limit its distribution. But I have found it in numbers on the Pennask Plateau west of Kelowna, where a winter snow accumulation of two or three metres is common, and its regular occurrence on the Canadian prairies belies a sensitivity to either snow or cold temperatures. Within its BC range, the Long-tailed Weasel regularly occurs in the Interior Cedar-Hemlock, Interior Douglas-fir, Montane Spruce and Bunchgrass biogeoclimatic zones, and adjacent upland habitats, but not in wet coastal zones or in the sub-boreal and boreal zones to the north. In second-growth Douglas-fir habitat in Washington, Andrew Carey and associates encountered Long-tailed Weasels almost exclusively in forest that had not been thinned by silviculture activities and retained abundant coarse woody debris.

Within its home range, a Long-tailed Weasel usually has one den that it returns to daily and others that it uses as occasional resting sites. Individuals are opportunistic in their selection of sites, as illustrated by reports from farmer-naturalists Norman and Stuart Criddle in Manitoba, who observed a male that had its primary den on a bed of straw inside a threshing machine and a female that regularly slept on a bag of feathers in a farmhouse basement. Of eight more natural dens and resting sites located in Andrew Carey's Washington study, three were in burrows under hollow stumps, three were in logging slash piles, and two were in Northern Flying Squirrel stick nests in trees 6 metres and 21 metres above the ground. Observers in other areas have found daily and natal dens in hollow trees and logs, rock piles, and pocket-gopher and ground-squirrel burrows. It was likely that the original occupants of the burrows had provided food for the weasels before they provided the housing. Burrow dens that have been excavated for examination have usually had a main nest chamber 10–20 cm in diameter lined with layers of chopped grasses and the fur of rodent prey, with side tunnels used as latrines and for prey storage.

Natural History

Feeding Ecology

The Long-tailed Weasel has the least specialized diet of the three North American weasel species. Like the others, it frequently and often primarily eats voles and mice, but its larger size permits exploitation of a greater range of prey species. Other mammals it has been known to eat include Snowshoe Hares, Cottontails, woodrats, barn rats, Muskrats, ground squirrels, tree squirrels, moles, pocket gophers, chipmunks, shrews, Ermines, Least Weasels and bats. Although they primarily hunt small and medium-sized mammals, there are also records of their preying upon other animals, including birds and their eggs (quail, Horned Larks, sparrows, ducks, domestic chickens), snakes and insects (especially grasshoppers). Eric Walters and Edward Miller found evidence of Long-tailed Weasel predation on nesting Red-naped Sapsuckers, Northern Flickers and Hairy Woodpeckers in the Cariboo region of central BC.

A Long-tailed Weasel requires up to 40 per cent of its body weight in food every day. A young captive male weighing 145 grams completely consumed (fur, bones and all) an 85-gram chipmunk and a 105-gram lab rat in 48 hours. Over a 37-day period in

1919, a female feeding six young in northern California brought at least 148 small mammals back to the den – 78 mice and voles, 34 chipmunks, 27 pocket gophers, 4 ground squirrels, 3 woodrats and 2 moles. The effect of such activity on the prey species is difficult to determine, but the Criddle brothers believed that Long-tailed Weasels significantly reduced the numbers of barn rats on their Manitoba farm, and were important predators of pocket gophers. They observed one weasel delivering seven dead gophers to its den in one day. Weasel specialist E. Raymond Hall cited other observations that suggest a regular predator-prey relationship between Long-tailed Weasels and pocket gophers. For example, a technician hired to trap gophers in Nevada captured 22 Long-tailed Weasels in his traps, all set underground in gopher burrows. In a long-term Washington study in the 1990s, biologist Andrew Carey and associates found that Long-tailed Weasels were major predators of Northern Flying Squirrels.

The Long-tailed Weasel is an opportunistic hunter, using its long, thin body to enter its prey's burrow and catch it there, or to tunnel through snow after prey; its endurance and perseverance enables it to pursue prey for long periods. Such pursuits are not limited to ground level. Once while camped under a large pine tree, I awoke to a commotion near my head. I watched a Red Squirrel dash up the tree with a Long-tailed Weasel in hot pursuit. Cornered on a branch more than 20 metres up, the squirrel escaped for that moment by jumping out of the tree. Minutes later the weasel, which had descended in a more conventional manner, picked up the scent where the squirrel had landed, and followed in that direction.

Those who have seen and examined kills report that Long-tailed Weasels often wrap themselves around their prey, snake-like, while biting at the neck and head. The killing bite is usually to the brain or the spinal cord in the neck, often damaging the major blood vessels in the neck in the process. Long-tailed Weasels have a reputation for ferocity and single-mindedness in pursuit of food, and for not stopping to consider the wisdom of an attack. A 200-gram weasel in Iowa had to be shot to stop its attack on a three-day old, one-kilogram domestic pig, and it was later discovered that it had already killed and cached one of the pig's littermates. There are also records of attempts to steal prey from a Snowy Owl and a Swainson's Hawk, both species known to prey on Long-tailed Weasels occasionally. But one of the few published observations on the species in BC illustrates that weasel perseverance is not absolute: in 1987, in Manning Park, Simon Fraser University professor Alton Harestad watched a Long-tailed Weasel withdraw from an attack on a juvenile Columbian Ground Squirrel when it was mobbed by other ground squirrels.

Home Range and Social Behaviour

Most of what is known on the social behaviour of Long-tailed Weasels has come from documenting the daily excursions of individuals by following tracks in fresh snow from and back to their dens. "Daily" in this sense actually means "nightly", since most weasel activity away from the den takes place between dusk and dawn. Winter home-range sizes varied from 80–160 hectares in a 1950s Colorado study to 10–18 hectares in a Kentucky study area in the early 1980s. The differences probably relate to differences in prey density – lower in the Colorado area, requiring the animals to range farther when hunting. The snow-tracking studies have confirmed that Long-tailed Weasels remain solitary for most of the year. Social arrangements are probably similar to those observed for other mustelids, with males having larger ranges than females and resident individuals maintaining territories that overlap little with those of other animals of the same sex, but the literature provides no detailed information on those points.

Activity and Movements

In a snow-tracking study based on five Long-tailed Weasels of undetermined sex on agricultural land in Iowa, daily movements away from home dens averaged just 125 metres and never exceeded 225 metres. But the trails of 11 males and 10 females in Pennsylvania averaged 215 metres and 105 metres, with maximums of 800 metres and 450 metres, respectively. A large male tracked on several occasions by Horace Quick in Michigan travelled an average of 4.4 km per day, with a maximum of 7.2 km.

There have been few studies that have involved marking Long-tailed Weasels so that individuals could be recognized and, accordingly, there appear to be no data on the nature or extent of dispersal. Their dispersal pattern is likely similar to that of most other solitary carnivores, with young males making the longest movements.

Reproduction

Much of what is known about the reproductive biology of Long-tailed Weasels in North America was established in intensive research by Professor Philip Wright of the University of Montana in the 1940s. He observed the species breeding in summer but not giving birth until the following April or May, after a gestation period averaging 9.3 months. The long gestation is due to delayed implantation. Once the embryo has attached to the uterine wall (implanted), the actual gestation is 21–27 days. Litters of up to nine young have been reported, but the average is probably closer to five or six. Newborn Long-tailed Weasels are naked, pink, blind and very small, weighing about 3 grams, but they grow quickly. By three weeks they are fairly well-furred and weigh 20 to 30 grams, and by six weeks their eyes are open and they are nearing one-third of adult body weight. Females reach adult size and maturity at three or four months; most mate shortly thereafter and bear their first litter at about one year. Young males attain full size by fall, but are not sexually mature until the following summer, at about age 15 months. The combination of low age at first reproduction among females, and the relatively large litter size, combine to produce a fairly high reproductive potential.

The mother raises her young alone. For more than 40 years researchers believed that both parents were involved but the early conclusion was apparently based on incorrectly interpreting observations of adult males near known or suspected natal dens.

Health and Mortality

In a comprehensive review paper, Steven Sheffield and Howard Thomas compiled a list of 19 species of fleas, 3 lice, 12 ticks, and 16 chiggers and mites that have been found as ectoparasites on Long-tailed Weasels. It is clear that many of those were contracted from prey species, and the large list reflects the diversity of weasel prey over its broad geographic range. In contrast, the number of internal parasites that have been identified is surprisingly small – only one species of fluke and three roundworms. One of the latter, the Sinus Worm, is also found in the frontal sinuses of other mustelids (see the American Mink account, page 286), and may cause severe damage to the skull of its hosts and possibly reduce the viability of a population. I found obvious worm-mediated skull damage in an adult male from the Pennask Plateau area in winter 1999. Not many diseases have been diagnosed in Long-tailed Weasels, but distemper is known to occur and individuals have been found positive for the bubonic-plague organism in some areas of the United States.

Documentation of mortality has been largely incidental, more often in studies on other species rather than in direct study of the weasels themselves. For example, during research on various hawk and owl species, Long-tailed Weasels have been found among the foods used. Biologists studying foxes have often found dead weasels (all species) at fox dens and, in the 1950s, biologist Roger Latham presented evidence suggesting that foxes were controlling weasel populations in Pennsylvania. He noted that the weasel carcasses were never eaten, but it appeared that foxes killed weasels "at every opportunity". More recently, naturalist Dick Dekker speculated that the "threatened" status of the northern prairie subspecies of Long-tailed Weasel in Alberta, as applied by COSEWIC, is due in part to predation by wild canids, especially the Coyote. Other species known to prey upon Long-tailed Weasels, but without population implications, are the Bobcat, American Marten and Water Moccasin, as well as an unspecified rattlesnake. Human causes of mortality include trapping, shooting for predator control purposes and vehicle incidents. Of 25 road-killed Long-tailed Weasels examined by Joseph Buchanan in western Washington, all but one were males, and all in summer. He concluded that increased activity and travel in search of mates made males particularly vulnerable to road accidents.

There is no specific information on longevity of Long-tailed Weasels but, based on records for Ermines, it is likely that some individuals may live seven to eight years. Marked wild animals have been known to live at least three years.

Abundance

Estimated population densities of Long-tailed Weasels range from less than one animal per km^2 in a western Colorado study area, to 7–9 per km^2 in a forested area of Pennsylvania, to 20–30 per km^2 in a marsh in Ontario. Those figures show that Long-Tailed Weasels can be fairly abundant at some times in some habitats. There have been no formal studies in BC, and fur-harvest statistics gathered to date have lumped all three weasel species together. Accordingly, there is no clearly documented information on populations in the province. From the perspective of BC trappers, the group of people who most regularly make contact with weasels, this species has been increasing in both local abundance and range over the past few decades. For example, Bob Gibbard who has lived in Naramata since the early 1940s, advises that Long-tailed Weasels were rarely encountered in the Penticton area when he was a boy, but are common there now. The general pattern of increase apparently does not apply to the subspecies *altifrontalis*, in the lower Fraser River valley, as there have been no specimens taken or reported since 1937.

Human Uses
Some First Nations, especially those on the prairies, used white weasel tails ornamentally. Writing about one of his exploratory expeditions for the US Geological and Geographical Survey in the 1870s, Elliott Coues noted that skins of Long-tailed Weasels "were in demand by the men of our party for the manufacture of tobacco pouches; they made very pretty ones and many were killed for this purpose." For hundreds of years the primary human use of the species has been the fur trade, primarily for garment trim.

Taxonomy
E. Raymond Hall recognized 21 subspecies in North America, of which 3 are found in British Columbia. No genetic studies have been done to verify their validity.

Mustela frenata altifrontalis Hall – a coastal form ranging from Oregon to the lower Fraser River valley of BC. This is a weakly defined subspecies based on minor skull differences and pelage features such as the extent of colour in the under parts and the presence of a dark spot at the angle of the mouth. The northern and eastern limits of its range in BC are not clear. Hall noted that several specimens from the east slopes of the Coast Mountains near Lillooet were intermediate between *altifrontalis* and *oribasus*.

Mustela frenata nevadensis Hall – the western United States and south-central BC from the eastern slopes of the coastal mountain ranges to Kootenay Lake. It differs from *oribasus* by its smaller size.

Mustela frenata oribasus Bangs – Montana, northwestern Wyoming, western Alberta and southeastern BC, where it inhabits the southern Rocky, Purcell, Selkirk and Monashee mountains.

Conservation Status and Management
In British Columbia, the five known specimens assigned to the subspecies *altifrontalis* all came from the lower Fraser River valley, within an area of about 12 km in diameter. The first of those was obtained near Chilliwack in 1896 by Major Allan Brooks and the last at Vedder Crossing in 1937 by D. Leavens. Based on this limited information, the subspecies appeared on the provincial Red List (as extirpated, endangered or threatened) for several years; but in 2007 its status was changed to unknown.

In the rest of BC, there are no management or conservation issues other than the need to deal with occasional depredations, particularly on poultry. The Long-tailed Weasel is managed as a furbearer, which can be harvested only by licensed trappers and only within prescribed seasons, and its populations appear to be increasing.

Remarks

Unlike Ermines and Least Weasels, the Long-tailed Weasel has no counterparts elsewhere in the world, and is therefore uniquely a New World weasel.

E. Raymond Hall determined that the colour change to white in northern populations is genetic and not an environmental response. Northern weasels kept in captivity in southern climates still turned white during their fall moult, and southern weasels moved into snow country maintained their brown coloration in winter.

The Long-tailed Weasel has three known vocalizations: trill, screech and squeal. The trill is a low intensity sound used around other weasels, especially associated with mating behaviour, and may be equivalent to a cat's purr. The weasel screeches when surprised or disturbed, and squeals when under stress, such as being handled by a human. In my experience, both squeal and screech have been accompanied by a strong-smelling emission from the anal glands.

In general, it appears that Long-tailed Weasels are less common or absent where American Martens are abundant, thus competition with and predation by American Martens may be factors limiting the spread of the species in northern BC. Extensive forest harvesting, which has resulted in reduced Marten populations in many areas of southern and central BC, may have contributed to the spread and increase in Long-tailed Weasels in those areas.

In the early 1940s, Horace Quick estimated a total population of 8000 Long-tailed Weasels in Gunnison County, Colorado. Based on kill rates conservatively estimated at four voles and mice per weasel per day, he calculated that those weasels accounted for the deaths of at least 10 million small rodents annually.

Selected References: Criddle and Criddle 1925, Harestad 1990, Quick 1951, Sheffield and Thomas 1997, Svendsen 1982, Walters and Miller 2001, Wilson and Carey 1996.

Least Weasel   *Mustela nivalis*

Other Common Names: Weasel; *Belette, Belette Pygmée.*
Provincial Species Code: M-MUNI.

Description

The Least Weasel is the smallest of all carnivores, about the size of a chipmunk. Like the other two weasel species, it has a thin body with short legs, a long neck, and short, dense fur. Unlike those species, its tail is short, less than one-fourth the length of the head and body, and is slender and pointed without a conspicuous black brush at the tip. Its head is flat in profile, and the ears are rounded and fairly prominent. In summer, the Least Weasel is bicoloured, with tan to dark brown above contrasting sharply with white to yellowish on the underparts (chin, throat, chest and belly). In winter, Least Weasels in British Columbia and other northern portions of their range are pure white, except for their bright, black eyes and occasional scattered black hairs on the tail. Males are generally larger than females, up to twice as large in some areas, but there have been too few specimens to ascertain the presence or extent of the size difference in BC.

Measurements:
total length (mm):
    male: 180 (149-194) n=6
    female: 174 (158-196) n=3

tail vertebrae (mm):
    male: 30 (23-35) n=6
    female: 27 (26-30) n=3

hind foot (mm):
    male: 22 (18-25) n=6
    female: 19 (18-20) n=2

ear (mm):
    male: None
    female: 10 (9-11) n=2

weight (g):
    male: None
    female: 42 (25-59) n=2

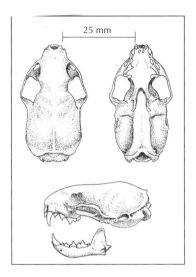

25 mm

Dental Formula:
    incisors: 3/3
    canines: 1/1
    premolars: 3/3
    molars: 1/2

Identification

Like the other two weasel species in British Columbia (Ermine and Long-tailed Weasel), the Least Weasel is white in winter and is bicoloured in summer. Its best distinguishing feature is the short tail (less than 20% of the animal's total length as compared to 40% or more for the other two species) that lacks a conspicuous black brush at the tip. The Least Weasel's small size is also characteristic, but there is some size overlap with female Ermines, especially juveniles.

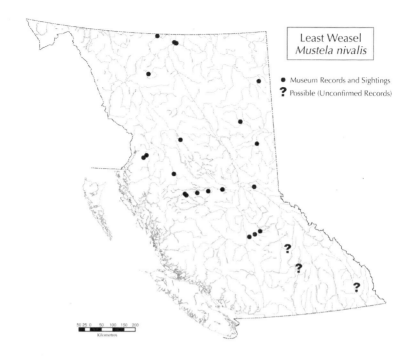

Distribution and Habitat

The Least Weasel is a circumpolar species, occurring widely in Europe and Asia and across northern North America. It has been introduced to New Zealand, and to the islands of Malta, Crete and the Azores. In North America, it has been recorded at many locations on the northern Prairies and central plains, but apparently does not occur west of the Rocky Mountains in the United States. It is believed to occupy most of Canada, with the largest gap being the coast and southern half of British Columbia. The southernmost known occurrence in this province is a specimen collected at Williams Lake in 1983, and there is a cluster of location records from the parkland and grassland habitats of the general Cariboo Region. The westernmost records are of cat-killed specimens observed by both myself (two) and biologist Allan Eadie in the Bulkley Valley area just east of Telkwa in the 1980s. Nevertheless, the Least Weasel is inconspicuous and probably occurs more widely in the province than current data indicate. Possible records (shown on the distribution map) include reports of one-time captures by trappers Valerie Gerber (Vavenby area), Robert Hooker (upland near Enderby) and Stan Smith (near Wasa, East Kootenays). Although all are reliable observers, and their verbal descriptions were compelling, it is not possible to clearly document a larger range for this species, extending into the south, without specimens or photographs.

Least Weasels are most commonly associated with sparsely forested or open habitats such as parklands, grasslands and meadows, including those at and above the timberline in northern mountains. The presence of suitable prey appears to be a more important aspect of habitat than are actual vegetation or terrain features. It is also likely that Least Weasels are excluded from some habitats, particularly the more heavily forested ones, by competition and predation from larger carnivores, especially American Martens and the other two weasel species.

Dens used by Least Weasels are often burrow systems commandeered from their small mammal prey. One such system, found under a haystack by farmer-naturalist Stuart Criddle in Manitoba, contained a central nest chamber that was lined, bottom and sides, with a 2.5-cm-thick, tightly packed mass of hair plucked from voles the weasel had killed. It is apparent that such a den is used repeatedly over a period of time, probably as long as a local source of prey is available. Criddle believed that the weasel in the above case had been living at that site for at least three months.

Natural History

Feeding Ecology

Least Weasels are small mammal specialists focusing mostly on rodents. Of 245 prey items consumed by Least Weasels in a British study area, approximately 80 per cent were rodents and most of the remainder were small birds and their eggs. Prey species recorded in North America include several species of field voles (genus *Microtus*), two species of lemmings, Red-backed Voles, Deer Mice and Harvest Mice. The Least Weasel's small, long, thin body allows it to hunt in the runways and burrows of its rodent prey, but it can also climb trees and swim. As observed in the laboratory, this weasel attacks prey with several quick bites to the back of the head, piercing the brain case or neck vertebrae, usually causing death in seconds.

A negative consequence of the Least Weasel's size and body shape is inefficiency in conservation of body heat. It compensates for that, in part, with a high metabolic rate. Studies in captivity indicate that it needs to eat frequently, ten or more times a day, consuming 40 per cent or more of its body weight just to maintain its weight. It hunts at intervals in daylight and darkness, and the frequency and duration of activity increases when the animal is deprived of food for a time. Captive animals studied by biologist Bruce Gillingham went up to 16 hours between feedings, but he doubted that they could have survived much longer than 24 hours without food.

Because Least Weasels can hunt their primary prey at any time and in almost any location, they are particularly efficient predators. When the opportunity presents itself, they commonly kill more than they need at the moment, and cache the extra prey items in their permanent or temporary dens. Stuart Criddle found five intact voles and parts of more than 40 others in a Least Weasel den. The combination of high hunting efficiency and the ability to quickly increase in numbers enables Least Weasels to exert considerable pressure on local prey populations, and they have been implicated as important contributors to the population cycles of voles and lemmings in some northern areas.

Home Range and Social Behaviour

I am aware of no detailed North American studies on home ranges of Least Weasels. Observers following tracks in snow in Iowa reported that none of four animals used areas of more than about 0.8 hectares, but that was over a short period. In Britain, weasel specialist Carolyn King concluded that most males had home ranges of 7–15 hectares and females had smaller home ranges of 1–4 hectares. Resident adults of the same sex appear to maintain ranges separate from each other, probably mostly by scent-mark signals, but rapid population turnover and high proportions of transient juveniles in populations makes the study of social organization difficult.

Activity and Movements

Daily movements are small. In a study in Poland most Least Weasels travelled less than 400 metres a day but up to 930 metres during a year of low prey abundance.

Reproduction

Reproduction and development in the Least Weasel are geared to respond quickly to the presence of abundant prey. It is the only one of the weasel species that does not exhibit delayed implantation, giving birth about 35 days after breeding. In temperate areas, females produce litters of up to 10 kits, with an average of 5 or 6, but litters of up to 15 have been recorded in the high arctic during a high in the lemming cycle. The young, weighing about 4 grams at birth, develop quickly and are weaned by about seven weeks of age, when 70 to 80 per cent of adult size. Given adequate nutrition (i.e., when prey populations are high), females reach sexual maturity at 12–16 weeks and may produce young before they are half a year old. The short time to weaning also makes it possible, in times of prey abundance, for females to produce two litters a year. A single female producing the first of two litters in the spring could end the year having produced 10 to 12 of her own and providing opportunity for an additional 15 or more young produced by her daughters from the first litter, a theoretical increase approaching 30-fold. In contrast, in years when food supplies are low, there may be little or no successful reproduction.

Least Weasels have no distinct breeding season but, although litters have been documented in most months of the year, winter breeding is rare. On May 13, 1983, I found seven well-developed fetuses in a Least Weasel killed by a house cat in the Bulkley River valley east of Smithers. That appears to be the only record of Least Weasel reproduction in British Columbia, suggesting breeding in the latter part of April that year.

Parental care is provided solely by the female, and anecdotal accounts suggest that mothers look after their young with considerable diligence. As noted by Elliott Coues in 1877, she will "defend her young with the utmost desperation against any assailant, and sacrifice her own life rather than desert them; and even when the nest is torn up by a dog, rushing out with great fury, and fastening upon his nose or lips."

Health and Mortality
Throughout its range, the Least Weasel is a regular host to the Sinus Worm, a species of roundworm that causes severe disfigurement and in some cases perforations of the skull (see American Mink account, page 286). Documented rates of infection have varied from about 5 per cent in one North American sample to a maximum of 81 per cent in a large sample of animals from Britain. The extent of damage in some specimens, which causes skull distortion sufficient to put pressure on the brain, has prompted speculation that the Sinus Worm must affect the viability of individuals and perhaps of local populations. But Carolyn King found no differ-

ences in age, size, or general condition of infested versus uninfested specimens in Britain, thus the nature and extent of effects of this parasite remains a mystery.

Least Weasels have been found to support at least five other kinds of roundworms, one fluke and one tapeworm, and several species of fleas, mites and ticks. None of those are known to have serious effects on populations and no chronic disease problems have been identified, except for those associated with the Sinus Worm. The most regularly documented cause of death for Least Weasels is predation by other carnivores (the other two weasel species, American Marten, Red Fox and Domestic Cat), raptors (Great Horned Owl, Barn Owl and various hawks including American Kestrel), and at least two species of snakes (Western Rattlesnake and Racer). Other mortality factors are difficult to detect, but it is likely that given this weasel's high daily nutritional demands, starvation is common. In Britain, Carolyn King identified declining body condition and weight loss as factors preceding the deaths or disappearance of several animals in her study.

Least Weasels may live three to four years in captivity, but few live past two years in the wild. In Carolyn King's British study, the average life span was about 11 months.

Abundance

There is no information on population densities and trends in British Columbia. The species is only rarely encountered, generally keeping to cover in the daytime and not coming into conflict with humans over livestock or poultry because of its small size. It rarely appears in the trapper's bag because it is not pursued and it is too small to spring the modern, humane traps used for other species. Observations elsewhere indicate that Least Weasel populations can increase quickly and dramatically when prey is abundant, and local densities of 0.2 to 1.0 animals per hectare have been recorded.

Human Uses

There are no known human uses, past or present.

Taxonomy

Of four subspecies recognized by E. Raymond Hall, one occurs in British Columbia. No genetic studies have been done to assess the validity of Hall's designations. Phil Youngman suggested that *M. n. eskimo*, a large, short-tailed form found in the Yukon Territory and Alaska, may occur in extreme northwestern BC, but there are too few specimens available to assess the taxonomy of Least Weasels in that region.

Mustela nivalis rixosa Bangs – eastern Canada and the north-central United States, west to the Northwest Territories and northern and central BC.

Conservation Status and Management

British gamekeepers have a long tradition of vigorously controlling species they consider vermin, including weasels (Least Weasels) and stoats (Ermines) on the estates they manage. Despite some high body counts, there has been little demonstrable effect either on the weasel populations or on the wildlife populations that the control efforts purport to protect. In North America, the Least Weasel is an inconspicuous animal, yet people who are aware of its presence, especially farmers, tend to appreciate the anti-rodent services it provides.

In British Columbia, the Least Weasel is technically classified as a furbearer, although there is no market demand and therefore no harvest effort expended. It has recently been assigned to the province's Blue List, a reflection of our very limited knowledge of the species in this province. We encourage readers to report sightings and deliver any specimens or photographs to wildlife authorities, along with dates and locations.

Remarks

A laboratory analysis identified four vocalizations made by Least Weasels. They chirp or hiss (interpreted as threat sounds) when suddenly disturbed; they trill during mating or in friendly encounters between individuals; and they squeal when in extreme fear, stress or pain.

The pelage of North American Least Weasels is reported to fluoresce to a vivid lavender colour under ultraviolet light.

The species name *nivalis* means "snowy", and is believed by some to refer to the animal's white coat in winter and by others to refer to its mostly northern habitat.

Selected References: Criddle 1947; Gillingham 1984; Heidt et al. 1968; King 1980, 1989; Osgood 1901; Sheffield and King 1994; Youngman 1975.

American Mink   *Neovison vison*

Other Common Names: Common Mink, Mink, Water Weasel; *Vison.* Provincial Species Code: M-NEVI.

Description
Although larger than the three weasel species in BC to which it is related, the American Mink is still a fairly small animal at less than half the size of an average house cat. It has the typical long-and-thin weasel shape, with short legs, a long neck, and short, dense fur. The head is flat in profile, with a narrow nose-pad, dark eyes, and short rounded ears that are usually visible but not prominent. The pelage is generally some shade of brown, but most specimens also have irregular white patches on one or more ventral areas including the chin, throat, chest and groin. The white markings are distinctive and have been used to identify individuals in biological studies. The tail is about half the length of the head and body combined, and is usually a darker shade than the rest of the body, often appearing black toward the tip; while not really bushy, it is well-furred and appears thick when dry. Males average about 20 per cent larger than females in body measurements and nearly twice the weight.

Measurements:

	All Specimens	Subspecies	
		energumenos	*evagor*
total length (mm):			
male:	609 (561-687) n=93	608 (n=42)	610 (n=51)
female:	524 (489-561) n=43	525 (n=14)	524 (n=29)
tail vertebrae (mm):			
male:	188 (158-216) n=93	194 (n=43)	184 (n=50)
female:	164 (148-187) n=43	171 (n=14)	160 (n=29)
hind foot (mm):			
male:	71 (60-86) n=92	69 (n=41)	73 (n=51)
female:	60 (51-66) n=43	59 (n=14)	61 (n=29)
ear (mm):			
male:	26 (22-29) n=76	26 (n=36)	26 (n=40)
female:	24 (21-27) n=35	24 (n=10)	24 (n=25)
weight (g):			
male:	1134 (717-1575) n=75	1054 (n=35)	1203 (n=40)
female:	735 (525-1025) n=37	600 (n=9)	778 (n=28)

Dental Formula:
incisors: 3/3
canines: 1/1
premolars: 3/3
molars: 1/2

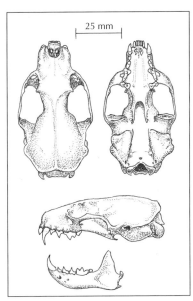

Identification

The American Mink (referred to as "Mink" in the rest of this account) is readily distinguished from the weasels by its solid brown colour in all seasons; weasels are bicoloured (brown above and white or yellow below) in the snow-free season, and most are white with black-tipped tails in the winter. The species with which the Mink is most likely to be confused is the American Marten, which is similar in size and is often generally brown, but has longer fur, larger feet, a bushier tail, ears that are more prominent and triangular rather than rounded in profile, and bib patches that are usually yellow or orange rather than white. The two species also differ in habitat and behaviour, with Minks almost always seen near or in water bodies (marshes, streams, ponds, lakes), where they swim and dive readily. Martens are primarily forest animals, climbing with great agility; they rarely enter water, and only reluctantly, and never dive beneath the surface.

Distribution and Habitat

The Mink occurs widely in North America, in all of the United States except the dry southwest and all of Canada except the extreme Arctic. It has also been introduced, mostly inadvertently through escapes from fur farms, in many European countries and in parts of South America. In British Columbia, Minks are present in at least the lower ends of watersheds over the entire mainland, but may be rare or absent at higher elevations over much of the province. They are also present on Vancouver Island and all but a few of the most remote offshore islands, but not on the Queen Charlotte Islands.

Minks are semiaquatic animals, rarely found far from water. There are two subspecies in BC, one occurring along marine shores on the islands and mainland coast, and the other in freshwater habitats in the rest of the province. Although physically and behaviourally similar, the two subspecies exhibit biological differences, labelled as "coastal" and "inland" in the rest of this account.

The best foraging habitats for coastal Minks are shorelines that are protected from heavy wave action and have an abundance of intertidal structure (boulders, rocky crevices, marine plants) that provides cover for prey: bays, inlets, lagoons, the lee sides of islands, and some rocky headlands in more exposed situations. Large expanses of sand or gravel beach, especially those regularly exposed to surf, are poor Mink habitat, although transient animals may subsist in those areas for a time by scavenging among beach debris. Coastal Minks den in suitable sites above the high tide line, and rarely venture more than a few hundred metres from the shoreline or beyond tidal influence in local freshwater streams.

Inland Minks also concentrate most foraging and resting activity near shorelines, but may be more dispersed over landscapes with abundant water features. They forage along ditches, streams and

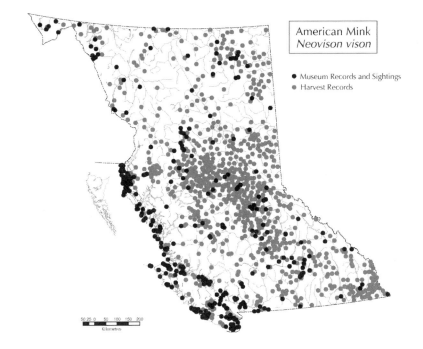

American Mink
Neovison vison

● Museum Records and Sightings
● Harvest Records

rivers, both in the water and in the riparian zone, and are common in wetlands and along pond and lake shores featuring either good growth of aquatic and marsh plants or thick upland cover. Shores with complex structure such as boulders, undercut banks with exposed tree root systems, and woody debris generally provide better habitat than do expanses of shoreline composed of small particles (sand or gravel) and little nearby vegetative or structural cover. Given a suitable food supply, inland Minks may be found along entire watersheds, from the mouths of large rivers to the smallest headwater streams.

In both kinds of habitats, Minks occupy dens for protection from weather and predators, both for production and rearing of young and for daily rest periods. They rarely dig their own dens, using a variety of natural cavities in hollow logs and stumps, under tree roots, in debris piles, and among rock rubble. They also readily use man-made structures, denning in and under buildings, bridges and wharfs, and occasionally in boats. Inland Minks also den in abandoned Beaver lodges and in burrows dug by other species (especially Muskrats and marmots). An individual Mink may use several different dens within its home range over a period of time although, on the coast, the den closest to a particularly good hunting spot may be used continuously over a long period.

Natural History

Feeding Ecology

The Mink is strictly carnivorous but eats a variety of terrestrial and aquatic prey, including mammals, birds, amphibians, reptiles, fishes and invertebrates. Coastal Minks forage primarily in the water or the intertidal zone at low tide, feeding mostly upon crustaceans and fish. I identified nine species of crabs among Mink foods on Vancouver Island, but the most regularly used were the Red Crab (year round), the Helmet Crab (summer), and the Kelp Crab (winter). The fishes caught by Minks in that area were generally small, weighing 15 grams or less, and mostly involved intertidal species such as sculpins, gunnels and pricklebacks. But Minks are opportunistic, taking whatever they can handle, including some larger, more subtidal forms such as rockfish and greenlings. In one case I watched a small female (weighing about 650 grams) expend considerable effort to beach and transport a struggling Kelp Greenling (273 mm long, 375 grams) she had caught off a rocky headland. (Determined and persistent, she reclaimed the fish within minutes after my brief interruption to examine and weigh it.)

After crabs and fish, birds are the next group that Vancouver Island Minks most often prey on. In my study of coastal Minks, I identified many species of ducks, grebes, gulls, shorebirds, alcids and cormorants, but also some more land-based species, such as Yellow-bellied Sapsucker, Song Sparrow, Fox Sparrow and Ruffed Grouse. In most cases, it was not possible to determine if the birds involved had been killed by the Minks or taken as carrion. That said, in winter 1971, island resident Joe Wilkowski watched a Mink stalk, catch and kill a roosting Pelagic Cormorant on a Barkley Sound islet. Charles Guiguet believed that the absence of Minks is a primary requisite for an offshore island in British Columbia to support nesting seabirds. That is very likely true, although individual Minks occasionally reach some of the closer islands. In June 1973, while monitoring gull and cormorant nests on the Chain Islets, near Victoria, Wayne Campbell discovered an active Mink midden (feeding site) and latrine on one of the islets. Cached under a nearby piece of beach debris plywood, he found the carcasses of a Glaucous-winged Gull, a Pelagic Cormorant and several shore-birds, including three Ruddy Turnstones, four Black Turnstones and a Black Oystercatcher. The Mink resided on that islet for the rest of the summer and, although its diet appeared to be primarily crabs, there was evidence that it also preyed upon Glaucous-winged Gull chicks.

Mammals were uncommon in the diets of Vancouver Island Minks. I had a few records of Mink preying on Norway Rats along the Tofino waterfront and at a nearby dump, but in examination of more than 1750 scats I found only three containing mammal remains, two with Townsend's Voles and one with a Deer Mouse. Other foods eaten by Vancouver Island Minks, in order of frequency, included the Sea Slater (a large sowbug that resides in rocky crevices above high tide), beach flea (an amphipod sometimes found in high numbers in wave-cast marine vegetation), and molluscs such as mussels, Basket Cockles and Northern Abalones.

Inland Minks have more variable diets than those on the coast, hunting in or near fresh water for fish, frogs, snakes, crayfish, Muskrats, insects such as water beetles, ducks, geese, coots, rails, shorebirds and blackbirds, and in more upland habitats for voles, mice, ground squirrels, marmots, moles, hares, rabbits and a variety of land birds. Whether involving cases of extreme hunger or just characteristic boldness, Minks often tackle prey larger than themselves. Among published records is one account of a Mink attacking a two-thirds-grown Trumpeter Swan cygnet. The Mink had attached itself to the cygnet's neck and would likely have succeeded in making the kill had the parent birds not intervened.

From metabolic studies of captive animals, I calculated that a male Vancouver Island Mink would require 12 average sized crabs or 100 small fish to meet its minimum daily requirement, while a female without maternal demands could subsist on about one-third fewer of either food. Based on conservative interpretation of hunting success in 452 direct observations of foraging Minks, I concluded that many would have met their requirements in about two to three hours of hunting per day (often much less). Such successes aside, Minks that fail to obtain sufficient food can lose weight at a rate of five to seven per cent daily, losing condition rapidly.

Efficient hunters that they are, and possibly as an innate hedge against periods of food deprivation, Minks often kill beyond their immediate needs. They regularly cache surplus items, a habit that is no doubt beneficial in some seasons in inland and northern areas, but which results in spoilage and waste in others. The minks did not eat any of the birds that Wayne Campbell found in the cache on Chain Islets, and I often found spoiled crabs and fish in caches in my study area. Perhaps more importantly, Minks can have a large impact on a prey population in a short time by returning again and again to a location where it is concentrated or vulnerable. In one case, a single Mink exterminated an introduced rodent population on a small island in Georgia, killing 41 of 42 animals in less than a month (and spoiling a biology experiment). Near Cordova, Alaska, biologist Pete Mickelson saw a female Mink "hanging by her hind feet (in sedge and grass) over a cutbank to reach in and enlarge Bank Swallow burrows. In about 1.5 hours, she cleaned out the entire colony (30+ adults)". Biologists Andre Breault and Kimberly Cheng reported a Mink killing 50 nesting Eared Grebes and 1 American Coot on a wetland near Williams Lake, eating portions of only 3 of them. At least 39 of the grebes were killed within three days, perhaps on the same night. Approximately half of the nests in that grebe colony were affected. The effect of coastal Mink predation on populations of intertidal organisms is difficult to assess, but I documented one case in which a Mink caught a minimum of 276 Kelp Crabs from a very small islet (less than 75 metres diameter) in less than two months.

Home Range and Social Behaviour

The ranges of male Minks are generally larger than those of females, but there is considerable individual variation, particularly as related to habitat characteristics (mostly distribution and abundance of prey) and probably also to population density. Minks foraging in the rich intertidal zone along Vancouver Island met their daily needs along very small sections of shore, and observed home ranges there are among the smallest known (averages of 0.56 km for 44 adult males, 0.41 km for 11 adult females, and 0.65 km for 95 animals of all sex and age classes). Ranges of feral coastal Minks in southern Scotland were somewhat larger at 1.5 km of shoreline for males and 1.1 km for females.

Average home range lengths documented for inland minks were about 2.5 km for males and 2.0 km for females in southern England and Sweden, and 6.8 km for males and 2.7 km for females in north England. Three males radio-tracked in eastern Tennessee had average range lengths of 7.5 km, with a maximum of 11 km. In the Prairie Pothole region of southwestern Manitoba, home ranges of six males were more two-dimensional than in other areas, covering 3.9 to 16.3 km^2, and averaging 6.5 km^2. All of the above figures relate to time periods that do not include the breeding season, when males travel farther and expand their ranges.

A Mink population consists of a relatively stable and productive core of residents and an often larger component of transients. Transients are usually juveniles dispersing from their birthplaces, but also include adults displaced from previously secure home ranges. Such displacement occasionally occurs when a Mink is unable to defend its home range against a larger or more aggressive intruder, but is most often a case of forced departure because the local food supply has failed or some other important feature of the home range has been lost.

Adult Minks are solitary except for short periods during the mating season, and when females are accompanied by dependent young. Adults remain solitary and separate as a result of a territorial social system, in which there is little home range overlap between neighbouring residents of the same sex. Minks mark their territories with feces, urine and musks from scent glands, and sometimes defend them with aggressive behaviour, most often expressed in bluff and ritual displays rather than serious fighting. I observed a number of encounters on Vancouver Island, and the usual result was for one animal to yield and flee before contact. When contact did occur, it was brief and often noisy, but caused little or no apparent damage to either participant.

Activity and Movements

Minks remain active throughout the year. In inland areas where water bodies freeze over and snow accumulates, they usually travel and hunt beneath the ice and in tunnels under the snow and may leave little visible sign of their presence during most of the winter. An exception occurs during the mating season, in February and March, when males search for females and may travel extensively on the surface. In most areas and seasons, Mink activity usually occurs between dusk and dawn, mostly in darkness. But the activity pattern of coastal Minks is also strongly influenced by the tidal cycle; they often travel and forage during daylight hours in periods of low tide.

Minks with home ranges that support a rich and concentrated supply of prey, a situation most common on the coast, may move little throughout most of the year. Such individuals may spend 20 hours or more each day resting, making only occasional short forays to obtain food or to mark and defend their territories, and their daily movements may amount to only a few hundred metres. On the other hand, transient animals and those with poorer home ranges, both on the coast and inland, may require daily movements of 10 km or more to satisfy daily food requirements. The largest annual movements are made by males during the mating season, when they travel well outside their territories, and by transient animals of both sexes. Juveniles are known to have dispersed up to 45 km in inland habitats, but longer distance movements likely occur. The longest move I documented on the coast was 9 km, undertaken by a young male that was tagged on Vargas Island in March 1969 and subsequently killed on the highway southeast of Tofino in July. That journey involved a substantial water crossing, at least 1.5 km if taken directly toward Tofino, and at least 0.8 km if island-hopping to the north.

Reproduction

Inland Minks breed in mid February through March in most areas throughout the species' broad North American range, but the mating season for coastal animals in BC is primarily in late May through mid June. Pregnancy is characterized by delayed implantation, up to five months inland, but as little as seven days on the coast. The actual gestation period, once the embryos have implanted, is about 30 days. The timing of birth has not been well-documented for inland Minks, but it occurs in the spring, probably in late April through early June. Coastal Minks are born primarily in mid to late July.

Mating is vigorous and prolonged to induce ovulation, and the beginning stages look more like fighting. The male pursues and catches the female, and secures a grip on the back of her neck with his teeth to hold her in position. The mating chase features much vocalization, mostly screeching and squealing by the female, and I saw a number of attempts that failed when the male did not get the right grip and was bitten for his trouble. In my only observation of a successful mating in full sequence, from chase to release, the female struggled and vocalized frequently in the first five minutes, but did so only occasionally as time went on. Each time the female struggled, the male tightened his grip on her neck, and at times he visibly chewed. He maintained the neck hold for the duration of the mating – an hour and 50 minutes – and both animals showed signs of fatigue afterward. Most females on my Vancouver Island study areas had open or healing neck wounds in June and July (figure 53). Two or more males can mate the same female resulting in a litter of kits with different fathers. That is likely the reason why males do not appear to compete or fight for mating privileges. In one case, at least three different males attended the home range of a female on a small islet at the same time, and largely ignored each other.

Most juvenile females are sexually mature by the first breeding season after birth (at 9 to 10 months inland and 11 to 12 months on the coast), but will likely be successful in producing young

Figure 53. Female Mink showing a neck wound incurred during mating.

only if they have obtained a secure home range with a sufficient food supply. In general, the size and vigour of a litter is related to the age and nutritional condition of the female. In the Vancouver Island study area, males were more abundant than females and appeared to impede production of young, directly by interrupting female foraging activity and causing injury during mating, and indirectly by dominating the best hunting spots. This may be a natural mechanism for regulating the population, suppressing production with an excess of males when the habitat is full and providing a more favourable environment for females when the male component is reduced by mortality.

Litters of up to eight young have been reported, but the average is usually about four. Newborn Minks, covered with a sparse coat of fine, silvery hair, are blind and tiny, weighing about 6 grams. They grow quickly, doubling their weight in the first week; by three weeks they are fairly well furred and weigh 40 to 50 grams. By five weeks their eyes open and they start to be weaned. And by eight weeks they follow the mother outside the den and begin hunting. The mother provides all parental care, and she may move the kits to fresh dens one or more times when they are fully dependent. I apparently interrupted such a movement in early August 1969, when I found a tiny kit on a trail near a den (figure 54). The mother must have dropped it there as I approached, then returned shortly afterward, because it was gone when I checked 20 minutes later.

Health and Mortality

The inland Mink has been semi-domesticated and is the most important species for fur-farming worldwide. As a result, much is known of its physiology and factors relating to its health. Diseases (anthrax, distemper, botulism, tularemia, tuberculosis, septicemia, enteritis and others) and parasites that affect Minks on fur farms have been identified, but the extent and importance of most of them in the wild are not known. In examinations of coastal Mink specimens for parasites, I found various fleas, ticks and mites externally and only two species of roundworms and one species of spiny-headed worm internally. Most infestations were light and clearly of no consequence, but the effects of two of those parasites warrant some discussion.

The Sinus Worm, a roundworm also found in the frontal sinuses of other small carnivores (especially weasels and skunks), may cause severe damage to the skull of the host. Of more than 100 coastal Minks necropsied, 59 per cent were infected with Sinus Worms and 15 per cent had skull damage evident externally. Swelling in the sinus area, readily visible in most of those, is known to cause pressure on the brain; in the most severe cases, the worms had perforated the top of the skull, and could be removed without further dissection. Although Sinus Worm infection likely affects individual survival, its population impacts have not been demonstrated.

The other common pathology I observed in coastal Minks was a tail wound, seen most often in June and July and apparently caused by mites. It appears that the mites cause an irritation at the base of the tail and the Mink wounds itself while attempting to relieve the irritation. I saw animals licking and nibbling at tail wounds on a few occasions and, although I saw a few with healed scars on and near their tails later in the year, all known (tagged) animals with large wounds and most with smaller wounds disappeared from the study area within a few weeks.

Figure 54. Mink kit found near its den on Vancouver Island in August 1969.

With their ability to hunt effectively on both land and water, Minks have a wider range of prey to choose from and generally maintain larger reserves of body fat than the other small mustelids in BC. But they have a high metabolic rate and their condition can decline quickly if food becomes scarce. Inadequate nutrition can lead to death either directly, by starvation, or indirectly, through reduced resistance to disease and parasites or increased exposure to predation, as animals spend more time searching for food. Transient animals, especially dispersing juveniles, are the least likely to have consistent access to productive hunting spots, and are therefore the most vulnerable. Adult females may also be relegated to a position of nutritional deprivation when the ratio of males in a population is high.

Among known Mink predators are the Bobcat, Canada Lynx, Coyote, Red Fox, Grey Wolf, Fisher, River Otter, and various raptors, including eagles and the larger hawks and owls. Most probably prey on individuals that are weakened or otherwise insecure for food-related reasons. Direct human-caused mortality of wild Minks occurs through legal trapper harvest, road kill, and the occasional killing of individuals preying on poultry or being a nuisance in boats and shore-side buildings. Given the Mink's proclivity for surplus killing, just one in the chicken house can destroy an entire flock in a single night.

As with River Otters, Minks are near the top of aquatic food chains and can have contaminants, such as mercury and PCBs, concentrated in their fatty tissues; for that reason they are commonly used as bio-indicators in pollution monitoring studies. Minks have died from mercury poisoning, and it has been postulated that populations in some areas in Europe and the United States have declined both through direct mortality and reduced reproductive performance. In their examination of Mink specimens from the upper Fraser River drainage near Prince George and from the Columbia River system in the east Kootenay area, Dr. J. Elliott and associates found contaminants to be at low levels. The cause of a die-off in Clayoquot Sound in 1970 was not determined, but there was strong circumstantial evidence that it was due to Paralytic Shellfish Poisoning (Red Tide), which also apparently affected the local Northern Raccoon population.

Minks are not particularly long-lived. Those in captivity may reach eight years, but wild Minks rarely live longer than five years and the average life span in most areas is believed to be about three years.

Abundance

Where there is an abundant and predictable food supply, Minks may occur in high numbers. Researchers express the density of most species in relation to area (e.g., the number of individuals per hectare), but because most Minks live next to water it is customary to express their density in relation to the distance along a shoreline or waterway. In my Vancouver Island studies, I found densities of from one to seven animals per km, with the highest along protected rocky shores where intertidal prey was abundant and the lowest along surf-washed sand beaches where prey was sparse and often difficult to access. Studies of inland Minks have produced similar numbers for small study areas, again relating to local prey abundance. Nevertheless, it is a certainty that coastal habitats support higher population densities and overall higher numbers of Minks than do those inland.

There is no evidence of significant population changes on the coast in recent years, but trappers, the people most regularly in contact with the species, reported a general and apparently widespread decline in inland Minks in central and northern BC from about the late 1980s to the late 1990s. While the cause was not determined, it was speculated that some disease was involved. Recent (2004–6) reports suggest that populations in those areas are now recovering.

Human Uses

The rich, shiny brown fur of the Mink has long been valued for its warmth, suppleness and durability. Influenced by fashion trends and world economics, the wild Mink harvest in BC peaked at 46,284 in 1933 (worth $1.1 million in the otherwise floundering economy of the Depression Era), but has averaged less than 1000 animals annually since the early 1990s. Increased world production of ranch Minks, with current annual pelt sales numbering in the millions, may be one of the factors involved.

Taxonomy

Of 15 subspecies recognized in North America, 2 occur in British Columbia. No genetic studies have been done to assess subspecies validity.

Neovison vison energumenos Bangs – most of western North America, from Alaska to California and New Mexico. In BC it is known from the entire mainland and adjacent islands.

Neovison vison evagor Hall – restricted to Vancouver Island and associated islands along its coast. The subspecies affinities of

Minks inhabiting islands that lie closer to the mainland, in the Strait of Georgia and Johnstone Strait, are unknown. This subspecies was described on the basis of its large body size, large broad skull with robust teeth, and reddish pelage.

Conservation Status and Management

In British Columbia, the Mink is classified and managed as a furbearer that can be harvested only by licensed trappers within prescribed areas and seasons. There are no management or sustainability issues and the species' occurrence in the province is considered secure. In terms of human conflict situations, the most common complaints are on the coast and involve messes and smells made by Minks establishing latrines and prey storage areas on docks and in buildings and boats. Occasionally a Mink develops a taste for small livestock, particularly chickens and ducks, and needs to be removed.

As reported by G. Clifford Carl and his associates in a 1951 provincial museum report, an unauthorized introduction of Minks to Lanz Island (Scott Island Group, northwest end of Vancouver Island) in the late 1930s resulted in the elimination of seabird nesting there within 12 years. As outlined in the feeding ecology section, Minks occasionally arrive and take up temporary residence on nearshore islands supporting seabirds, with some negative results, but the Lanz Island account is the only one from BC involving Mink occupation of a formerly secure seabird habitat. In Europe there are numerous accounts of serious feral Mink depredations in bird colonies, Minks causing local population declines of rare species such as the European Water Vole in Great Britain, and introduced American Minks outcompeting and threatening native species such as the European Mink (a smaller species). Vandalism and raids on Mink farms by animal rights groups set the animals free contributes to those problems.

Remarks

Despite their semiaquatic lifestyle, Minks are not well-adapted for life in the water, having small feet, toes that are only partially webbed, a rather small lung capacity, and a tail that drags rather than propels. Minks dive for prey in shallow water, mostly less than three metres in my own observations, although I recorded one adult male swimming between two islets and diving for about 30 seconds to come up with a Helmet Crab (necessarily, from the bottom, about 7.4 metres deep). Of 264 dives for which I recorded times, 205 (77.5 %) were 20 seconds or less and only 5 exceeded 40 seconds (the maximum was 48 seconds).

In an analysis of ranch Mink vocalizations, Fred Gilbert categorized them into four groups: chuckling, mostly attributable to males during mating; hissing and screaming and low- and high-level defensive threats, respectively; and squeaking, most commonly associated with fear or pain. Screams (or squeals), and squeaks are often accompanied by the release of musk from the anal glands, in a manner similar to skunks. Indeed, in his 1877 treatise on mustelids, the naturalist Elliott Coues declared: "No animal of this country, except the skunk, possesses so powerful, penetrating, and lasting an effluvium."

Selected References: Ben-David et al. 1996, Breault and Cheng 1988, Craik 1997, Eagle and Whitman 1987, Elliott et al. 1999, Errington 1967, Gerell 1971, Hatler 1976, Hatler and Beal 2003b, Lariviere 1999, Mitchell 1961.

American Badger   *Taxidea taxus*

Other Common Names: Badger, North American Badger, Yellow Badger; *Blaireau*. Provincial Species Code: M-TATA.

Description
The American Badger's body is about the size of a Springer Spaniel, but its legs are so short that it appears flat to the ground, and that appearance is enhanced by its pelage hanging over the flanks like the shell on a turtle. With its broad, compact and powerful body, thickset front legs, and long, curved front claws, the American Badger is built for digging. Its head is wedge-shaped, with a sharp-pointed snout, short, well-furred ears, and small, dark eyes. The pelage is long and thick, particularly on the sides where it hangs down; it is often grizzled on the back, ranging from silver-grey to yellow-brown, and the shorter fur on the belly is usually paler, in cream, buff or grey. The feet and lower legs are dark brown to black, and there are distinctive black-and-white markings on the face and head. Typically, the face has a central white stripe, starting at about the nose-pad and extending between the eyes, over the head and down the back of the neck to about the shoulders. The stripe is flanked by black from muzzle to forehead, and the cheeks and throat are light, often pure white, forming a conspicuous contrast to

black, sideburn-like markings in front of the ear on each side. The tail is short, stout and bushy, but not prominent. The sexes look alike, but males are larger than females by about 20 per cent.

Measurements:

total length (mm):
male:	810 (730-900) n=34
female:	745 (643-830) n=14

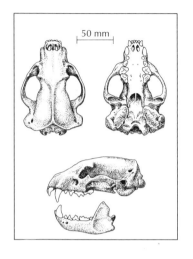

tail vertebrae (mm):
male:	146 (120-170) n=35
female:	142 (115-165) n=14

hind foot (mm):
male:	117 (110-130) n=16
female:	107 (100-117) n=7

ear (mm):
male:	44 (40-45) n=6
female:	39 (37-40) n=2

weight (kg):
male:	9.8 (7.2-13.5) n=30
female:	6.7 (5.9-8.6) n=12

Dental Formula:
  incisors: 3/3
  canines: 1/1
  premolars: 3/3
  molars: 1/2

Identification:
The heavyset, flat body shape of the American Badger (called "Badger" in the rest of this account) is unique, making it difficult to confuse with other species. Probably the closest is the Northern Raccoon, similar in size and also with a grizzled pelage. But unlike the Badger, the Raccoon does not have a white head stripe or white cheeks, its black facial markings form a mask over the eyes rather than sideburns, and it has a series of black rings along the length of its tail. Some early North American explorers apparently confused Badgers with Wolverines, having had little experience with either, but Wolverines are dark brown rather than grey or tawny, do not

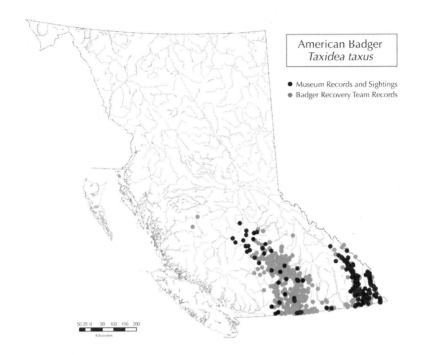

American Badger
Taxidea taxus

● Museum Records and Sightings
● Badger Recovery Team Records

have a white head stripe, and usually have prominent pale stripes along the sides.

Distribution and Habitat

The Badger occurs throughout the southern half of North America from the Great Lakes westward throughout the United States and south to Baja, California and central Mexico. In Canada, it occurs in a separate pocket in southwestern Ontario, near Lake Erie, and then almost continuously from southern Manitoba across the Prairies, except for the mountainous and northern boreal forest portions of Alberta. The Badgers in BC are part of a separate distribution extending northward from the United States (western Montana to eastern Washington).

Badger distribution in BC is discontinuous, with a few centres of occurrence: the East Kootenays, Okanagan Valley, and the drainages of the Thompson and Fraser rivers north to the vicinity of Williams Lake. Records from the Kootenays are mostly from lowland locations in the major river valleys, from Cranbrook and Kimberly south and eastward, and along the Rocky Mountain Trench north to at least Spillimacheen. The most northerly record from the Kootenays is in the Blaeberry River area, near Golden, in 1975. Records from the Okanagan have been sparse in recent years, but in a 1995 review, Ann Rahme, Alton Harestad, and Fred Bunnell documented historic sightings from Anarchist Mountain near Osoyoos north to about Armstrong. In the dry south-central interior, Rich Weir and his associates have documented sightings from areas around Princeton, Merritt, Cache Creek, and Kamloops, up the Thompson River Valley to at least Clearwater, and westward across the Fraser Plateau to 100 Mile House. Corinna Hoodicoff and Roger Packham have recently documented larger numbers and a wider distribution in the 100 Mile House area than had been known previously, with records from 70 Mile House in the south to Soda Creek in the north (27 km north of Williams Lake), and occurrences west of the Fraser River in the Chilcotin River and Big Creek drainages and on the Gang Ranch.

The wide open spaces of North American Prairies, plains, and parklands constitute the primary habitat of the Badger throughout its range. In British Columbia, such habitat is coincident with the productive cattle ranching country in the Interior Douglas-fir, Bunchgrass and Ponderosa Pine biogeoclimatic zones in the dry south-central and southeastern portions of the province. The occasional records of Badger occurrence in forested areas may primarily involve dispersing young, although it is possible that forest openings produced by agriculture, forestry, and certain other developments may provide new, usable habitat for Badgers and their prey in some areas. BC researchers have found Badgers occupying golf courses in the Kootenays and at Marmot Ridge near 100 Mile House, and Corinna Hoodicoff documented one Badger foraging frequently in unpaved patches at an urban shopping centre. A report from Nancy Greene and Al Raine indicates that in summer 2005, one or more Badgers had discovered the increased ground squirrel population on the ski runs at Sun Peaks, northeast of Kamloops. The fossorial lifestyle of the species dictates a depen-

Figure 55. Badger burrow.

dence upon particular soil conditions. Thus, areas with shallow or excessively rocky soils may not be suitable.

Dens used by Badgers for daily shelter and protection are almost exclusively burrows that they have dug themselves (figure 55), either specifically for that purpose or in pursuit of prey. Of 127 dens examined in a Utah study, only 19 (15%) were dug on the day of use. Individual Badgers maintain a number of such dens in their home ranges, rarely using any one on two consecutive days except in winter when they may confine themselves to one site for a long period. In the winter of 2000–2001, a radio-collared male monitored near Kamloops by Rich Weir and his associates appeared to stay in its den for four months (November through March). The multiple den sites in use during summer (e.g., 46 dens used in August and September 1963 by a radio-collared female in Minnesota) give the Badger a network of escape options in its area of activity, and there is evidence that they also serve as a kind of trap for prey. Badgers regularly visit and explore burrows they or others have used previously, and often catch burrowing prey such as small mammals, reptiles and insects.

Unlike day dens, natal and maternal dens are used for considerably more than a day or two at a time, up to several weeks in spring and early summer. Those examined by biologist Fred Lindzey in Utah and Idaho were up to 2.3 metres deep and 5 metres long, each generally consisting of a main tunnel, pairs of secondary tunnels that branched from and then rejoined the main tunnel, numbers of blind pockets and short (dead-end) tunnels in which scats were found, and one or two nest chambers about 0.5 metres in diameter. None of the dens contained nesting material other than loose soil.

Natural History

Feeding Ecology

Badgers are particularly adapted for predation on burrowing rodents such as ground squirrels, prairie dogs, marmots, chipmunks and pocket gophers, and local species in one or more of those groups usually constitute the bulk of their food. But Badgers also eat a large variety of non-burrowing animals, many of them also located and cornered in burrows. The list of known foods includes various voles, mice and rats, cottontails, Red Squirrels, Muskrats, Striped and Spotted skunks, Coyote pups, several kinds of birds and eggs, snakes, lizards, fishes, arthropods (scorpions, spiders, centipedes and various insects), and all kinds of carrion. In 1949, observers in Wyoming watched a Badger catch a 35–40 cm Carp at the edge of a reservoir, and found other evidence that it or other Badgers had fished in that area previously. In the East Kootenay area of BC, Badger food items documented by Nancy Newhouse and Trevor Kinley included two fish (small salmonid, large sucker), and they also found Common Loon in two samples. One of the Badgers radio-collared for the Thompson-Okanagan study was originally located by a biologist working on a Western Rattlesnake study. The Badger, a young male, had killed and was eating one of his radio-tagged snakes. Badgers sometimes also eat vegetable matter. Ten of seventeen Badgers collected in the fall in Iowa contained corn in their stomachs, and four of the stomachs were packed full. Corinna Hoodicoff found heavy use of Saskatoon berries by a Badger in the Okanagan region.

Although Badgers are occasionally observed in daylight, radio-tracking studies have confirmed that they are mostly nocturnal. Thus, most of their predation on diurnal species, such as ground squirrels and marmots, is accomplished by digging them from their burrows. Badgers are particularly adept at locating natal burrows (those occupied by females with young), thus maximizing their return on burrowing effort. During an intensive study of Columbia Ground Squirrels in eastern Washington, University of Alberta biologist Jan O. Murie counted 183 Badger excavations of squirrel burrows over a six-year period and 129 of those (70%) were in natal burrows; Murie estimated that Badgers killed up to 56 per cent of the juvenile ground squirrels in that population each year.

Metabolic studies have shown that adult Badgers in captivity require about 350 grams of prey per day for basic maintenance, but the energetic demands of movement and foraging activity among free-ranging individuals raises that daily requirement to at least 575 grams (the equivalent of one to one-and-a-half adult Columbian Ground Squirrels, or three or four Northern Pocket Gophers). Although the energy cost of digging by Badgers has not been determined, it is probably large. Individuals may make two or three major excavations per night, and biologist R. Lampe found that Badgers moved an average of 180 litres of soil at excavation sites examined in Minnesota.

Badgers have several adaptations for winter survival, when food is less available. As with bears, the primary strategy appears to involve building up body weight and fat reserves by fall, and then retiring to a den for extended periods of fasting. The den enclosure plus an insulating layer of snow, when present, helps conserve body heat. Badgers can also enter a state of mild torpor, reducing the metabolic rate and thereby slowing the depletion of body stores. Physiological studies at the University of Wyoming by biologist Henry Harlow showed Badgers decreasing their metabolism by 54 per cent over a 30-day fasting period, and increasing the efficiency of nutrient absorption after the fast by 8 to 10 per cent.

In 13 years of detailed studies on Richardson's Ground Squirrels, University of Lethbridge researcher Gail Michener observed 27 instances of Badgers caching intact (uneaten) ground squirrel carcasses for later use. They stored each carcass on its own, either in burrows or on the ground covered by loose soil. The Badgers cached all but three of the squirrels in the fall, killing several while they were hibernating. The Badgers retrieved and presumably ate all of the cached squirrels by early December, suggesting that they cache food to assist in the fall fattening rather than to provide an overwinter food source. In an interesting side study, Michener and Andrew Iwaniuk found that most of the cached ground squirrels showed little external trauma. They appeared to have been killed by a clamping bite to the thorax which did not pierce the skin. That technique may reduce odours from the carcass that might attract other mammalian predators, such as Coyotes, and may also reduce the entry of microbes and invertebrates that would initiate or hasten decay.

Home Range and Social Behaviour

In good habitat supporting abundant prey, Badger home ranges are small. In his study area in Idaho, John Messick documented densities of up to 790 ground squirrels per km², and the average home-range size for Badgers was 2.4 km² for resident males and 1.6 km² for resident females. Studies in Utah and Wyoming documented average home ranges of 5.8 and 12.3 km² for males and 2.4 and 3.4 km² for females, and a single adult female radio-tracked in Minnesota by Alan Sargeant and Dwain Warner used an area of 8.4 km² over a 165-day period. In contrast, recently documented home ranges in British Columbia (at the northern edge of the species' range) were much larger, averaging 35 km² for seven females and 301 km² for nine males in an East Kootenay study, and 78.8 km² for seven males in a Thompson-Okanagan study. A female with young near Kamloops used an area of 8.1 km² between early June and late August 2000.

Even in study areas where home ranges are small and population densities high, all evidence indicates that Badgers are solitary for most of the year, except when mating, when females have young, and when siblings associate briefly during the early stages of dispersal. Despite their solitary inclination, Badgers do not seem to routinely engage in rigid defense or marking of territories, and biologists in most areas have found considerable home-range overlap both between and among the sexes. The only exception was in a Wyoming study, where females (but not males) maintained ranges that were almost completely exclusive. In that case, researchers John Goodrich and Steven Buskirk speculated that the spacing among females reflected the more even distribution of the local prairie-dog food base. It should be noted that the extensive home-range overlap among males in all areas and females in most areas studied does not mean that Badgers are not strongly attached to particular places. In the few studies that have been able to follow marked adults over more than one year, the animals have remained in the same general areas.

Activity and Movements

The daily movements by Badgers are difficult to determine, because the consistent location of radio-collared animals has been hampered by problems with signal transmission when the animals are underground. In general, movements vary by sex, season and prey availability, but Badgers are not built for travel and normal movements for foraging in summer are usually less than two kilometres per day. More extensive movements, particularly by adult males during the mating season, have been recorded. Probably in that category, a radio-collared male monitored by Rich Weir and his associates north of Kamloops moved at least 13.7 km in four hours (at night in late July 2000).

As with most carnivore species, the longest movements are made by juveniles during dispersal from their natal ranges in mid to late summer. Badgers of both sexes make such movements, which may take them through large areas of unsuitable habitat and across major waterways. A juvenile male tagged by John Messick in Idaho was caught by a trapper in Oregon five months later, having moved a minimum of 110 km, and a juvenile female from the same study area moved 52 km, crossing the Snake River in the process.

Reproduction

Badgers breed in the summer, generally in July and August, and after a lengthy gestation period featuring delayed implantation, the young are born early the following spring, mostly in March and

April. Litter size ranges from 1 to 4, with averages of 1.7 to 2.2 in major studies in the western United States. In BC, records compiled by Trevor Kinley and Nancy Newhouse for the East Kootenay area (1996 to 2003) include 12 litters, ranging from 1 to 3 and averaging 1.6 young.

At birth, Badgers are fully furred, but are small, blind and wholly dependent upon the mother, which provides all parental care. Relatively little is known about their development, because they remain in underground dens for the first 4–6 weeks, until they are mobile enough to emerge on their own. By 10–12 weeks, young Badgers are near full size and family breakup and juvenile dispersal commences shortly thereafter. A third or more of the juvenile females may attain sexual maturity and be bred in their first summer, but juvenile males do not mature until the following year.

Health and Mortality

Only a few invertebrate parasites have been found on or in Badgers, including some species of fleas and ticks externally, and some common tapeworms, roundworms, and flukes internally. While those may contribute to poor condition of some individuals on occasion, no population effects relating to such parasites have been documented. Badgers have been listed as susceptible to two diseases relevant to human health, tularemia and rabies, but I found no details or case histories in support of that contention. The disease of greatest interest in relation to the Badger is bubonic plague, because it can affect humans. This disease causes occasional major die-offs of rodents, such as ground squirrels, in the American West. There is a possibility that Badgers, which have tested positive for antibodies of the disease, may contribute to its transmission among rodent populations.

Studies to date have not documented much natural mortality among adult Badgers, but juveniles are known to have been preyed upon by Golden Eagles, Common Ravens, Coyotes, Domestic Dogs, and other Badgers, and the Kootenay study identified probable predation by Cougars, Bobcats and Black Bears. In addition, dispersing juveniles typically lose weight in their search for suitable habitat and many do not survive. In John Messick's Idaho study, two of seven juveniles that died during dispersal starved to death, one drowned in an irrigation ditch, and the other four were killed by predators.

For most Badger populations studied to date, the majority of documented mortality was caused by humans. In Idaho, 94 (60%)

of 157 mortalities were due to trapping and shooting (including problem wildlife control activities) and incidents involving farm machinery, 52 (33%) were road kills, and the remaining 11 (7%) were attributed to natural or unknown causes. As reported by researcher Ronald Case, highway crews monitoring road-killed animals along a Nebraska highway recorded 615 dead Badgers over a six-year period. In BC, the legal harvest of Badgers by trapping and hunting was terminated in 1967, but deaths documented in the past decade are still primarily human-caused, particularly road kills. Of thirteen animals radio-collared in the Thompson Okanagan region, five of the seven that died were killed by highway traffic and one by a train; and of ten collared animals for which cause of death was known in the East Kootenay study area, five were killed by vehicles (one by a train), three were taken by predators, one starved, and one died of "age or weather related factors" in a burrow.

The oldest known Badger lived to 15 years and 5 months in captivity and, although 352 of 354 wild specimens aged by John Messick in Idaho were 7 years or younger, the other two (sex not reported) had attained ages of 12 and 14 years, respectively. The oldest BC study animal, a female in the East Kootenay study, died at 13.6 years of age.

Abundance

In the few Badger population studies undertaken to date, densities varied from 0.4/km² in Lindzey's Utah study area to 5/km² in the Snake River Birds of Prey Natural Area in Idaho. The latter area, studied intensively by researcher John Messick in the late 1970s, was characterized by large populations of ground squirrels and minimal human development, and was probably optimum habitat. In British Columbia, recent and ongoing Badger studies have not yet generated data sufficient to calculate meaningful population densities. A government estimate in 1987 set the provincial population at 500–1000, but following the intensive field studies undertaken in the 1990s, Nancy Newhouse and Trevor Kinley revised that estimate to 250–600.

Although the Badger population in BC was believed to be in decline when the intensive studies commenced, recent findings suggest a somewhat brighter prognosis. Based on accumulated reproduction and survival information from 1996 to 2005, Newhouse and Kinley report that the overall Kootenay population may now be stable. In the southern half of the study area, Badger numbers are increasing, and in the northern half, translocation of

16 Badgers from Montana since 2002 appears to have been success-ful in re-establishing a productive population in that area. In the results of her research in the 100 Mile House area, Corinna Hoodicoff reported another hopeful sign: "There may be a stable source population of Badgers in the Cariboo Region despite it being in the extreme northwestern limit of their range."

Human Uses
Badger pelts are too thick and heavy for use in production of gar-ments, but the fur has received some use in the fur trade, primarily as trim material. The best known and most regular use of Badger fur over the years has been as the durable bristles in brushes for shaving and painting.

Taxonomy
Four subspecies are recognized, with one occurring in BC. Christopher Kyle and his colleagues assessed DNA in samples from BC, Alberta and Montana. Their results were somewhat ambiguous, indicating that some Badgers inhabiting the Rocky Mountain Trench in extreme southeastern BC may be genetically linked with the Prairie subspecies *Taxidea taxus taxus*.

Taxidea taxus jeffersonii Harlan – the western United States and southern BC. It is distinguished from *taxus* by minor skull traits and its more reddish pelage.

Conservation Status and Management
Historically, the primary management consideration for the Badger throughout its range was in the context of "problem wildlife". Despite its predation on rodent pests, the net public attitude was negative, because of the potential for injuries to livestock from stepping in Badger holes. Concern about the conservation status of Badgers in BC was first formalized in 1967, when the legal harvest by trapping and hunting was terminated. Although that action was probably necessary, it was obviously not sufficient as the Badger's risk status has continued to move from threatened (provincial Blue List) in the late 1970s through the mid 1990s to officially imperilled (provincial Red List) in 1999. Most recently, in 2000, the subspecies that occurs in BC was also declared endangered nationally by COSEWIC.

The challenge of maintaining viable Badger populations in British Columbia is substantial, because humans have taken most of the best habitat for such purposes as agriculture, urban development and transportation corridors, or flooded it for hydroelectric developments. Appropriately, the East Kootenay, Thompson-Okanagan and Cariboo research projects (mentioned above) feature significant public education and participation components, both of which will be needed as much as the research findings to ensure success. Up-to-date information on Badger research and recovery efforts in BC can be found on the website www.badgers.bc.ca.

Remarks

The Badger's digging proclivities and abilities are regular features of the literature on the species. Eliott Coues, writing in 1877, described a captive animal that "for sport" regularly dug itself out of sight in a few minutes, and to the end of its 3.5-metre chain in only a few minutes more. A captive animal kept by Mary Louise Perry in Washington in the 1930s "raised havoc with a firmly packed automobile road", and also worked its way through the 2.5-cm concrete layer of her basement floor.

The literature also contains several accounts of "Badger-Coyote partnerships" – that is, Coyotes catching rodents frightened from their burrows by digging Badgers. But opinions differ as to whether the Badgers realize any benefit from such associations. That would occur if the presence of the Coyote above ground nearby caused some prey to remain underground and vulnerable to the Badger attack. In any case, the two species have been seen hunting within metres of each other without any aggression towards each other.

Regarding what he termed "The cruel sport of Badger baiting" (a historic practice setting one or more large dogs against a cornered Badger), Elliott Coues said, "The fighting qualities of the Badger and stubborn resistance it offers at whatever unfair odds, have supplied our language with a word of particular significance: to 'badger' is to beset on all sides and harass and worry."

Selected References: Adams et al. 2003; Apps et al. 2002; Hoodicoff 2003, 2005; Kinley and Newhouse 2005; Kyle et al. 2004; Messick 1987; Messick and Hornocker 1981; Minta et al. 1992; Newhouse and Kinley 2000, 2001; Rahme et al. 1991; Weir et al. 2003.

Striped Skunk   *Mephitis mephitis*

Other Common Names: Large Striped Skunk, Polecat, Skunk, Woods Pussy; *Bête Puante, Moufette, Moufette Rayée, Putois, Sconse.*
Provincial Species Code: M-MEME.

Description
The Striped Skunk is a compact, stocky animal about the size and shape of a well-fed house cat. The head, viewed from above, is triangular, tapering to a somewhat bulbous nose-pad on the otherwise pointed snout. It has small, round ears and dark, beady eyes. The pelage on the back and sides is thick and long, and the tail, which is composed of long, coarse hairs, is prominent and bushy. The tail hairs show mixtures of black and white on most individuals, but appear mostly black on some. The overall body colouration is the characteristic glossy black with conspicuous white markings. The general pattern for the white markings is a narrow line between the eyes, from forehead to nose, and a broad patch that extends back from the top of the head and neck and separates at the shoulders into two lateral stripes, one along each side of the back. But there is considerable variation on that theme, particularly in the length, width, and conspicuousness of the lateral white stripes. Some individuals are almost completely black. The measurements shown below for BC specimens are within the size range known for the species elsewhere, but the averages indicated are

probably not accurate because of the small sample sizes. Skunks can be both time-consuming and unpleasant to process, so apparently have not been popular subjects for museum collection and preparation. Throughout the species' range, males average about 10 per cent larger than females.

Measurements:

	All Specimens	*hudsonica*	*spissigrada*
		Subspecies	
total length (mm):			
male:	669 (612-755) n=9	684 (n=7)	616 (n=2)
female:	609 (565-640) n=6	610 (n=1)	608 (n=5)
tail vertebrae (mm):			
male:	239 (192-305) n=10	224 (n=7)	273 (n=3)
female:	233 (215-260) n=8	242 (n=3)	227 (n=5)
hind foot (mm):			
male:	84 (73-94) n=11	84 (n=8)	84 (n=3)
female:	74 (70-77) n=8	76 (n=3)	73 (n=5)
ear (mm):			
male:	33 (31-37) n=5	33 (n=5)	none
female:	30 (30) n=2	30 (n=1)	30 (n=1)
weight (g):			
male:	3120 (2110-4581) n=9	3309 (n=7)	2460 (n=2)
female:	2287 (1790-2948) n=4	2381 (n=3)	2006 (n=1)

Dental Formula:
incisors: 3/3
canines: 1/1
premolars: 3/3
molars: 1/2

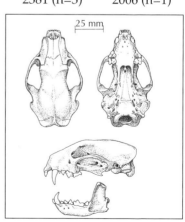

25 mm

Identification
The conspicuous black-and-white colouration is unique to skunks, thus the potential for confusion in the identification of the Striped Skunk in British Columbia is

limited to a small area in the southwestern corner of the province, where it shares range with its smaller relative, the Western Spotted Skunk. Those two species are best separated by the patterns of white markings, which are simpler and less variable in the Striped Skunk (see figure 16). Both species usually have white on the face, between the eyes, but that occurs as a narrow line in the Striped Skunk and as a broad triangular to elliptical patch in the Western Spotted Skunk. The Striped Skunk typically has a broad white stripe from nape to shoulder which forks there and extends as single stripes along both sides of the body to (and sometimes onto) the tail. It, therefore, shows only two horizontal stripes, in comparison to six on the Western Spotted Skunk and, unlike that species, the Striped Skunk has no vertical stripes or spots on the rear half of its body and does not have a white-tipped tail.

Distribution and Habitat

The Striped Skunk is widely distributed over North America, occurring throughout most of Canada to the extreme southern Northwest Territories, in all of the continental United States except the large deserts of Nevada and southern California, and in northern Mexico. In British Columbia, it is most common in the southern half of the province, and does not occur on coastal islands. During species management planning exercises in the late 1970s, provincial biologists W.T. Munro and L. Jackson estimated that 75 per cent of the province's Striped Skunks occurred in the Lower Mainland Resource Management Region (mostly in the lower Fraser River drainage from about Hope to Vancouver), 13 per cent in the Thompson-Okanagan Region, 6 per cent in the Kootenay Region, and the remaining 6 per cent in the three regions that constitute the northern two-thirds of the province.

Previous distribution maps have indicated that the species does not occur west of the Coast Mountains. While that may be mostly true, correspondence from trapper Herb Hughan indicates that individual movements through the mountains along the major river valleys sometimes occurs. Hughan, a long-time resident of the upper Nass River valley, notes that Striped Skunks first appeared in that area in the 1930s (so new to the area that the local Nisga'a had no name for them), were locally abundant for a number of years, and then seemingly disappeared. During the period of abundance, he saw skunks that had been killed on the highway between Terrace and Prince Rupert (lower Skeena River area). His most recent coastal record is of an individual that "let go a blast"

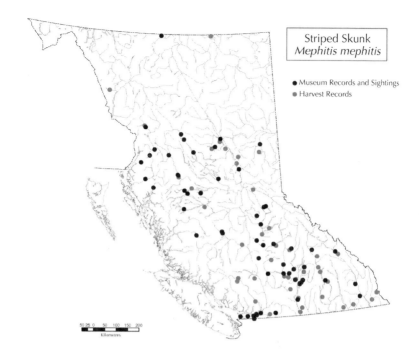

Striped Skunk
Mephitis mephitis

● Museum Records and Sightings
● Harvest Records

under a logging camp cookshack near Greenville (lower Nass River) in about 1991.

The Striped Skunk occurs and thrives in numerous habitats over its broad geographic range, but is not well adapted to the thick coniferous forests and rugged mountainous terrain that characterize much of British Columbia. The highest densities occur in the northern Prairies and Central Plains, both in the patches of natural habitat still remaining and in certain agricultural and urban environments that are common there now. In a detailed habitat study in a farming area of southern Saskatchewan, researchers Serge Lariviere and Francois Messier found radio-collared animals foraging most often in wetland areas, in uncultivated habitat patches, such as along fence lines and road or railway rights-of-way, and in open deciduous woodland.

Where Striped Skunks occur in coniferous forests, in most of their range in BC, they commonly forage along edges and in openings such as wetlands, meadows and riparian areas along and around streams, lakes and ponds. Biologists suggest that the extensive forest clearing that occurred post-European contact greatly increased Striped Skunk habitat in North America, providing for both expanded ranges and higher populations.

Den and daily resting sites used by Striped Skunks in the northern portions of their range (Canada and the northern United States) vary by sex and season. In spring through fall, a male or a female without den-bound young may use several different sites over a period of days. In most cases, the resting skunk is in or under some structure that provides protection from both weather and predators: hollow trees and logs, cavities in and under rock and debris piles, burrows made previously by skunks or other similar sized mammals, and a variety of man-made structures, such as buildings, haystacks and farm machinery. During one summer in Great Smoky Mountains National Park, Tennessee, all of 25 culverts that were either plugged with debris or not occupied by raccoons were used one or more times as day dens by Striped Skunks.

For about two months in early summer, adult females with small young use one den site more intensively, returning to it daily. Natal dens are similar to the summer sites used by other skunks, but probably more secure. In Minnesota, mother skunks appeared to prefer underground burrows, while in the southern Saskatchewan study Lariviere and Messier located a high proportion of natal dens in or under farm buildings. Once the young are mobile and able to follow the mother, the family may adopt the more nomadic summer lifestyle of other Striped Skunks, changing den sites every one or two days.

Dens used in winter by all sex and age classes are often underground burrows, but 109 of 117 winter dens located in the early to mid 1970s by biologists John Gunson and Ronald Bjorge in Alberta and Saskatchewan were under farm buildings. All accounts indicate that Striped Skunks are particularly diligent in preparation of winter dens, lining the nest chambers with thick layers of leaves and grasses. Although Striped Skunks do not truly hibernate, they may spend several months (up to 150 days in the vicinity of Edmonton) in a relatively inactive state in the winter den. A peculiar and interesting feature of winter denning is that it is often done communally, with up to 20 adults occupying the same den. In most cases, the animals involved are all females, or females with one male. Males often den alone, and may venture outside the den in mid winter during periods of thaw.

Natural History

Feeding Ecology

During the snow-free season, Striped Skunks prey mostly on insects, particularly grasshoppers, crickets, and both the adults and larvae of beetles, moths and butterflies. But Striped Skunks are omnivorous and opportunistic, eating a wide variety of other foods, including mice, voles, birds eggs and nestlings, frogs, crayfish, carrion, various fruits and, when available, human garbage and grain crops such as corn. In the northern portions of their range, skunks may spend three to four months in their winter dens, subsisting largely by metabolizing stored fat. Overwinter weight losses of 32 to 55 per cent for females and 14 to 48 per cent for males have been reported.

During active seasons, most foraging activity occurs between dusk and dawn, and involves behaviour more like gathering than hunting. Individuals wander back and forth extensively, locating potential food items mostly by smell and sound. They usually catch mobile prey, such as mice and frogs, with a cat-like pounce, trapping them under the forefeet. Most foraging is done above ground and, although Striped Skunks may wade in shallow water, they rarely swim when searching for food. Equipped for defense as they are, skunks apparently do not need to keep as alert for danger as most other species do, and they may concentrate on foraging quite oblivious to their surroundings. In October 1985, while resting on a log in a wetland near Smithers, I became aware of a rustling beside me and moments later a Striped Skunk appeared less than a metre away. Highly motivated not to startle it, I held still for several minutes while it explored the area around me, during which time it sniffed my feet twice without exhibiting either recognition or alarm.

Home Range and Social Behaviour

As determined from movements of radio-collared animals, the home ranges of Striped Skunks in natural habitats cover 2–5 km² in most areas, but in the Saskatchewan area studied by Lariviere and Messier they averaged 11.6 km² for males and 3.7 km² for females. On the other extreme, biologist Richard Rosatte documented much smaller ranges, averaging 0.6 km², in urban Toronto, and surmised that the small ranges reflected a higher availability of den sites and food sources in that area. Lactating females generally have the smallest home ranges, due to their need to focus foraging activities near the natal den.

The evidence from research to date indicates that Striped Skunk home ranges are not sharply defined and stable except over short periods and, although individuals remain solitary for most of the year, they are not strongly territorial. Males appear to wander widely, and most of the collared males in the Saskatchewan study area moved out of radio-contact range within a few weeks of capture. The ranges of females overlap considerably and that, together with the incidence of communal denning, indicates that neighbouring females are much more tolerant of each other than are the females of most other carnivores.

Activity and Movements
Striped Skunks tracked in snow or with radio transmitters have travelled two to three kilometres daily, but they usually stayed within a kilometre or less from the previous day's den. Over longer periods, long distance movements appear to be common, particularly for males. In North Dakota, Alan Sargeant and his colleagues documented minimum spring movements of 10, 51, 76 and 119 km for four adult males and 54 km for one adult female. Known fall dispersal movements by juveniles have ranged up to about 70 km and, unlike the case for most other carnivore species, long dispersals by both sexes have been reported. In a southern Alberta study area, Richard Rosatte and John Gunson observed only small juvenile dispersal movements, to a maximum of 9.7 km for one female and averaging 3.0 km for 28 animals of both sexes. Their study was in an area undergoing intensive population control following a rabies outbreak, and they believed the short dispersals reflected a low level of competition among remaining skunks and therefore little need for extensive travel to locate suitable habitat.

Reproduction
Striped Skunks in the north breed primarily in February and March, in the winter dens. The several females in a communal den at that time will likely all be bred by the same male, often one that has denned with them, but during the breeding season males may travel extensively in search of females. Gestation ranges from about 59 to 77 days, including a variable period of delayed implantation, but is most often 62 to 66 days. The young are usually born in May or early June. Litter sizes are fairly large, ranging from two to ten and averaging five or six in most areas. In a study area near Edmonton, Ronald Bjorge and his associates found reduced litter sizes (average 2.5) following an unusually long and severe winter in the early 1970s.

Striped Skunks weigh approximately 30 grams at birth, and are blind, sparsely furred and helpless. They remain in the natal den or a maternal den until weaned at about six or seven weeks, after which they begin following the mother on her foraging trips. The mother provides all parental care. The family group travels together until the fall when the young, at the age of four or five months, begin to disperse. Both sexes attain sexual maturity within their first year and most yearling females mate in their first winter (i.e., at about age ten months). A male in captivity exhibited rut behaviour in his first winter, but it is not known how common it is for yearling males to breed in the wild.

Health and Mortality
Certain features of Striped Skunk life history are conducive to the spread of disease, particularly the sharing of dens, the high degree of overlap in home ranges, and the tendency and capacity for adults and juveniles to disperse. Those traits are compounded in urban environments, where the abundance of food and den sites result in higher than normal skunk densities. Striped Skunks carry zoonoses, including rabies, leptospirosis, listeriosis and tularemia. Rabies appears to be the most serious, and is the disease for which Striped Skunks are most notorious. In a recent review paper, researcher John Krebs and associates noted that the Striped Skunk was the number one wildlife species reported with rabies in the United States from 1960 to 1989, and in the years since has been second only to the Raccoon. Although there is no comparable compilation of Canadian records, it is evident that Skunk rabies has also been of concern in some areas of this country. The southern Alberta study by Rosatte and Gunson was initiated in response to reports of 77 rabid skunks in a small area between 1979 and 1981. To date, the situation in British Columbia is less imposing, with only one recorded incidence. In Spring 2004, four young Striped Skunks from Stanley Park (Vancouver) were diagnosed with a strain of rabies from bats. There were no known effects on humans, and subsequent monitoring indicated that other animals in the area were not infected.

Largely because of the rabies connection, Striped Skunks have been subject to numerous pathology investigations, and the list of parasites and diseases identified across the species' geographic range is large. Among the parasites are at least 3 species of protozoans, 17 species of fleas, lice, mites and ticks, and more than 50 species of internal worms. In an investigation of Striped Skunk skulls in museums across North America, G. and C. Kirkland found cranial damage due to Sinus Worms (see the American Mink account, page 286) in 20 of 30 specimens from BC. Diseases identified in addition to the zoonoses listed above include histoplasmosis, canine distemper and canine hepatitis.

Despite an effective defense with its well-known anal gland discharges, the Striped Skunk is subject to occasional predation by a number of larger carnivores and raptors. Documented predators include the Cougar, Bobcat, Coyote, Red Fox, American Badger, Golden Eagle and Great Horned Owl. Although it is believed that most such encounters involve inexperienced skunks, inexperienced predators, or both, some observers believe that the raptors are less affected by or deterred by skunk odour. In one case, the remains of 57 Striped Skunks were found beneath a Great Horned Owl nest in New Mexico. But the relationship is not strictly one-sided, as indicated by a report of emaciated and injured owls "smelling strongly of skunk" delivered to wildlife rehabilitation centres in California; most of them died, which is likely also the fate of others in the wild that were not found by humans. American Badgers and, occasionally, adult male Striped Skunks sometimes prey on juvenile Striped Skunks in the natal den. Another common source of natural mortality is starvation, particularly for young orphaned during the nursing period and adults confined too early or too long to the winter den.

Human causes of Striped Skunk mortality include road accidents (e.g., 2417 along one highway in Nebraska over a seven-year period), incidents involving farm machinery, shooting and trapping by individuals protecting their property (sometimes from poultry depredations, but most often from the odour), occasional (mostly unintentional) kills by fur trappers, and public health-related removals during rabies outbreaks. As reported by Richard Rosatte, two rabies control operations of 20–24 months duration in southern Alberta killed 1450 Striped Skunks in one area and 838 in another, and those deaths were in addition to the animals that died directly from the disease.

Population studies and examination of carcasses from the large-scale disease control operations all indicate that Striped Skunk populations have a high rate of turnover. Among radio-collared skunks in a North Dakota study by Alan Sargeant and his associates, an average of only 43 per cent of the adult females, 11 per cent of the adult males and 9 per cent of the juveniles (both sexes) were present in the area in two successive years. Although some disap-

pearances were due to dispersal, studies of population age structure indicate that the high reproductive capacity of the species
(high pregnancy rates, low age of first reproduction, and large litters) is regularly balanced by high mortality, especially of juveniles.
Based on analysis of Striped Skunk carcasses from a boreal forest
area of northern Minnesota, Todd Fuller and David Kuehn calculated that a theoretical spring population of 100 animals (65 juveniles and 35 adults) would be reduced to 15 (5 yearlings and 10
adults) a year later. In a sample of 750 specimens from Ontario,
Manitoba and Quebec, more than half were less than one year old,
and fewer than 75 (10%) were older than three years. The oldest
were two females, both between five and six years.

Abundance
In Striped Skunk studies across North America, population densities vary from 0.5–2.4 animals per km^2 in parkland habitat of central Alberta to 13–26 per km^2 in a study area in Illinois. There have
been no population studies in British Columbia, but provincial
wildlife managers W.T. Munro and L. Jackson estimated a total
provincial population of 40,000 to 60,000 animals in 1979, with the
highest densities in the Lower Mainland and the Thompson-
Okanagan Regions.

Human Uses
Skunk meat is said to be palatable when obtained under the right
circumstances, and was reportedly eaten (though not regularly) in
some aboriginal societies and by early pioneers. The fur of the
Striped Skunk is thick, glossy and durable, and it has a long history
of use in the fur trade. From the mid 1800s through the 1930s, the
value of skunk pelts was sufficient to make their pursuit worth the
risk of "odour incidents", but that has not been the case for most
trappers for the past several decades. Striped Skunk fur was once
marketed as "Alaska Sable" or "Russian Sable" for, as noted by
Eliott Coues in 1877, "our elegant dames would surely not deck
themselves in obscene skunk skins if they were not permitted to
call the rose by some other name".

A more unusual use for Striped Skunk products was application
of its musk to provide relief from asthma. Quoting from a 1953
treatise by the naturalists John James Audubon and John Bachman,
Eliott Coues told of "an asthmatic clergyman who procured the
glands of a skunk, which he kept tightly corked in a smelling bottle,
to be applied to his nose when his symptoms appeared. He

believed he had discovered a specific for his distressing malady, and rejoiced thereat; but on one occasion he uncorked his bottle in the pulpit, and drove his congregation out of church."

Taxonomy
Of thirteen recognized subspecies, two occur in British Columbia. The validity of these subspecies has not been verified by modern study of structural or genetic variation.

Mephitis mephitis hudsonica Richardson – a vast range across western Canada and the western United States. In BC, it occupies mainland regions east of the coastal mountains.

Mephitis mephitis spissigrada Bangs – the Pacific coast of Oregon and Washington, north to the lower Fraser River valley in BC. This appears to be a weakly defined subspecies distinguished from *hudsonica* by broader stripes and some minor differences in skull structure.

Conservation Status and Management
Throughout its continental range, the Striped Skunk has received most press in relation to its association with rabies, its role in the predation of waterfowl nests, and its role in domestic incidents involving actual or feared discharge of its scent glands. Fortunately, in British Columbia only the last has been an issue to date. In this province, the species is classified and managed as a furbearer, with a specified harvesting season, but there is little market demand and very little trapping effort expended. The reported Striped Skunk harvest in BC has averaged fewer than ten animals per year since the Fur Harvest Database system was initiated in 1985.

Remarks
The Striped Skunk's scientific name *mephitis* is from the Latin *mephit*, meaning "bad odour", and its chemical defense speciality is a central theme in literary accounts of the species. For example, J. Turner-Turner, an English gentleman who spent a season on a trapline in north-central BC in the late 1880s, said: "The smell of a skunk I consider one of the few things in the description of which exaggeration is impossible...."

When provoked the animal discharges a butylmercaptan compound from its anal glands. The spray can cover a distance of up to five metres and can cause temporary blindness, nervous system depression and vomiting in the recipient of a direct hit. The odour itself, which is detectable by the human nose at concentrations of about 6 one/billionths of one mg per ml of air, reaches far beyond the actual spray distance and, from my personal experience, two layers of closed plastic bags do not completely contain it. In the event of an encounter which leaves clothing, skin, or one's dog contaminated with skunk musk, the following recipe provided by animal control specialist Pete Wise may be of assistance:

> *Ingredients: 1 litre of household hydrogen peroxide*
> *+ ½ cup of baking soda + 1 teaspoon of liquid dish soap*
>
> *1) Mix well in a sufficiently large container.*
> *2) Saturate the affected area thoroughly.*
> *3) Rinse immediately and well.*
> *4) Do not cap or prepare ahead of time as this solution is*
> *   volatile.*
> *5) Unused liquid should be discarded immediately.*
> *6) This operation should be conducted outdoors.*

Young Striped Skunks are known to produce the anal gland musk shortly after birth, but are not able to control and direct it until they are about four weeks old. The distinctive black and white colour of skunks, plus ritualized behaviour that precedes spraying (foot stamping, charging, hissing, raising of the tail) probably evolved to warn would-be predators, especially those that have already had a spray experience. In central Saskatchewan, biologists Lyle Walton and Serge Lariviere witnessed a Striped Skunk successfully repel two Coyotes by threat alone, without spraying. Thus, the warning colour and behaviour reduce the number of local predators willing to risk making an attack and the skunk's need to actually use its ultimate defense.

In a discussion about domestication of the Striped Skunk, written in 1877, the naturalist Eliott Coues noted:

> *Writers speak of the removal of the anal glands in early life,*
> *to the better adaptation of the animal to human society, and*
> *such would appear to be an eminently judicious procedure.*
> *For though skunks may habitually spare their favors when*
> *accustomed to the presence of man, yet I should think that*
> *their companionship would give rise to a certain sense of*
> *insecurity, unfavorable to peace of mind. To depend upon*
> *the good will of so irritable and so formidable a beast, whose*
> *temper may be ruffled in a moment, is hazardous – like the*
> *enjoyment of a cigar in a powder-magazine.*

Skunks are quiet animals, and even those in captivity rarely utter any sounds. Those that have been described include the hiss, a short squeal and a growl, all made when under stress, and a churring or bird-like twittering that occurs when the animal is in a more contented frame of mind.

Selected References: Bjorge et al. 1981; Dragoo et al. 1993; Gunson and Bjorge 1979; Krebs et al. 1995; Lariviere and Messier 1997, 1998, 2000; Lariviere et al. 1999; Rosatte 1987; Turner-Turner 1888; Verts 1967; Wade-Smith and Verts 1982.

Western Spotted Skunk — *Spilogale gracilis*

Other Common Names: Civet, Civet Cat, Polecat, Tree Skunk, Weasel Skunk; *Mouffette Tachetée*. Provincial Species Code: M-SPGR.

Description
The Western Spotted Skunk is small, about half the size of a Domestic Cat. Its general body shape is long and slender, though less so than a weasel's. The head is roughly triangular, with a pointed snout and rounded ears that are small but prominent. The pelage is short and soft, although the tail is bushy with long scraggly hairs. The Western Spotted Skunk follows the classic skunk theme in colouration – black with contrasting white markings. The white markings vary, but generally include a large patch from forehead to nose, three pairs of stripes parallel to the spine on the front half of the body, three pairs of vertical (back to belly) stripes or irregular spots on the rear half of the body, a line on the underside of the tail, and a large brush at the tip of the tail. On some animals, some of the white stripes are broken into series of spots or, more accurately, dashes. We have no weights for specimens in British Columbia, but data from areas farther south show females at 300 to 450 grams and males at 400 to 750 grams.

Measurements:

total length (mm):
male:	399 (362-434) n=17
female:	376 (365-389) n=8

tail vertebrae (mm):
male:	117 (100-144) n=19
female:	114 (90-148) n=11

hind foot (mm):
male:	47 (42-53) n=20
female:	44 (40-48) n=11

ear (mm):
male:	none
female:	none

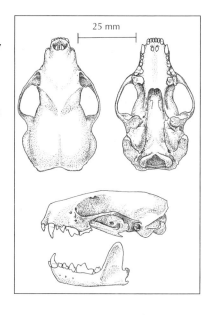

25 mm

weight (g):
    male: none
    female: none

Dental Formula:
    incisors: 3/3
    canines: 1/1
    premolars:3/3
    molars: 1/2

Identification

The only species in British Columbia that the Western Spotted Skunk (called the "Spotted Skunk" in the rest of this account) might be confused with is the Striped Skunk, which is considerably larger (more than twice the size). The two species are best separated by the patterns of white markings, which are much more extensive on the Spotted Skunk (see figure 17, page 47). Both species have a

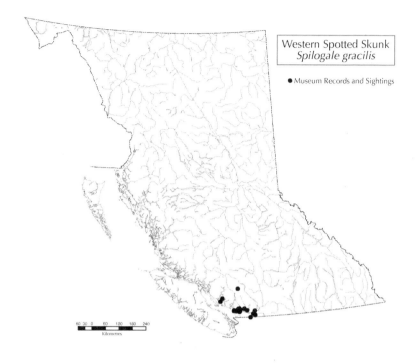

white patch between the eyes, but the Spotted Skunk's is triangular or elliptical and is only a narrow line. The Spotted Skunk has six horizontal stripes along the back, compared to only two on the Striped Skunk, and also has vertical stripes or spots on the rear half of its body while the Striped Skunk does not. Finally, the tails of most Spotted Skunks have a conspicuous bushy white tip while those of Striped Skunks are usually all black or show a mixture of black and white hairs but do not have a white tip. A behavioural trait that will serve to distinguish the species is the Spotted Skunk's distinctive threat posture, appropriately referred to as a handstand. With hind feet up and tail arched forward, the animal exposes both barrels of its business end.

Distribution and Habitat

The Western Spotted Skunk has southern roots. Its range extends from southwestern British Columbia and western Washington in the north, southward west of the Rockies to southern Guanajuato in central Mexico. All of the 83 known museum specimens from BC were collected between 1892 and 1950, and came from the south side of the Fraser River valley as far east as Cultus Lake, the north side of the lower Fraser (New Westminster, Port Moody), Vancouver city (Marine Drive, Point Grey), and the north shore (North Vancouver, Deep Cove, Mount Seymour). More recent records include a family group (five animals, presumed to be a mother with four nearly grown young) that shared the premises of naturalist Myrtle Winchester in Pender Harbour for several months over the 1986–87 winter. Winchester advises that Spotted Skunks occur over the whole of the Sechelt Peninsula and are fairly common over most of the lower Sunshine Coast. My most recent records for the province are of a male trapped by Al Starkey at Pitt Lake (near Maple Ridge) in January 2001, and a sighting of one skunk near Egmont by resident D.G. Stephenson; the latter, at the north end of the Sechelt Peninsula, is the northern and western-most continental record of the species to date.

Although there have been few studies, the Spotted Skunk appears to avoid wetlands and timbered areas over most of its range, reportedly preferring drier lowland habitats with thick shrub or herbaceous cover and often rocky terrain. But biologists Andrew Carey and Janet Kershner found the species widely distributed in coniferous forests of the southern Cascade Mountains in Washington and Oregon. Most documented occurrences were in lowland forests (below 500 metres), and those higher were usually in shrubby riparian areas. Spotted Skunks were captured in forests of all ages, particularly in Washington, but more often in old growth in Oregon.

In BC, the Spotted Skunk occurs only on the southwest mainland, where it is primarily associated with urban and agricultural environments in the Coastal Western Hemlock Biogeoclimatic Zone. In 1902, naturalist Allan Brooks noted that this species was fairly common in the Chilliwack District, with occurrences as high as 1300 metres in surrounding mountains.

Spotted Skunks use a variety of sites for denning, including natural cavities among rocks, in hollow trees or stumps and in woody debris piles, burrows excavated by themselves or other animals, and man-made structures (in or under buildings, culverts, haystacks). The species is an agile climber. Its tree dens have been found as high as seven metres above the ground. One Spotted Skunk may use several different dens during its travels over its home range. On a study area in California, two individuals each used 12 dens in three to six months. Spotted Skunks do not hibernate, but may be den-bound and inactive for weeks at a time during the winter. Occasionally, more than one adult animal occupies a den, especially in winter; the relationship of the animals sharing a den is not known.

Natural History

Feeding Ecology
There are no food habits data for British Columbia, but insects often dominate the diets of Spotted Skunks elsewhere during most of the year, and small mammals (mice and voles) are primary prey when insects are not available, especially in winter. Invertebrates eaten by Spotted Skunks include grasshoppers, adult and larval beetles, crickets, worms, caterpillars, and centipedes. Spotted Skunks are far from being specialists, though, and consume a large variety of other foods including bird eggs and nestlings, frogs, salamanders, lizards and various fruits. They also readily eat carrion.

All accounts indicate that Spotted Skunks are primarily nocturnal, and sightings during daylight are rare. In a Florida study of the Eastern Spotted Skunk, there was evidence that the animals even avoided moonlight, delaying the night's activity until after the moon had set. There is little published information on hunting behaviour, but the species appears to be a more versatile hunter than the Striped Skunk, using its smaller size and weasel-like body to invade underground burrows and tunnels and its agility to hunt above ground in trees and shrubs.

Home Range and Social Behaviour
In a study of individually marked Spotted Skunks on Santa Cruz Island in California, in the early 1990s, home ranges averaged 29.6 hectares in the wet season (December to April) and 61.1 hectares over the rest of the year. The few studies done to date produced no evidence that Spotted Skunks attempt to maintain exclusive use of their home ranges and defend them against others. The current view is that they are largely nomadic, remaining in and intensively using distinct areas as long as their nutritional and cover needs are met, and moving on when that is no longer the case. An Eastern Spotted Skunk in Iowa moved up to 4.8 km per night during its hunting excursions.

Reproduction
As with other aspects of the species' life history, there is no information on the reproductive biology of the Spotted Skunk in British

Columbia. Elsewhere, based primarily on laboratory studies, Spotted Skunks breed in the fall and give birth in the spring, the six-to-seven month pregnancy made possible by delayed implantation. In areas studied, implantation usually occurred in April and the young were born about 30 days later, in May. Litter sizes are small, ranging from 2 to 5 and averaging 3.1 to 3.8.

Newborn Spotted Skunks are small, weighing 9–12 grams, and are blind and helpless. Although only sparsely furred, the pattern of white markings on their backs and sides is already discernible. As with most small carnivores, young Spotted Skunks grow and develop quickly. Their eyes open at about 32 days, they are able to expel scent from their anal glands by 46 days, and they are nearly full grown at three months. Given adequate nutrition, young animals of both sexes can become sexually mature and breed in their first fall, at 4–5 months of age. The mother provides all parental care.

Health and Mortality
In their examination of Spotted Skunk skulls in museums across North America, G. and C. Kirkland found cranial damage due to Sinus Worms in all 31 specimens from British Columbia. The population implications of that finding are unknown, but it is consistent with a general pattern in which the parasite occurs most frequently in specimens from wet coastal climates. There is no other published information on parasites or diseases of the species in BC, but studies elsewhere have identified a number of fleas, lice and ticks, and a few intestinal tapeworms and roundworms. Although individuals in captivity have contracted diseases such as distemper, pneumonia, and coccidiosis, wild populations have shown no disease problems. There have been a few cases of rabies in Spotted Skunks, but the incidence is negligible compared to that in the Striped Skunk.

The strong-smelling discharge from the anal glands of the Spotted Skunk, said by some observers to be worse than that of the more common Striped Skunk, is apparently used only in defense. Prior to engaging in its particular brand of chemical warfare, the Spotted Skunk exhibits warning behaviours: a rapid stamping of the forefeet and, most conspicuously, the handstand stance. It may discharge scent during the handstand, but it can also spray from other positions. Despite this effective defense, Spotted Skunks are killed by predators such as Coyotes, Bobcats, Domestic Dogs, Domestic Cats and Great Horned Owls. It is presumed that, at least among the mammal predators, skunk kills are mostly by inexperienced individuals. The unique black-and-white colour pattern of skunks plus the various warning behaviours help to remind potential predators of previous bad experiences.

Most Spotted Skunk mortality from human causes occurs in the form of road accidents, but a few animals are killed when there is concern about odour (rather than livestock depredation). Myrtle Winchester notes that in the Pender Harbour area, Spotted Skunks found in yards or buildings are usually trapped and killed, but "kinder souls capture them live and relocate them to an unpopulated area or a disliked acquaintance's property". In British Columbia, trappers occasionally catch Spotted Skunks, providing important information on the continued existence of the species in the province over the past few decades.

There is no published information on the lifespan of Spotted Skunks in the wild, but there is a record of a captive individual that lived to 9 years and 10 months.

Abundance

Population densities of 9–20 animals per km^2 have been documented for the closely related Eastern Spotted Skunk (in Iowa and Florida). But there have been few field studies of the Western Spotted Skunk and I found no published information on population densities or trends for this species. From all accounts, it occurs sparsely in northwestern North America, with an average capture rate of less than 1 animal per 1000 trap nights in live-trapping studies in Washington and Oregon as compared to about 6 per 1000 trap nights on Santa Cruz Island, California. There have been no Spotted Skunk studies in British Columbia, and the most recent museum specimen is an animal collected in New Westminster in September 1950. Jack Lay, an animal control officer in the Lower Mainland for many years, notes that when he started work in that area in the early 1940s, Spotted Skunks were common and there were no Striped Skunks. As people removed local forests for agricultural, residential and industrial development, the larger Striped Skunks, which do well in urban and agricultural settings, gradually replaced the Spotted Skunks.

Human Uses

The fur trade has had some interest in Spotted Skunks over the years, but the species has never been in high demand for its fur and, because of the odour risk, it is not popular with trappers.

Taxonomy

Although originally lumped together as a single species, the Eastern Spotted Skunk and the Western Spotted Skunk are now recognized by most taxonomists as two distinct species, *Spilogale putorius* and *Spilogale gracilis*, respectively. Based on skull and pelage features, Richard Van Gelder recognized seven subspecies of Western Spotted Skunk, with one occurring in British Columbia.

Spilogale gracilis latifrons Merriam – along the coast from Oregon and Washington to Garibaldi Provincial Park and the Sechelt Peninsula in BC. Richard Van Gelder noted that BC specimens of this subspecies are distinctly smaller than those from Washington and Oregon, but he concluded that the size differences were insufficient to classify the BC population separately.

Conservation Status and Management

Despite its limited distribution and apparent rarity in British Columbia, the Spotted Skunk is not considered to be at risk, and there are no specific conservation concerns or issues. The species is managed as a furbearer, although there is no market demand and, therefore, no harvest effort. Because there has been little reported contact with the species in this province for over half a century, we encourage readers to report any sightings and deliver any specimens or photographs to wildlife authorities, noting the dates and locations.

Remarks

The small BC range represents the only occurrence of either species of spotted skunk in Canada.

Selected References: Brooks 1902, Carey and Kershner 1996, Johnson 1921, Van Gelder 1959, Verts et al. 2001.

Cougar   *Puma concolor*

Other Common Names: Catamount, Mountain Lion, Painter, Panther, Puma; *Couguar, Panthere*; Provincial Species Code: M-PUCO.

Description

The Cougar is the largest cat in British Columbia, most adults larger than a German Shepherd dog. The Latin species name *concolor* means "one colour", and is applied in reference to the generally uniform appearance of adults. But not all Cougars are the same colour. They vary from a light greyish brown to tawny to a rich reddish brown that, on any individual, covers the back, sides and legs. The belly is usually paler, but it doesn't show from most angles. The pelage is relatively short and coarse. The body is long and slender, and the head appears disproportionately small. The tail is prominent, thick but not bushy and tipped with darker hairs; it is usually more than one-third of the total length. The ears are short, somewhat rounded in profile and dark (to black) on the back. The muzzle is dark on the sides, mustache like, contrasting sharply with a white to cream area on the upper lip and chin. Cougar kittens have dark brown to black spots that are conspicuous to about four months of age, but gradually fade and are faint to

absent at one year. The measurements and weights given below for BC specimens are within the general range for the species, but do not reflect known extremes such as a 125 kg male taken in Arizona.

Measurements:

total length (mm):
    male: 2200 (1867-2794) n=32
    female: 1972 (1753-2159) n=23

tail vertebrae (mm):
    male: 821 (660-914) n=32
    female: 754 (584-838) n=23

hind foot (mm):
    male: 301 (265-343) n=32
    female: 275 (241-305) n=23

ear (mm):
    male: 102 (102-102) n=1
    female: 86 (85-87) n=2

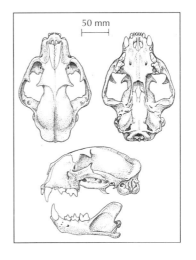

50 mm

weight (kg):
    male: 62.9 (40.8-80.7) n=30
    female: 44.3 (28.6-54.5) n=20

Dental Formula:
    incisors: 3/3
    canines: 1/1
    premolars: 3/2
    molars: 1/1

Identification:
The large size, uniform colouration and long tail serve to distinguish an adult Cougar from any other cat in British Columbia. A young, spotted Cougar kitten could possibly be confused with a Bobcat, but unlike the Bobcat, it has a conspicuous long tail.

Distribution and Habitat
The Cougar was once one of the most widely distributed of North American mammals, but it was reportedly extirpated in much of the central and eastern portions of the continent by shooting and habitat loss during early agricultural and urban expansions.

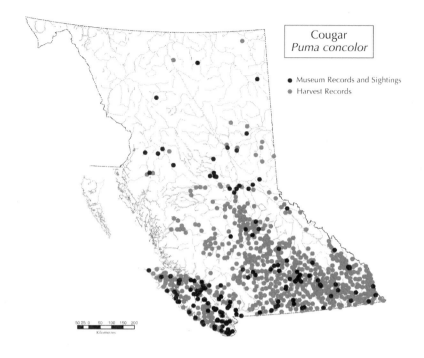

Notwithstanding occasional records of individuals along the eastern seaboard north to at least Maine, the only extant population east of Texas is in Florida. The Cougar's stronghold has probably always been the mountainous west, where its occurrence has been recorded from southern Yukon Territory through Mexico and Central America to Chile and Argentina. The distribution in BC is almost province-wide in terms of reported sightings, although there is not a continuous, viable Cougar population over the whole province. The areas of primary occurrence are Vancouver Island, the mainland coast in and adjacent to the Coast Mountain Range north to about Bella Coola, and most of the interior in the southern half of the province.

Both Mule Deer and White-tailed Deer have been expanding range northward over the past few decades, and it appears that Cougars have followed. There were only occasional records from areas along Highway 16, from Prince Rupert to Prince George, through the mid 1980s, but Cougar sightings have been fairly common both there and in more northerly areas since. Trappers Ron Timothy (Fort St James) and Don Wilkins (Prince George) advised me of several sightings and incidents in their respective areas in the late 1990s, and biologist Brian Churchill told me that sightings in the Peace River area, around Fort St John, have been more common in recent years. Trapper Mike Green was so surprised to see a Cougar near the Alaska Highway in Muncho Lake Park that he "stared too long and went in the ditch".

Cougars are strong swimmers, readily crossing large freshwater bodies (figure 56) and moving between and among many of our coastal islands. Occasional misjudgements occur, and when I resided on the west coast of Vancouver Island in the late 1960s and early 1970s, I heard of two incidents of swimming Cougars encountered by commercial fishermen 15 km or more out to sea. In one of those, the obviously very weary cat tried to get into the boat and had to be shot.

The Cougar is widely distributed in both North America and South America, and it is found in a variety of habitats over that broad range, including western mountains, eastern deciduous woodlands, the desert country of the southwestern United States, and tropical jungles of South America. In most areas, it appears that the habitats used by Cougars are primarily occupied by the local ungulate prey species. In British Columbia, that includes mature and second-growth coastal forests supporting Black-tailed Deer in the west, the forested mountains and valleys supporting Mule Deer and White-tailed Deer in the southern and central interior, and local patches of relatively dry microclimate supporting Mule Deer in the north. During the snow-free season, Cougars may use all local habitats, but in winter they avoid areas of deep snow both because they are not well adapted to travel in snow, and because ungulates usually move from those areas during winter.

Steep, broken, often rocky terrain and thick clumps of vegetation afford good cover for stalking and ambush, and are often used for those purposes by hunting Cougars. Habitats used for day beds and

Figure 56. A Cougar swimming across Horsefly Lake in central BC.

resting sites are variable, depending on local conditions, and usually change from day to day. Included are open knolls and rock outcrops exposed to the sun on cool days or wind on hot days, and various protected locations such as under rock overhangs and in the hollows under low tree branches during periods of rain or snow. A Cougar may use the same bed site repeatedly for several days when it is consuming the carcass of a large animal. Females with young may occupy the same site for several weeks after the young are born. I have seen no description of natal dens in British Columbia, but records from elsewhere indicate use of rock caves and crevices, hollow stumps, and more simple shelters under overhanging banks or trees, or in thickets. Whatever the nature of the site, the female apparently provides no supplementary nesting material.

Natural History

Feeding Ecology

Cougars are not specialists in the strict sense, as they are known to kill and consume a large variety of prey, but they are particularly adapted to hunt and kill medium-sized ungulates, especially deer, which comprise the bulk of the diet in most areas. In Cougar stomachs collected from the Okanagan region in winter between 1958 and 1967, biologist David Spalding and Conservation Officer John Lesowski found Mule Deer remains in 74 per cent of the samples. Porcupines (11%) and Beavers (5%) were the second and third most common prey in that study, with several other species in minor amounts, including Coyote, Ruffed Grouse and livestock (sheep, cattle and horses). During the same time period, samples from near Williams Lake in the Cariboo region contained equal occurrences of Mule Deer and Snowshoe Hares (27% each), followed by Moose and livestock (11% each), and Porcupine (8%). The differences between areas likely reflect availability and opportunism more than selection. Spalding and Lesowski noted that White-tailed Deer were common in the Okanagan region at the time, but did not appear in any of the samples. They took that to indicate that Cougars hunted mostly in local mountain areas where Mule Deer were wintering rather than in the wooded lowlands occupied by White-tails in that season. More recently, Hugh Robinson and his co-workers also demonstrated higher rates of Cougar predation on Mule Deer (17%) than on White-tailed Deer (9%) in the West Kootenay area, and concluded that the slow recovery of the local Mule Deer population after the harsh winter of 1996–97 was likely related to continuing Cougar predation.

White-tailed Deer are on the list of Cougar prey species identified in other areas, as are Mule Deer and Black-tailed Deer, Elk, Moose, Bighorn Sheep, Mountain Goat, Caribou, Pronghorn, Peccary, other carnivores (including Black Bear, Coyote, Bobcat, Red Fox, Raccoon, American Badger, Striped Skunk and other Cougars), most of the smaller mammals (marmot size and smaller) that occur within its range, as well as fish, insects and several bird species. Cougar predation on Porcupines has been recorded in several areas, and it is apparent that at least some individuals become proficient at killing that species. Oregon researchers found "flesh, bones, and large amounts of flaccid quills" in Cougar stomachs and noted "no damage ... attributable to the ingestion of quills".

Cougars usually kill their own food, but will scavenge when necessary, and there are several accounts of Cougars taking prey from other carnivores, especially Coyotes and Bobcats, often killing them in the process. Like other cats, Cougars hunt with patience, stealth (figure 57) and a quick, short pursuit. When hunting large prey, a Cougar often ends up riding its quarry for a time, usually bringing it down with a bite through the neck that severs the spine. But a bull-Elk kill in Banff National Park indicated that the cat had grasped the Elk's muzzle from behind and pulled its head back until the neck broke, an impressive display of the predator's power considering the Elk outweighed the Cougar by a factor of at least three to one. Once the Cougar brings down a large animal, it may drag the carcass several hundred metres to a more sheltered or secure area.

A Cougar may feed on a large carcass for three weeks or more, either remaining near it or returning to it repeatedly after excursions elsewhere. In the case of the Banff Elk, the Cougar devoured the kill for over

Figure 57. Cougar and Black-tailed Deer tracks after a rare snowfall on west Vancouver Island.

a month. Between feeding bouts, apparently to discourage scavenging by smaller carnivores and birds, Cougars typically cover the carcass at least partially with snow, soil, leaves, sticks or other debris scratched up from the ground nearby (figure 58). The evidence from both food habits and tracking studies indicates that even though carcasses of large prey may provide meals for several days, Cougars may not eat on many days during the year. Alberta researchers Ian Ross and Martin Jalkotsky documented intervals of up to 23 days between an animal leaving one kill and making another. In the western United States, biologist John Laundre estimated that the number of deer killed by adult Cougars each year ranges from 20 by males to 40 by females with kittens.

Home Range and Social Behaviour

Cougar populations consist of residents and transients. Most residents are mature adults, while transients consist of newly independent young or displaced older adults. Cougars travel, hunt and live alone, except briefly during the mating season, when adult females have dependent young, or when dispersing siblings associate temporarily. The degree of overlap between the home ranges of adjacent resident females appears to vary, from almost non-existent among radio-collared animals in a southeastern BC study, to fairly considerable (up to 45%), among study animals in Utah and Idaho. The ranges of adjacent resident males have overlapped very little in studies to date, but they have extensively overlapped those of local resident females. Areas of overlap between adjacent residents are usually not occupied by both animals at the same time, and the same is true for most incidents of trespassing by transients on resident ranges. Maurice Hornocker used the phrase "mutual avoidance" to describe the behaviour which helps the animals maintain spatial separation without having to resort to violent conflict. Territorial marking, which helps facilitate avoidance by serving as notification of current occupancy, is done with

Figure 58. A female Mule Deer killed and partially covered with debris by a Cougar.

feces, urine and scrapes (conspicuous piles of soil and debris 3–5 cm deep and up to 45 cm long and 30 cm wide). Scrapes, usually made by scratching up the substrate with the hind feet, are mostly done by resident males.

Among residents, the annual home ranges of males are generally larger than those of females, but vary between areas. In two separate studies in southeastern BC, average home-range sizes in the Elk and Fording river valleys for males and females, respectively, were 152 km^2 and 55 km^2, as compared to 782 km^2 and 628 km^2 for Cougars in the south Selkirk Mountains. Comparable figures from study areas in the United States include 187 km^2 and 74 km^2 in the San Andres Mountains of New Mexico, and 453 km^2 and 268 km^2 in Idaho. In a Utah study, the only male range recorded was 826 km^2, while the average for four females was 626 km^2 (396 to 1,454 km^2). The above differences probably reflect differences in local prey distribution and abundance.

Activity and Movements

> "Active as a cat" is an established measure of physical fitness. There is not on earth a creature ... that is better equipped in point of agility, muscular control, weapons, and – shall I add? – personal prowess, and sheer fortitude.... Let us then remember that the Cougar is in all physical respects a Cat, simply a Cat multiplied by 20.
>
> – Ernest Thompson Seton

In an intensive study on the daily activity and movement patterns of Cougars in southern California, researcher Paul Beier and his associates found that when hunting (which was done almost exclusively at night). Cougars typically stalked or sat in ambush for periods averaging 42 minutes, then moved an average distance of 1.4 km over the following hour or so. They repeated the pattern up to six times on nights when they killed no prey. When a Cougar made a kill it usually took four to six hours to eat a small animal and two to five days for a large one. Actual distances moved during a night's hunting period averaged 9.9 km for females and 11.0 km for males.

Two other types of movement of particular interest are seasonal migration and dispersal. Although it was known that Cougars in mountainous areas relocate their activities up and down slope in response to seasonal movements of their prey, recent studies in

California, led by Becky Pierce, have demonstrated that such migrations are more extreme in some areas than previously reported, resulting in individuals occupying seasonal ranges that are widely separated, and interacting in two completely different Cougar sub-populations over the course of one year.

Young Cougars dispersing from their natal areas replace residents in other areas that have died and recolonize temporarily vacant habitat, thereby providing gene flow between areas. As illustrated in a detailed study in New Mexico by Linda Sweanor and her colleagues, the general pattern is for considerably longer movements by males than by females. In that study, males moved an average of 8.1 times farther than females after becoming independent from their mothers. Further, none of 13 males established activity centres within their natal ranges, as compared to 13 of 21 females that did. In southeastern BC, no marked kittens were known to have established adult home ranges within the study area. Most of the documented dispersal movements of males in the BC study were 30–60 km, including one into adjacent Alberta, and the longest was 163 km southwest, to Montana. The longest dispersal on record was 1067 km, by a young male that was radio-collared in the Black Hills of South Dakota and was killed by a train nine months later in north-central Oklahoma.

Reproduction

Cougars are generally polygamous, with a male potentially servicing several females living in or near its home range. Although mature Cougars are capable of breeding and producing young throughout the year, a more restricted breeding schedule that results in summer births (June through September) appears to be the rule in the northern hemisphere. Based on evidence collected by Brian Spreadbury and associates in 1985–87, and on the known gestation period of about 90 days, successful matings (producing litters) occurred in March, May, August and November in southeastern BC. Females in estrus are very vocal, producing what has been variously described as yowling or caterwauling sounds like those of domestic cats, but louder and harsher. Responding males also yowl, but not as frequently or as loud. Consort pairs usually remain together for only a few days, mating repeatedly during that period (as many as nine times in an hour and 50 to 70 times in a day have been documented).

Most female Cougars are not sexually mature until they are about two years old, and few produce litters until they have become resident on a stable home range. Among males, which also become mature at about two years, most mating is probably also accomplished by residents. Cougar litters range from one to six kittens, averaging three in most areas. In southeastern BC, seven litters documented by Brian Spreadbury and his co-workers in the Elk and Fording river valleys (late 1980s) ranged from two to four cubs and averaged 3.1, while in the South Selkirk Mountains the range for seven litters observed by Ross Clarke (1998 through 2002) was two to three with an average of 2.4. Productivity was lower on Vancouver Island in the 1990s, with a range of one to three and an average of 1.9 for 21 litters documented by Steven Wilson and his co-workers. They suspected that reduced prey availability resulting from severe declines in Black-tailed Deer populations on the island was the primary factor involved.

Newborn kittens weigh about 500 grams and, with their eyes and ear canals closed for about their first ten days of life, are quite helpless. The mother provides all parental care, reducing the size of her hunting range while the kittens are small, but gradually expanding it as they get older and more mobile. Kittens may stay with their mother for up to two years, during which time she will not have another litter. The interval between births for four females studied in southeastern BC ranged from 15 to 23 months, and that appears to be typical for the species.

Health and Mortality

Cougars are said to be remarkably free of ectoparasites such as fleas and lice, but they are host to at least six species of ticks. Like most carnivores, they harbour a number of different internal parasites. The most commonly encountered are tapeworms, especially *Taenia omissa*, which exists in its larval stages in the internal organs of deer. Various roundworms have also been identified in Cougars including the Trichina Worm (the causative agent of trichinosis), which occurred in more than 50 per cent of specimens examined in a study in the western United States. A number of diseases, including distemper, rabies and anthrax have been reported in Cougars, but incidence is low and has little or no effect on populations. The Cougar's solitary, mobile existence at low density is not conducive to the spread of disease.

Cougars can be killed in their attempts to bring down large prey. Utah researchers Jay Gashwiler and Leslie Robinette, reported adult Elk killing Cougars by trampling them with their forefeet, and accidents that apparently occurred when Cougars were clinging to fleeing prey. In one such case, a Cougar was impaled by a tree branch and, in another, both Cougar and deer died when they collided with a tree, the Cougar from a broken neck. Ian Ross and his associates in Alberta documented an adult male and a Bighorn Sheep falling off a cliff while struggling. Cougars not killed outright in such encounters, or in fights with other Cougars, may be injured seriously enough so that they are unable to provide for themselves and die later from starvation or predation.

Most documented cases of predation on Cougars has been by other Cougars, mostly adult males killing younger animals. The only records of predation by other species on Cougars, other than small kittens have involved Grey Wolves, two in Glacier National Park, Montana, as reported by wolf researcher Diane Boyd and her associates. In one case, a three to four month old kitten that had treed near a deer kill was killed and partly eaten by a pack of four or five wolves when it came down; in the other an adult female was caught and killed, but not eaten, by a pack of eight wolves.

Other natural deaths that have been reported include starvation not precipitated by injuries, and the debilitating effects of Porcupine quills in the face.

The most common human-caused deaths of Cougars in North America are vehicle collisions on roadways, predator control activities and hunter harvest. In a southeastern BC study in the mid 1980s, Brian Spreadbury and his associates documented the deaths of seven Cougars, four (three adult males and one female kitten) by road accidents and three (one young adult and two male kittens) by other Cougars. Ross Clarke documented a quite different scenario in the South Selkirk study area, where eight of twelve known deaths were due to hunting, one to animal control, and three to natural causes (starvation, from an apparent injury by prey and probably from a fight with another Cougar).

Accurate determination of Cougar age is difficult, and there is, therefore, little data on that subject. It is believed that few live longer than 13 years in the wild, although ages of at least 20 years have been attained in captivity.

Abundance

The official provincial population estimate for Cougars in British Columbia in 1987 was about 3000 animals. Patterns of harvest and sightings since then suggest that BC Cougar populations are stable in most areas. A dramatic increase in sightings and conflict situations with humans in southern BC in the mid to late 1990s may have reflected an increase in Cougar numbers in that region. But it was probably due to significant deer population declines from winter kill during that period, resulting in unusual numbers of hungry cats that were more prone to activity in daylight and near human settlements, making them more conspicuous.

Based primarily on the number killed for bounty in the first half of the 20th century (up to 456 in 1933), Vancouver Island was once considered to support the densest Cougar population in North America. More recently, in the 1990s, Steven Wilson and his co-workers documented minimum densities of 1.4 to 2.0 Cougars per 100 km^2 on the north end of the island, near Kelsey Bay, and 2.6 to 7.3 per 100 km^2 near Parksville. Most mainland populations occur at lower densities year-round; for example, up to 1.4 Cougars per 100 km^2 documented over a nine-year period on a large (1900 km^2) study area in Utah. In the south Selkirk Mountains, near Nelson, Cougar densities reported by Ross Clarke are among the lowest known, averaging 0.55 animals per 100 km^2 between 1998 and 2002. Higher local and seasonal densities occur, as illustrated by Brian Spreadbury's study on a 540 km^2 area in the East Kootenays, where he estimated winter densities of 3.5 to 3.7 Cougars per 100 km^2 over a three-year period in the mid 1980s.

Human Uses

The fur of the Cougar is relatively coarse and dull, and has never attracted attention for use in making or trimming garments. Its meat is quite palatable, often compared to pork by those who have tried it, but has never been and will never be a staple in anyone's diet. The primary human use for Cougars for at least a century has been as a quarry for hunters, especially those using trailing hounds, and its value for that purpose has been important in offsetting the negative effects of its depredations and maintaining its positive big-game status.

Taxonomy

In their classic book *The Puma, Mysterious American Cat*, Stanley Young and Edward Goldman recognized 15 subspecies in North America, with three occurring in British Columbia, and that classification was adopted by Cowan and Guiguet in *The Mammals of British Columbia*. Many of those subspecies were defined from subjective skull traits and minor differences in pelage colour. More recently, based on a study of DNA samples from throughout the species' range, M. Culver and colleagues demonstrated that there are six distinct genetic groups of *Puma concolor*, one in North

America and five in South America. They found no evidence for distinct genetic groups among North American Cougars, and classified all continental populations as a single subspecies:

Puma concolor cougar Kerr – applicable to the entire North American range of the Cougar, including all populations on the islands and mainland of BC.

Conservation Status and Management
The Cougar in British Columbia is on the Yellow List, considered not at risk. Since the 1960s it has been managed as a big-game species, its harvest by hunting restricted by seasons, bag limits and local closures to ensure sustainability. But for more than 50 years before then, the species was considered vermin, and had a bounty on its head. Writer Lyn Hancock studied the history of human-Cougar relationships in the province, describing legendary characters such as Cougar Brown and "Bring 'em Back" Cougar Holcombe, professional bounty hunters who allegedly killed thousands of Cougars during their careers. Provincial Game Commission reports indicate that an average of 460 BC Cougars were killed for bounty each year between 1930 and 1956, with the take exceeding 500 in 11 of those years and 700 twice. Between 1980 and 2003 the annual average legal harvest was 278 animals; the harvest increased dramatically between 1996 and 2000 (average 467; maximum 505), in concert with increased abundance or vulnerability of Cougars.

Most of the predator-control effort directed at Cougars throughout the west in the first half of the 20th century was undertaken both to protect and enhance wild ungulate populations and to prevent or reduce depredations on livestock. In most studies since, predation effects on local populations of deer, the primary prey, have not been apparent. But that is not uniformly the case for all prey species, especially Bighorn Sheep. California researchers reported that, in recent decades, Cougar predation caused serious declines in two sheep populations, and was impeding the recovery of a third population that had been declared endangered. Ian Ross and his co-workers in Alberta found that some Cougars appeared to specialize on Bighorn Sheep, and had the potential to seriously affect small populations. One female Cougar in their study area killed 9 per cent of the population and 26 per cent of the lambs in a single winter.

In BC, Scott Harrison and Darryl Hebert reported selective predation by Cougars on California Bighorn Sheep in the Junction Herd west of Williams Lake. In that case, of 40 kills attributed to two radio-collared female Cougars between 1986 and 1988, 16 (40%) were adult rams. Most of those were killed in fall or early winter, and the researchers believed that "poor condition following the rigours of the rut" was the primary factor in their apparent vulnerability. Considering incidents reported by biologist Andrew Bryant, Cougars also pose some threat to recovery of the endangered Vancouver Island Marmot and, based on studies since the mid 1990s, Trevor Kinley and Clayton Apps have identified Cougar predation as a major factor in a continuing decline of an endangered Caribou population in the southern Purcell Mountains of BC. A radio-collared male Cougar in the south Selkirk Mountains was removed in an animal control action after it killed at least two and probably three of the Caribou from the endangered population in that area.

As for livestock, Cougars have a long and continuing history of preying on domestic sheep, goats, pigs, cattle, horses and poultry in rural areas, and domestic pets in urban areas. Another of BC's legendary figures, Cougar Annie, reportedly killed at least 70 Cougars defending her west coast homestead (northern Vancouver Island) over a number of years. In addition, and in increasing frequency in recent years, Cougars appear to be the most disposed of all our carnivores to attempt predation on humans. Although the frequency is still extremely low in comparison to automobile or boating accidents, there have been a number of such attacks, some fatal, in western provinces and states in the last two decades, and especially in California where Cougars are fully protected. In BC, three humans were killed and 32 injured in Cougar attacks between 1980 and 2005.

Remarks

Although rare, both albino and melanic Cougars have been documented. Trapper Bob Hooker told me that on two occasions in 2000, an acquaintance saw a black Cougar near Adams Lake in southern BC. Museum curator Gavin Hanke provided two other records of black cougars, one that he observed in Manitoba and a recent (May 2006) sighting by Bernice Wheeler in Victoria.

Selected References: Beier 1991, Clarke 2003, Culver et al. 2000, Hancock 1987, Harrison and Hebert 1988, Hornocker 1970, Laundre 2005, Logan and Sweanor 2000, Robinson et al. 2002, Ross et al. 1997, Seidensticker et al. 1973, Seton 1909, Spalding and Lesowski 1971, Spreadbury et al. 1996, Wilson et al. 2004.

Canada Lynx   *Lynx canadensis*

Other Common Names: Grey Lynx, Grey Wildcat, Link, Lucivee, Lynx; *Loup-Cervier, Pichu*. Provincial Species Code: M-LYCA.

Description

The Canada Lynx is a short-tailed cat about the size of a Brittany Spaniel. It has long legs, and conspicuously large, well-furred paws. Its thick, luxurious pelage appears silver-grey, especially in winter; undertones of buff to red-brown sometimes show through where the fur is shortest (head, tail and lower legs), more so during moult and in summer. Black spots or bars, if present, are usually confined to the paler and longer belly fur, but in some cases show faintly on the legs. Up close, the pelage is bicoloured, with the brown-toned underfur and the guard hairs brownish for about 80 per cent of their length, but tipped with the greys, silvers and blacks that compose the overall outward appearance. The face is rimmed with a furry ruff, including beard-like clumps of white-and-black hairs beneath a white chin. The large yellow eyes are rimmed by black eyelids surrounded by an edge of white fur. The ears are edged with black, and tipped with conspicuous 4–5 cm long black tufts, and the tail has a solid black tip (figure 15, page 46). The sexes are similar in colour and overall appearance, but males are about 15 to

20 per cent larger than females. Body weights of Canada Lynxes are less than would be expected, given the species' height and apparent bulk, and those listed below for some BC specimens are typical. But female weights to 13.5 kg have been recorded elsewhere, and I weighed a well-fed (fat) male in Alaska at 19.1 kg.

Measurements:

total length (mm):
    male: 902 (825-945) n=18
    female: 862 (813-925) n=18

tail vertebrae (mm):
    male: 106 (92-116) n=18
    female: 102 (91-118) n=17

hind foot (mm):
    male: 241 (215-261) n=18
    female: 227 (212-245) n=18

ear (mm):
    male: 81 (76-86) n=16
    female: 77 (73-81) n=14

weight (kg):
    male: 11.1 (9.1-15.0) n=22
    female: 9.1 (6.6-11.8) n=23

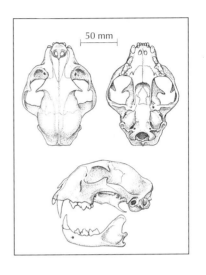

50 mm

Dental Formula:
    incisors: 3/3
    canines: 1/1
    premolars: 2/2
    molars: 1/1

Identification:
The only other short-tailed wild cat in British Columbia, the Bobcat, is similar in size to the Canada Lynx (called "Lynx" in the rest of this account), but has shorter fur and much smaller feet. Although light-tipped guard hairs on the back and upper sides may present a silver to greyish appearance from some angles, the dominant colour of most Bobcats is usually buff to reddish. Unlike Lynxes, Bobcats usually show some spotting on the back and sides, and are marked with rich black spots and bars on the lighter coloured face, chest,

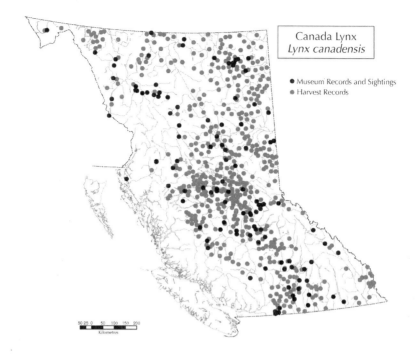

Canada Lynx
Lynx canadensis

● Museum Records and Sightings
● Harvest Records

50 25 0   50   100 150 200
Kilometres

legs and belly. Bobcats also differ from Lynxes in having black on only the dorsal surface of the tail tip rather than all the way around (figure 15), black ears with a prominent white spot (viewed from behind), and smaller ear tufts (less than 3 cm in length, as opposed to 4–6 cm for Lynxes).

Distribution and Habitat

The broad range of the North American Lynx includes much of Alaska, most of the forested portions of Canada from coast to coast, and scattered locations in the northern portions of the continental United States. In British Columbia, Lynxes regularly occur in all regions except the coastal islands, the wet forests west of the Coast Mountains, and the Lower Mainland west of Hope. There are centres of abundance in the boreal and sub-boreal forests in the northern third of the province, and in the dry forests along the vast central and southern interior uplands of the Nechako, Fraser and Thompson plateaus.

The relationship between the Lynx and its primary prey, the Snowshoe Hare (figure 59), is well known and is reflected in most aspects of the Lynx's life history. As expressed by the naturalist

Ernest Thompson Seton, "The Lynx lives on rabbits, follows the rabbits, thinks rabbits, tastes like rabbits, increases with them, and on their failure dies of starvation in the unrabbited woods."

The best hunting habitats for Lynxes are in the dense shrub and conifer cover where hares thrive, particularly in young forests (15 to 40 year-old) regenerating after a fire (burns), high elevation riparian thickets, and older forests that support a sufficiently thick undergrowth of shrubs and young conifers. Regenerating forest cutblocks, especially those featuring pine, may also be good for Snowshoe Hares and Lynxes, particularly if they develop naturally rather than under intensive silviculture.

Figure 59. Snowshoe Hare, the primary prey of the Lynx.

Lynxes usually establish day beds and resting sites in or adjacent to hunting habitat and, depending upon weather conditions, either out in the open or under shelter. When they need shelter Lynxes often select sites such as the hollows at the bases of thickly branched trees. Natal and maternal dens used by females are more structurally complex, and usually a cavity in or under a pile of large woody debris. In a Yukon study area, 70 per cent of which was a 40-year-old burn, Brian Slough documented 39 den sites, 35 in the burn and all but 2 of those associated with tangles of fire-killed trees. In unburned forest there and elsewhere, most of the maternal sites have been associated with deadfalls and blowdowns. In the early summer of 1981, near Houston in the central interior, biologist Grant Hazelwood observed a small Lynx kitten at a den in a snarl of logging debris and blowdown at the edge of a forest cutblock.

Natural History

Feeding Ecology
Lynx food habits studies in Alaska, Yukon, Northwest Territories, Alberta, Newfoundland, Nova Scotia, Montana, Washington and

southern British Columbia have all found Snowshoe Hares predominating in all seasons, and even in areas or years with low hare populations. When they are abundant, Snowshoe Hares often constitute 80 per cent or more of the diet, especially in winter. In summer and at low hare densities in the north, other common prey species are Red Squirrels, grouse and voles, and Lynx predation on ungulates (Dall's Sheep, Caribou, Mule Deer) and on other carnivores (Red Fox, American Marten, American Mink, other Lynxes) has been recorded. In the south, the diet appears to be more varied and includes marmots, ground squirrels and several species of birds. On Clayton Apps' study area in southeastern BC, Snowshoe Hares predominated in the Lynx winter diet, but only slightly, at 56 per cent; Red Squirrels, Northern Flying Squirrels and grouse made up about 80 per cent of the remaining (non-hare) component. In addition, the researchers in that area documented an adult male Lynx killing and eating two Mule Deer, a fawn and a doe, feeding on the latter for 28 days in January 1999.

Although there have been few applicable studies, Lynxes probably hunt more opportunistically during the snow-free season. One fall in the Spatsizi area of northern BC, outfitter Reg Collingwood watched a Lynx catch a duck by leaping upon it in a backwater off the Spatsizi River. The Lynx displayed the typical cat disdain for water upon emerging, violently shaking the water from each of its feet before moving off with its prey.

Figure 60. Adult male Lynx

A dead Caribou calf found by biologist John Elliott in northern BC had apparently been killed by a Lynx. Such an occurrence may not be unusual, since A.T. Bergerud found Lynxes to be significant predators of Caribou calves in Newfoundland. Lynxes appear to prefer fresh kills, but will eat carrion if alternatives are not available. In addition to published records of Lynxes eating Moose, Caribou and Mule Deer carcasses, I observed a Lynx in northern BC that had claimed the carcass of a hunter-killed Grizzly Bear. The Snowshoe Hare cycle was at a low, and the Lynx was continuously at or near that carcass for at least 15 days in October 1992 (figure 60).

Several biologists have monitored Lynx hunting behaviour and success by following tracks in snow. Most Lynxes hunt alone by either stalking or waiting at a strategic location in ambush, and then attacking with a short, explosive pursuit. In family groups (females with mobile young), they typically "fan out" while moving through their hunting habitat, and that probably increases the chances of flushing out a prey animal. Lynx hunting behaviour and strategies appear to be attuned to Snowshoe Hares, because success rates in pursuits are highest for hares. In one Alberta study, success rates in different years averaged 16 per cent for hares (range: 9–24%), 12 per cent for grouse (8–19%) and 8 per cent for Red Squirrels (0–12%). Individual Lynxes averaged 0.42 kills per night in that study, and about 1 kill per night in a Nova Scotian study.

Home Range and Social Behaviour

Given adequate food, resident adults occupy home ranges that overlap little with those of neighbouring Lynxes of the same sex, except for the occasional larger overlap between neighbouring females that are related (e.g., mother-daughter). They appear to maintain separate territories more by mutual avoidance, perhaps facilitated by scent marking, than by aggressive confrontations. Individuals may use the same core areas and general home ranges over several years. The documented sizes of Lynx home ranges have varied from about 8 km² to 783 km², with differences depending upon geographic area, habitat features, sex and age classes, seasons, and stages of the Snowshoe Hare cycle. Males generally have larger ranges than females and, in most cases, ranges are smallest when hares are abundant. In the Rocky Mountains of southeastern British Columbia, where hares do not attain the high densities characteristic of northern populations, Clayton Apps documented large home ranges, averaging 369 km² for four resident adult males and 313 km² for seven resident females.

Activity and Movements

The daily movements of Lynxes are also related to the abundance and availability of prey. During periods of food scarcity, such as during the low phase of the Snowshoe Hare cycle, a Lynx may need to travel extensively to meet its nutritional requirements. Studies in the Yukon have documented average daily movements ranging from 2.4 km to 20 km at different hare densities. In southeastern British Columbia, at low hare densities, the minimum daily movements of radio-collared males and females averaged 3.4 and 3.2 km, respectively.

Most daily movements are within home ranges, but Lynxes also undertake longer dispersal movements. Young animals leave their mother's home range (juvenile dispersal), and resident adults move out of their own home ranges if the local food supply fails (environmental dispersal). In both types, the animal wanders until it locates and establishes a new home range, or until it perishes. Studies in the Snowshoe Hare boom-and-bust habitats of the north have documented many Lynx dispersals of more than 500 km, up to a maximum of 1100 km. Tagged and collared Lynxes from both the Yukon and Northwest Territories have been recovered in BC. Lynx populations in the contiguous United States have been augmented by dispersing animals from Canada, sometimes in great numbers. But dispersing Lynxes do not always travel southward. Two radio-collared Lynxes from Washington were recovered in BC, one near Prince George (travelling at least 616 km in 202 days) and the other near Sorrento (210 km, time not recorded). In addition, a resident male in Clayton Apps's study left its home range in Yoho National Park and travelled at least 498 km to a location northwest of Edmonton, Alberta.

Reproduction

The breeding season for Lynxes in most areas is in March and April, during which time individuals of both sexes can be heard making what biologists Garth Mowat and Brian Slough have described as a wailing call. Most information on breeding is from studies in the north, although Clayton Apps documented the presence of pairs and observed a pair mating on his southeastern BC study area in March.

Most young are born in late May or early June, after a 65–70 day gestation period. Like domestic kittens, Lynx kittens are blind and helpless at birth, but they grow and develop fairly quickly. Their eyes are open in about two weeks, they take solid food by about six weeks and, although not yet fully mobile, they leave the maternal den shortly thereafter. All parental care is by the mother, which maintains a small activity radius around the den until the young are old enough to follow. Kittens remain with the mother through the first winter and occasionally into the following fall before dispersing.

Lynxes have a high reproductive potential, which is realized only in the north when Snowshoe Hare populations are high. Under conditions of adequate nutrition, most adult and many yearling females (9–10 months old) can conceive, producing litters of one to six young, with averages as high as five. In March 1980, during a high in the hare cycle, I observed a family group of six obviously well-fed Lynxes moving across a subalpine slope on Level Mountain, north of Telegraph Creek. During the low phase of the hare cycle, yearlings do not breed, adults produce smaller litters, and most kittens do not survive long enough to emerge from the natal den. In southeastern British Columbia, where hare numbers were low, Clayton Apps documented no litters larger than two, and detected no litters at all in the last year of his study.

Health and Mortality

Lynxes are host to a number of different parasites, including several species of fleas and at least one species of louse on the outside, and numerous roundworms, tapeworms, flatworms and a spiny-headed worm inside. No chronic health problems associated with those infestations have been documented. Similarly, although cases of pneumonia, rabies and distemper have been reported for individuals, no major outbreaks that might have implications for populations have been described.

Among natural Lynx deaths for which cause could be determined in three northern radio-tracking studies, ten of sixteen were due to starvation, all following Snowshoe Hare population crashes. The other six deaths were from predation, including four by other Lynxes and two by Wolverines. Of three juvenile Lynxes that died during the first three years of a southeastern British Columbia study, two starved and the other was cannibalized by another Lynx. In the following year, after a decline in both Snowshoe Hares and Red Squirrels, no kittens survived to winter and six of eight adults died, five to starvation and one to apparent Wolverine predation. In north-central Washington, three of four recorded deaths were believed to be due to starvation, and the other was a result of predation by either another Lynx or a Bobcat and, in a Montana study, three of five deaths were due to starvation and the other two to predation by Cougars.

Coyotes and Grey Wolves also prey on Lynxes. In March 1991, I followed a Lynx track along the Spatsizi River, in northern BC, to a location where it abruptly stopped, with only minimal signs of struggle, at its intersection with the tracks of two wolves. Following the wolf tracks into adjacent timber about 50 metres, I found a few tufts of Lynx fur, a portion of a Lynx shoulder blade, and a blood stain in the snow.

In areas closer to human settlements and access, Lynx mortality may be mostly due to human causes, including commercial trapping. During a period of vulnerability and widespread dispersal following a Snowshoe Hare population crash, many of the animals that would otherwise be destined to starve end up as trapper harvest statistics, and some are killed during attacks on pets and livestock. Lynxes have been known to live to 22 years old in captivity, but wild animals rarely reach 10 years and the oldest on record was 14.5 years.

Abundance

Records of North American Lynx pelt sales since the early 1800s indicate a distinct cycle of abundance, peaking approximately every 10 years, and that pattern has also been confirmed by population studies. Population peaks do not occur in all areas of the continent at the same time, and the numbers of Lynxes during peak times vary both regionally and between cycles. Further, local population peaks may be primarily the result of reproduction in some areas and mostly due to immigration in others.

In the remainder of this account, I distinguish between populations in the north (Alaska, Yukon Territory, Northwest Territories, and the northern halves of British Columbia and Alberta), and those farther south. The northern areas, where most studies have been undertaken, are characterized by extreme fluctuations in numbers and very high peak densities of both Snowshoe Hares and Lynxes, while fluctuations of both species are typically less extreme and densities consistently lower in the south.

Provincial authorities have estimated that the Lynx population in BC could be as high as 240,000 animals during the highest peak, and as low as 20,000 animals during the lowest low. But there is no practical way to confirm those numbers, and Lynx population levels and performance are monitored primarily by characteristics of the fur harvest. For all of BC, fur harvest peaks since the 1950s have been in the second or third year of each decade, with a high of 12,500 pelts in the winter of 1962–63. The most recent peak, in 2002–3, was much lower at about 1120 pelts, a result in part reflecting low demand and trapper effort at that time, but possibly also relating to diminished habitat at both the provincial and continental scale.

Despite local, regional and temporal differences, Lynx cycles have a single unifying characteristic, and that is their relationship to the ten-year cycle of the Snowshoe Hare. When hares are abundant, adult Lynxes are secure, maintaining good physical condition and high productivity within well-defined home ranges with minimal activity and effort. A year or two after hare numbers decline, Lynxes enter a period of stress that may be manifested by deteriorating physical condition, increased activity, increased home-range size, abandonment of home ranges, long-distance dispersal, decreased productivity, poor kitten survival, cannibalism and starvation. During the low phase, dispersing Lynxes may appear in unusual and unsuitable areas. From his many years as an animal control officer in the Lower Mainland, Jack Lay reports that Lynx tracks were occasionally seen in the Skagit River valley area in the 1950s, but that the most memorable occurrence of the species in the area was in 1963, following the crash of Snowshoe Hare populations in the north. In that year he had complaints of Lynxes killing livestock near cities, including Cloverdale, Fort Langley, Aldergrove and Chilliwack, and all of the cats that he handled or observed were in poor condition.

Expressing population highs and lows in terms of actual population numbers is difficult, and has been attempted only for small study areas. Densities of 30–45 Lynxes per 100 km^2 have been documented in optimal post-fire habitats in the north during cyclic highs, while peak densities in older forest habitats and those farther south are lower, at 8–20 Lynxes per 100 km^2. Densities typically drop to below 3 per 100 km^2 in all areas during the low phase, and in what may be marginal habitats in the extreme southern portions of the species' range. In British Columbia, Lynxes attain their highest numbers in the boreal and sub-boreal forests in the inland northern half of the province.

Human Uses

Use of Lynxes by First Peoples in British Columbia prior to European contact has not been documented. The meat of the animal is quite palatable, similar to turkey breast, and its fur is luxurious and warm so it is likely that First Peoples hunted Lynxes for both food and clothing. Some rural British Columbians still eat Lynx meat as a secondary product, but the primary human use for the Lynx in this province for more than 100 years has been in the fur trade.

Taxonomy

Several studies have shown that this species is remarkably uniform across its entire range, and taxonomists have not recognized subspecies. A DNA study from samples across the North American range, conducted by Ell Rueness and colleagues, confirmed that there are no genetically distinct geographic groups of Canada Lynx, although populations east and west of the Rocky Mountains show some minor genetic differences.

Conservation Status and Management

In northern and central British Columbia, Lynx populations fluctuate in response to the Snowshoe Hare cycle, and elicit no special conservation or management interest in those areas. In the extreme southern portions of the provincial range, where neither Snowshoe Hare nor Lynx densities are known to increase appreciably over time, studies in progress will help to determine whether a more intensive management focus is required. To date, management has been directed primarily to regulation of human uses, and not to the consideration of habitat needs. Over the past 50 to 60 years, since the advent of aggressive forest-fire suppression programs, the amount of the highest quality habitats for Snowshoe Hares and, therefore, also for Lynxes has diminished continent-wide. In many areas, forest cutblocks have the potential to supply some of the habitat that would previously have been provided by fire, but that potential may be reduced by silviculture practices (thinning, pruning, herbicides) designed to decrease tree density and remove competing vegetation (hare food and cover).

Remarks

Lynxes are particularly adapted to winter conditions, with thick fur, long legs that can negotiate deep snow, and large feet that act as effective snowshoes. The Lynx's snowshoe effect is about four times greater than that of a Bobcat and eight times greater than that of a Coyote, thus helping the Lynx maintain an ecological separation from those competitors during the critical winter period.

Some Lynxes show variations from the basic colours identified in the description. In the "blue" phase, the blacks of a normal Lynx (including ear tufts) are a steely blue-grey. I have not seen any records of full albinos, but trappers Dennis Brown and Paul Blackwell, live-capturing Lynx in the 100 Mile House area for translocations to Colorado, obtained a very striking specimen that may have been partially albino.

Selected References: Apps 2007; Breitenmoser et al. 1993; Hatler 1988; Hatler and Beal 2003c; Koehler 1990; Mowat and Slough 1998; Mowat et al 2000; Poole 1995, 1997, 2003; Poszig et al. 2004; Rueness et al. 2003; Slough 1999; Ward and Krebs 1985.

Bobcat   *Lynx rufus*

Other Common Names: Bay Lynx, Catamount, Lynx Cat, Red Lynx, Wildcat; *Chat Sauvage, Loup-Cervier, Lynx Roux, Pichou*. Provincial Species Code: M-LYRU.

Description
The Bobcat is a short-tailed cat about the size of a Brittany Spaniel. Like its close relative, the Canada Lynx, it has fairly long legs, but its feet are not disproportionately large. Bobcat fur is short and dense with an overall reddish to tawny cast above, especially in summer, and with shades of white on the insides of the legs and the belly. The winter pelage may appear greyer, owing to the grizzled effect of its black-tipped guard hairs. Most specimens are marked with dark spots and bars, which may show over the whole body, but are most conspicuous where they contrast with the lighter underparts. The Bobcat has a facial ruff of cheek whiskers, each side usually marked with one or more black bars extending down from below the eye. Most animals have a rim of white fur around each eye, and white areas around the muzzle and above the upper lip. The ears are large and pointed with prominent white spots on a black background in back, and with tufts of short black hairs at the tips. The characteristic featured in the species' common

name refers to its bobbed tail, which is white below and has a series of parallel black bars, including one on top, at the tip (figure 15). The sexes appear similar, but the measurements available for BC specimens suggest that, as elsewhere, males average 10 to 15 per cent larger in total length and 20 per cent or more heavier than females. Measurements provided below are within the range given in a continent-wide summary for the species, although weights of up to 28.5 kg for males and 15.0 kg for females have been recorded.

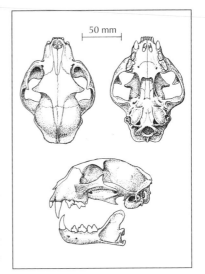

Measurements:

	All Specimens	Subspecies	
		fasciatus	*pallescens*
total length (mm):			
male:	917 (830-983) n=20	887 (n=5)	928 (n=15)
female:	837 (738-898) n=18	771 (n=3)	850 (n=15)
tail vertebrae (mm):			
male:	161 (145-184) n=17	160 (n=3)	161 (n=14)
female:	159 (140-181) n=16	145 (n=2)	160 (n=14)
hind foot (mm):			
male:	187 (175-195) n=19	189 (n=5)	186 (n=14)
female:	174 (164-185) n=17	171 (n=3)	175 (n=14)
ear (mm):			
male:	80 (70-85) n=13	73 (n=2)	81 (n=11)
female:	80 (72-85) n=14	72 (n=1)	81 (n=13)
weight (kg):			
male:	11.3 (7.3-16.6) n=25	8.8 (n=3)	11.7 (n=22)
female:	7.8 (5.0-9.8) n=23	none	7.8 (n=23)

Dental Formula:
  incisors: 3/3
  canines: 1/1
  premolars: 2/2
  molars: 1/1

Identification:

In British Columbia, the Bobcat is most likely to be confused with the Canada Lynx, but the Lynx is greyer, longer-furred and less conspicuously spotted, and has much larger feet. Other features that help distinguish between the two species are differences in ear coloration (only the Bobcat has the central white spot), differences in ear tuft length (more than 3 cm and up to 6 cm on the Canada Lynx, and usually less than 3 cm on the Bobcat), and differences in tail coloration and marking (buff with a solid black tip on the Canada Lynx, and bicoloured – rufous with black bars above and white below – with black only on the upper surface at the tip on the Bobcat; (see figure 15, page 46). Cougar kittens may be of similar size, rufous-coloured, and spotted but, unlike Bobcats, they have long tails and no facial ruff.

Distribution and Habitat

The Bobcat occurs throughout the continental United States, although it is now considered extirpated, by extensive forest clearing and agricultural development, in a large block of the southern Great Lakes region, including southern Minnesota, Wisconsin, Michigan, and all or portions of Iowa, Illinois, Indiana and Ohio. To the south, the Bobcat's range extends well into Mexico, to the State of Oaxaca just below the 18th parallel. The northern limit of its range is in southern Canada, in the southern interior of British Columbia (see below) and, east of the Rocky Mountains, a narrow band 50–100 km along the United States border from Alberta through Quebec, and in all of New Brunswick and Nova Scotia. Occurrence in much of that area is believed to be a recent range expansion in the generally warmer climate of the 20th century.

The largest expanse of Bobcat habitat in Canada is in BC, with the northernmost extreme of regular occurrence in the dry interior of the Cariboo region (Interior Douglas-fir Zone) near Williams Lake, including most of the Chilcotin River drainage. Transient individuals may appear elsewhere, and there is one museum record of a specimen from farther north, in the vicinity of Quesnel. Trapper and taxidermist Bryan Monroe reports that at least two

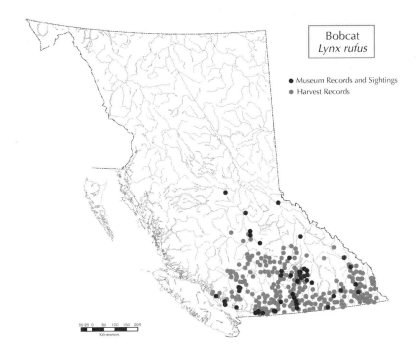

Bobcat
Lynx rufus

● Museum Records and Sightings
● Harvest Records

Bobcats have been caught in the McBride area, one in the mid 1960s and the other in about 1998. In addition, Conservation Officer Bill Richmond showed me a photograph of a Bobcat killed while raiding a chicken house in the early 1960s, near Burns Lake. It was the only one he observed in his long experience in that area. On the coast, Bobcats occur from the Fraser River valley west to the Sunshine Coast, and north to about Bute Inlet.

The Bobcat is at the northern extreme of its geographic range in BC. It is most commonly associated with dry, rocky terrain and does best in areas with little snow accumulation. The habitat relationships of this species in BC have been studied only in the Rocky Mountain Trench area of the East Kootenay region. There it is most strongly associated with the relatively dry Interior Douglas-fir Zone, primarily inhabiting lowland (< 1200 metres) mature forest habitats in winter though some (mostly males) venture into upland habitats in nearby mountains during the snow-free season. Large conifers in mature forests intercept winter snowfall, thereby reducing accumulation on the ground, and also provide shelter, such as hollows at the bases of trees and stumps and spaces under fallen and low-level branches. Hunting habitat is not uniformly distributed in such areas and consists primarily of patches of dense understorey vegetation in multi-layered mature forest, riparian thickets and meadowlands, or natural regeneration in small openings and at the edges of larger ones. A Bobcat requires the greater structural and vegetative complexity of such habitats to support the variety of its prey species and to provide the necessary cover for its hunting methods.

Bobcats probably select day beds and resting sites for their microclimate features, providing shade on hot days, radiating warmth on cold days, and overhead protection during periods of rain or snow. Another important consideration in some areas has to be security from predators, especially Cougars and Coyotes. Throughout their geographic range, Bobcats regularly use rocky terrain features, such as caves, ledges and boulder rubble, for resting and also for natal den sites. In BC, biologist Clayton Apps reported that all three Bobcat natal dens located in his East Kootenay study area were in or near steep, rocky slopes surrounded by thick vegetative cover.

Natural History

Feeding Ecology
Unlike Canada Lynxes, Bobcats do not normally rely on one prey species. Local food habits may vary considerably between seasons and from year to year (depending upon availability), and between individuals (depending upon availability of prey and upon the sex and age of the Bobcat). In general, the larger cats take larger prey species. In the southern and western United States, Bobcats prey upon cottontails, jackrabbits, cotton rats, woodrats, ground squirrels, marmots, pocket gophers, mice and voles, galliforme birds (grouse, pheasants, quail, turkeys), and the smaller ungulates (Mule Deer and White-tailed deer, Pronghorn, Peccary). They also occasionally prey on many other small and medium-sized mammals and birds, a few reptiles (including rattlesnakes), and some insects. In Arizona, a Bobcat was seen killing and eating at least 10 bats within a half an hour, catching or knocking them from the air while they were hunting insects over an artificial water source for Bighorn Sheep. In another case, an adult male Bobcat killed and ate a Red-tailed Hawk, apparently catching it on the ground shortly after the hawk had killed a cottontail (which the Bobcat probably ate for dessert). The Mountain Beaver, a relatively rare rodent in BC, was the most commonly recorded item in Bobcat prey remains

from wet forest habitats of western Washington, and Bobcats were the most frequent predators of American Martens in an Oregon study. A few instances of cannibalism have also been recorded.

There are fewer potential prey species in northern areas of the Bobcat's range. Studies in Maine and Nova Scotia found Snowshoe Hares the most common remains in winter scats and stomach samples (50% and 74%, respectively), and White-tailed Deer second (15% and 12%). The deer were sometimes eaten as carrion. Other foods included voles, squirrels, Porcupines, songbirds, grouse, and single records of Striped Skunk and River Otter. In British Columbia, the East Kootenay studies found Red Squirrels the predominant food for Bobcats in winter, occurring in 36 (52%) of 70 digestive tracts collected over a five-year period. Ungulates, mostly deer, were next at 18 (26%), followed by voles at 8 (11%). Those species are all common in the lowlands of the Rocky Mountain Trench where the Bobcats concentrate in winter, while upland species such as Snowshoe Hare (used commonly in other areas) were found in only 5 (7%) of the samples. Other foods identified in that study included five occurrences of native birds and two of domestic fowl. Based on local climate and animal metabolism, Trevor Kinley estimated that a 10-kilogram Bobcat would require the equivalent of about 1100 Red Squirrels, 15,000 mice or voles, 350 Snowshoe Hares, or 180 kg of deer meat to sustain its weight over one East Kootenay winter.

In the only other Bobcat food-habits study in BC, university student Connie Young examined 17 digestive tracts from the north Okanagan area, near Enderby. Prey found in the 11 that were not empty included Mule Deer (7 occurrences, some known to be carrion), Snowshoe Hare (5), Long-tailed Vole (2), and Red Squirrel, Bushy-tailed Woodrat, American Marten and Ruffed Grouse (1 each).

The hunting strategies and behaviour documented for Bobcats are similar to those of Canada Lynxes. They either ambush prey along trails or at specific locations where they are likely to appear (burrows, feeding and watering sites), or stalk them and then attack with a quick, short rush. Bobcats are built for speed, not endurance, and usually give up a pursuit if not successful within about 15 metres. They kill large prey, such as deer, by biting its throat or the base of its skull. Bobcats bite with great speed and power, as biologist C.M. McCord discovered first hand. A five-month old kitten that he was feeding suddenly attacked, biting his hand to the bone three or four times before he could withdraw it.

If prey is too large to be consumed in one meal, Bobcats some-times cache it by loosely covering the carcass with fine materials such as snow or leaves. Cached carcasses provide a significant win-ter resource, but they can be lost to other scavenging predators, such as Cougars or Coyotes. In the East Kootenay area, a radio-collared male Bobcat killed a female White-tailed Deer and retained possession for 14 days, apparently not moving more than 50 metres from the carcass during that period.

Home Range and Social Behaviour

Like most members of the cat family, Bobcats are generally solitary. The only times they are found together are as pairs during the short, late-winter breeding season, in family groups (females with depen-dent young) through the first winter until about March, and occa-sionally as groups of two or more siblings when dispersing. A local population is usually made up of established residents (mostly adults) and transients (mostly juveniles, but also some displaced adults). Residents maintain home ranges that are largely separate from those of their similarly established same-sex neighbours. As with most carnivores, violent territorial disputes are probably rare, but do occur. In California, a radio-collared resident male Bobcat was observed in its home range eating another adult male that it had apparently killed. In Colorado, an adult female in its home range repeatedly attacked another Bobcat for 26 minutes, drawing blood but apparently not causing serious injury, until the intruder retreated.

In most cases, separation between neighbours appears to be maintained by mutual avoidance, facilitated by scent and visual marking. Scent marking may involve depositing feces at elevated locations along trails, and backward squirting of urine or rubbing of anal glands on objects such as stumps and boulders (scent posts). Visual marks most often involve scrapes on the ground, made with the hind feet and often enhanced by scent from feces or urine. Scratch marks on trees and stumps, made by what some observers refer to as "claw sharpening", may also serve as visual signs that indicate an occupied territory.

Female home ranges are smaller than those of males, and may overlap completely with them. In the East Kootenay study area, the average annual ranges of five females and seven males were 56 km² and 139 km², respectively. Those are larger than home ranges observed in many Bobcat habitats in the United States, particularly in southern and coastal areas, and probably indicate less favourable conditions at the northern edge of the species' range.

In the Kootenay area, the winter ranges of males were about 40 per cent smaller than their annual ranges, reflecting the effects of increasing snow at higher elevations forcing them into more concentrated use of drier lowland sites. The ranges of females differed little between seasons, presumably because they were tied to lowland den sites and kitten care in the summer and were subject to the same snow-cover limitations faced by males in winter.

Activity and Movements

Daily movements made by Bobcats vary considerably, mostly in relation to prey availability. Longer movements may be required when prey densities are low, or where good hunting spots are widely dispersed or seasonal. For several studies in which movements over 24 hours were determined by snow-tracking, averages ranged from 4.9 km (Massachusetts) to 8.8 km (Minnesota), with extremes of 0.8 km and 18.5 km. More recently, radio-tracking studies have shown that daily movements of females are generally less than those of males (2.6 vs. 4.8 km in one study), suggesting that females use their smaller home ranges more intensively than males do.

In the East Kootenay radio-tracking study, Clayton Apps found that six of seven juvenile males undertook long-distance dispersal movements once independent of their mothers, while none of three juvenile females did so. The males travelled from 42 to 506 km (average 186 km), among the longest dispersal movements recorded, possibly reflecting a more scattered distribution of suitable habitat or prey at the northern limits of the species' range. Few other studies have reported dispersal movements longer than 50 km, and in one of the exceptional cases (straight-line distances of 182 and 158 km by two juvenile males in Idaho), a major prey decline was believed to have been at least partly responsible.

Reproduction

The breeding season in the northern portion of the Bobcat's range is generally in February and March, although there is evidence of later breeding (through July) in more southerly areas in the United States. There are no specific reports on the timing of breeding in British Columbia, but Clayton Apps observed signs of mating activity in about March in the East Kootenay area.

Females can breed as yearlings, but pregnancy rates are lower for such young cats than for adults. In a Nova Scotia study, pregnancy rates for Bobcats aged 1, 2 and 3 or more years were 26, 55 and 73 per cent, respectively. In Washington, pregnancy rates among yearlings were 40 per cent in the west and 48 per cent in the east. Males do not attain sexual maturity until their second year, but they apparently remain fecund year-round from then on. The gestation period averages about 63 days, and most females at northern latitudes produce litters in May or June. Although a female Bobcat in Florida gave birth to two litters in one year, that is rare and probably never occurs in the north. Litter sizes range from one to six kittens, varying with factors such as environmental effects on physical condition. In an Idaho study, Bobcats failed to produce young during a decline in prey populations. The age of the mother is also a factor. In Nova Scotia, average in-utero litter sizes for 1-year-old, 2-year-old and 3-or-more-year old females was 2.2, 2.4, and 2.7 respectively. Average litter sizes in Washington (as interpreted from placental scar counts) were 2.0 for yearlings and 2.5 for adults in the western subspecies and 2.8 for both yearlings and adults in the eastern subspecies.

Bobcat kittens are blind and helpless at birth, weighing 200 to 300 grams. They gain about 10 grams per day if well-fed, their eyes open at about 10 days, they begin exploring the den and use solid food by four to five weeks, and they are usually weaned by about eight to ten weeks. The mother provides all parental care, which restricts her hunting movements when the young are den-bound. Mothers sometimes move their kittens from the natal den to one or more maternal dens during the period of early dependency. Such moves may be primarily a security measure against potential kitten predators, but may also reflect the need to change hunting areas.

In northern areas, including British Columbia, most young males remain in family groups with their sisters and mother through the first winter until the onset of the next breeding season, about March, then begin to disperse. The situation for female kittens is less clear, but some remain on or near the mother's home range, dispersing only short distances if at all. In one case in the east Kootenay area and three cases in Idaho, female kittens took over their natal ranges after their mothers' deaths.

Health and Mortality

Bobcats are host to a long list of parasites, including fleas, lice, mites and ticks, various helminths (roundworms, tapeworms, flukes and spiny-headed worms) in intestines and other organs, and more microscopic hitch-hikers (protozoans and larval helminths in body fluids, including the blood). Cases of large, debilitating numbers of parasites in individual Bobcats are rare and may indicate poor physical condition rather than be its cause. A number of diseases have been reported, including rabies, feline distemper, feline infectious peritonitis, feline infectious anemia, salmonellosis and toxoplasmosis. Although individuals infected with some of those are known to have died as a result, major outbreaks affecting populations have not been documented. The potential for such occurrences is low due to the solitary nature of Bobcats, and their propensity to regularly change den sites and resting areas.

Starvation is one of the most common natural causes of Bobcat mortality, especially for transient animals but also for resident adults in winters with unusually deep snow. Kittens are susceptible to predation by a number of carnivores, including other Bobcats and large raptors such as Great Horned Owls, and there are several reports of adult Bobcats being killed by Coyotes and Cougars. Accidental deaths have also been recorded, including a kitten fracturing its skull in a fall from a rocky ledge. Although some Bobcats prey successfully on Porcupines, others (usually juveniles) make mistakes and there are at least two records of deaths directly attributable to wounds from imbedded quills. Bobcats sometimes sustain injuries when hunting larger prey, such as deer, and such injuries can lead to death if the Bobcat is unable to hunt efficiently afterward.

Bobcat range in North America overlaps with areas used intensively-by humans, and many are killed by humans or human activities. Legal, regulated harvest by hunting and trapping accounts for much of that in most years, but others are removed to protect livestock and pets, especially in deep-snow winters and during periods of scarcity in natural prey. Other human-related mortality includes poaching (illegal hunting), road accidents, and other accidents (in Idaho, six kittens were electrocuted while climbing powerline poles).

In the East Kootenay studies, of ten radio-collared animals for which cause of death was documented, four were shot, three starved, two were trapped and one was killed by a Cougar. Four of those deaths, including two from starvation, occurred during the most severe of five winters during the study period (1992–93). In that same winter, an adult female and a male kitten found in emaciated condition were restored to health in captivity before being returned to the wild (they probably would have perished). The digestive tracts were empty in 8 of 25 specimens (32%) that year, as compared to just 3 of 45 (7%) in the other four winters combined.

The negative effects of severe winter conditions on Bobcats, which result in reduced reproductive output as well as starvation, increased vulnerability to predation and human harvest, and increased nuisance interactions with humans, are a primary theme throughout the literature on the species, and it is likely that those effects are felt most strongly in northern populations such as those in British Columbia.

Bobcats have been known to live as long as 25 years in captivity, but wild animals rarely live half that long. Among 1105 specimens from five US states (Wyoming, Minnesota, South Dakota, Washington and Arkansas), only 5 were 10 years old or older. The oldest documented wild specimen, at 15.5 years, was a female from Washington.

Abundance

In the early 1980s, provincial wildlife authorities estimated the provincial Bobcat population at between 5000 and 10,000 animals, based on projections from published information on average densities elsewhere. The only specific population density information for the province is from an early 1990s study by Trevor Kinley in the East Kootenay area. The preliminary, conservative estimate of 0.6 Bobcats per 100 km^2 of annual range obtained there is low compared to 0.6–1.2 per 100 km^2 in Montana and 3.5–4.3 in Idaho. But it is likely that the lower snowpack areas farther west in BC Bobcat range (Okanagan and lower Fraser River valleys) support higher densities than were observed in the Kootenay study area.

Both fur-harvest statistics and anecdotal evidence suggest that the provincial Bobcat population is at least stable and, probably in response to increasingly mild winters, has apparently been increasing in numbers and in range since the mid 1990s.

Human Uses

Due to its largely nocturnal and secretive habits, the Bobcat is not a good candidate for wildlife viewing, although occasional glimpses of the animal and observations of its sign will continue to be of interest to both official and unofficial naturalists. The primary human use for the Bobcat throughout recorded history is trapping and hunting for its fur, which is used to make fine jackets and hats. The legal harvest in British Columbia is relatively small, ranging from a low of 62 animals (1993) to a high of 188 (2003) since the early 1990s.

Taxonomy

Defined from pelage colour and skull traits, twelve subspecies are recognized in North America and two of those occur in British Columbia. No modern genetic studies have been done to verify the validity of the currently recognized subspecies.

Lynx rufus fasciatus Rafinesque – along the coast from northern California to Bute Inlet, BC, this subspecies is characterized by its dark-reddish pelage with prominent dark markings. According to Cowan and Guiguet, it intergrades with *pallescens* along the eastern slopes of the Cascade and Coast mountains.

Lynx rufus pallescens – the western United States, southern Saskatchewan and Alberta, and south-central BC as far north (rarely) as Prince George. This subspecies differs from *fasciatus* in its softer and paler pelage, fewer dark markings, larger size and longer skull.

Conservation Status and Management

Although its distribution in this province is limited by climate, the Bobcat is a secure member of British Columbia's fauna and is managed as a game animal and furbearer. It is one of a few species that would probably benefit from global warming and, indeed, it appears to have been increasing in local numbers and expanding its range in the province in recent, mild-winter years, although a few severe winters could reverse that pattern. As with Bobcats elsewhere, many in BC occur in areas occupied by humans, and so the potential for conflict continues in the form of depredations on small livestock and pets; but such conflicts can be easily dealt with as they occur.

Overall, the most important issue in Bobcat conservation in BC may be maintaining winter habitat. In the Kootenay area in particular, there are many competing human uses for the lowland mature-forest habitats used by wintering Bobcats. As outlined by researcher Clayton Apps, the Interior Douglas-fir Zone is "more than 90 per cent fragmented and less than 1 per cent protected, placing it among the most threatened ecosystems in the province".

Remarks

There are records of both melanic (dark, to black) and albino (white) Bobcats although neither has been reported in British Columbia.

There is evidence that where Canada Lynx and Bobcat ranges overlap, and snow conditions are not limiting, the Bobcats will out-compete the Lynxes. On Cape Breton Island, Nova Scotia, the range of Canada Lynx has shrunk considerably since the 1960s when Bobcats arrived on the island. Meanwhile, the relationship of Bobcats with Coyotes appears to be negative for Bobcats. The Bobcat's selection of certain habitats and den sites and its reluctance to hunt and travel in the open are thought to be, in part, to avoid predation by Coyotes. There is also considerable diet overlap between the two species. It is not clear which factor is more important, but Bobcats have generally increased during periods when predator control efforts directed to Coyote populations have been successful.

In regard to the species' reputation, at least two automobiles, one small, but aggressive piece of earth-moving machinery, one WWII fighter plane, and numerous sports teams carry the name "Bobcat(s)" or "Wildcat(s)".

Selected References: Apps 1996; Bailey 1974; Hatler et al. 2003c; Kinley 1992; Knick 1990; Knick et al. 1984, 1985; Lariviere and Walton 1997; McCord and Cordoza 1982.

APPENDIX

Scientific Names of Organisms Mentioned in this Book

Trees and Shrubs

Balsam Poplar	*Populus balsamifera balsamifera*
Black Cottonwood	*Populus balsamifera trichocarpa*
Choke Cherry	*Prunus virginiana*
Crowberry	*Empetrum nigrum*
Devil's Club	*Oplopanax horridus*
Douglas-fir	*Pseudotsuga menziesii*
Hairy Manzanita	*Arctostaphylos columbiana*
Highbush-cranberry	*Viburnum edule*
Kinnikinnick	*Arctostaphylos uva-ursi*
Oregon-grape	*Mahonia aquifolium*
Saskatoon	*Amelanchier alnifolia*
Sitka Spruce	*Picea sitchensis*
Soopolallie	*Shepherdia canadensis*
Trembling Aspen	*Populus tremuloides*
Western Hemlock	*Tsuga heterophylla*
Western Redcedar	*Thuja plicata*
Whitebark Pine	*Pinus albicaulis*

Other Plants

Arrow-leaved Groundsel	*Senecio triangularis*
Bunchberry	*Cornus canadensis*
Common Dandelion	*Taraxacum officinale*
Cow-parsnip	*Heracleum lanatum*
Crowberry	*Empetrum nigrum*
Dwarf Mistletoe	*Arceuthobium americanum*
Glacier Lily	*Erythronium grandiflorum*
Lingonberry	*Vaccinium vitis-idaea*
Skunk Cabbage	*Lysichiton americanum*
Spring Beauty	*Claytonia lanceolata*

Invertebrates (including Parasites)

Army Cutworm Moth	*Chorizagrotis auxiliaris*
Basket Cockle	*Clinocardinum nuttali*

Butter Clam	*Saxidomus giganteus*
Guinea Worm	*Dracunculus insignis*
Helmet Crab	*Telmessus cheiragonus*
Hydatid Worm	*Echinococcus granulosus*
Kelp Crab	*Pugettia producta*
Kidney Worm	*Dioctophyma renale*
Mange Mite	*Sarcoptes scabiei*
Northern Abalone	*Haliotis kamschatkana*
Oregon Shore Crab	*Hemigrapsus oregonensis*
Pacific Razor Clam	*Siliqua patula*
Purple Shore Crab	*Hemigrapsus nudus*
Raccoon Roundworm	*Baylisascaris procyonis*
Red Crab	*Cancer productus*
Sea Slater	*Ligia pallasii*
Sinus Worm	*Skrjabingylus nasicola*
Soft-shell Clam	*Mya arenaria*
Trichina Worm	*Trichinella spiralis*

Fishes

Arctic Grayling	*Thymallus arcticus*
Cabezon	*Scorpaenichthys marmoratus*
Carp	*Cyprinus carpio*
Chum Salmon	*Oncorhynchus keta*
Cutthroat Trout	*Oncorhynchus clarkii*
Kelp Greenling	*Hexagrammos decagrammus*
Kokanee	*Oncorhynchus nerka*
Lingcod	*Ophiodon elongatus*
Northern Pikeminnow	*Ptychocheilus oregonensis*
Pacific Lamprey	*Lampetra tridentata*
Perch	*Perca flavescens*
Red Irish Lord	*Hemilepidotus hemilepidotus*
Sockeye Salmon	*Oncorhynchus nerka*
Wolf-eel	*Anarhichthys ocellatus*

Amphibians and Reptiles

Coastal Giant Salamander	*Dicamptodon tenebrosus*
Common Garter Snake	*Thamnophis sirtalis*
Oregon Spotted Frog	*Rana pretiosa*
Pacific Treefrog	*Hyla regilla*
Racer	*Coluber constrictor*
Red-legged Frog	*Rana aurora*
Water Moccasin	*Ancistrodon piscivorus*

Western Pond Turtle	*Clemmys marmorata*
Western Rattlesnake	*Crotalus oreganus*
Western Toad	*Bufo boreas*

Birds

American Coot	*Fulica americana*
American Kestrel	*Falco sparverius*
American Robin	*Turdus migratorius*
Ancient Murrelet	*Synthliboramphus antiquus*
Bald Eagle	*Haliaeetus leucocephalus*
Bank Swallow	*Riparia riparia*
Barn Owl	*Tyto alba*
Barn Swallow	*Hirundo rustica*
Black Oystercatcher	*Haematopus bachmani*
Black Turnstone	*Arenaria melanocephala*
Blue Grouse	*Dendragapus obscurus*
Canada Goose	*Branta canadensis*
Cassin's Auklet	*Ptychoramphus aleuticus*
Common Loon	*Gavia immer*
Common Raven	*Corvus corax*
Dark-eyed Junco	*Junco hyemalis*
Downy Woodpecker	*Picoides pubescens*
Eared Grebe	*Podiceps nigricollis*
Fox Sparrow	*Passerella iliaca*
Glaucous-winged Gull	*Larus glaucescens*
Golden-crowned Kinglet	*Regulus satrapa*
Golden Eagle	*Aquila chrysaetos*
Great Blue Heron	*Ardea herodias*
Great Horned Owl	*Bubo virginianus*
Hairy Woodpecker	*Picoides villosus*
Herring Gull	*Larus argentatus*
Horned Lark	*Eremophila alpestris*
Lapland Longspur	*Calcarius lapponicus*
Leach's Storm-petrel	*Oceanodroma leucorhoa*
Mew Gull	*Lapus canus*
Northern Flicker	*Colaptes auratus*
Northern Goshawk	*Accipter gentilis*
Pelagic Cormorant	*Phalacrocorax pelagicus*
Pigeon Guillemot	*Cepphus columba*
Pine Siskin	*Carduelis pinus*
Red-tailed Hawk	*Buteo jamaicensis*
Red-breasted Sapsucker	*Syphrapicus ruber*

Red-naped Sapsucker	*Syphrapicus nuchalis*
Rhinoceros Auklet	*Cerorhinca monocerata*
Ring-billed Gull	*Larus delawarensis*
Rock Ptarmigan	*Lagopus mutus*
Ruby-crowned Kinglet	*Regulus calendula*
Ruddy Turnstone	*Arenaria interpres*
Ruffed Grouse	*Bonasa umbellus*
Snow Bunting	*Plectrophenax nivalis*
Snowy Owl	*Nyctea scandiaca*
Song Sparrow	*Melospiza melodia*
Spotted Owl	*Strix occidentalis*
Spruce Grouse	*Falcipennis canadensis*
Steller's Jay	*Cyanocitta stelleri*
Swainson's Hawk	*Buteo swainsoni*
Trumpeter Swan	*Cygnus buccinator*
Varied Thrush	*Ixoreus naevius*
Williamson's Sapsucker	*Sphyrapicus thyrodeus*
Willow Ptarmigan	*Lagopus lagopus*
Winter Wren	*Troglodytes troglodytes*
Yellow-bellied Sapsucker	*Sphyrapicus varius*

Mammals

Arctic Fox	*Alopex lagopus*
Arctic Ground Squirrel	*Spermophilus parryii*
Beaver	*Castor canadensis*
Bighorn Sheep	*Ovis canadensis*
Bison	*Bison bison*
Black Rat	*Rattus rattus*
Black-tailed Deer	*Odocoileus hemionus columbianus* and *O.h. sitkensis*
Bushy-tailed Woodrat	*Neotoma cinerea*
Caribou	*Rangifer tarandus*
Columbia Ground Squirrell	*Spermophilus columbianus*
Dall's Sheep	*Ovis dalli dalli*
Deer Mouse	*Peromyscus maniculatus*
Domestic Cat	*Felis sylvestris*
Domestic Dog	*Canis familiaris*
Eastern Cottontail	*Sylvilagus floridanus*
Eastern Spotted Skunk	*Spilogale putorius*
Elk	*Cervus elaphus*
European Mink	*Mustela lutreola*
European Water Vole	*Arvicola terrestris*

Giant Short-faced Bear	*Arctodus simus*
Grey Squirrel	*Sciurus carolinensis*
Harvest Mouse	*Reithrodontomys megalotis*
Hoary Marmot	*Marmota caligata*
House Mouse	*Mus musculus*
Killer Whale	*Orcinus orca*
Long-tailed Vole	*Microtus longicaudus*
Meadow Jumping Mouse	*Zapus hudsonius*
Meadow Vole	*Microtus pennsylvanicus*
Moose	*Alces alces*
Mountain Beaver	*Aplodontia rufa*
Mountain Goat	*Oreamnos americanus*
Mule Deer	*Odocoileus hemionus hemionus*
Muskrat	*Ondatra zibethicus*
Noble Marten	*Martes nobilis*
North American Opossum	*Didelphis virginiana*
Northern Flying Squirrel	*Glaucomys sabrinus*
Northern Pocket Gopher	*Thomomys talpoides*
Norway Rat	*Rattus norvegicus*
Pacific Water Shrew	*Sorex bendirii*
Peccary	*Pecari tajacu*
Porcupine	*Erethizon dorsatum*
Pronghorn	*Antilocapra americana*
Red Squirrel	*Tamiasciurus hudsonicus*
Richardson's Ground Squirrel	*Spermophilus richardsoni*
Ringed Seal	*Pusa hispida*
Sabretooth	*Smilodon fatalis*
Snowshoe Hare	*Lepus americanus*
Southern Red-backed Vole	*Clethrionomys gapperi*
Stone's Sheep	*Ovis dalli stonei*
Townsend's Big-eared Bat	*Plecotus townsendii*
Townsend's Mole	*Scapanus townsendii*
Townsend's Vole	*Microtus townsendii*
Trowbridge's Shrew	*Sorex trowbridgii*
Vancouver Island Marmot	*Marmota vancouverensis*
Water Vole	*Microtus richardsoni*
White-tailed Deer	*Odocoileus virginiana*
Woodchuck	*Marmota monax*
Yellow-bellied Marmot	*Marmota flaviventris*

GLOSSARY

Albino   Having a congenital pigment deficiency resulting in the absence of colour to the skin, hair and eyes.

Auditory bulla   Bony capsule that covers the middle and inner ear (see figure 41, page 56).

Biogeoclimatic zone   An area with a relatively homogeneous climate and characteristic vegetation; BC has 14 biogeoclimatic zones.

Blowdown   An accumulation of fallen trees and branches resulting from wind action.

Body length   Total length of an animal excluding the tail (minus the tail-vertebrae length).

Burn (noun)   An area of forest regenerating after a fire.

Canid   A member of the dog family.

Cheek tooth   A premolar or molar.

Coarse woody debris   Dead or dying wood in the form of snags, stumps, logs, branches and associated litter; may occur naturally or as a by-product of logging.

COSEWIC   Committee on the Status of Endangered Wildlife in Canada.

Crepuscular   Active during the period of half light that occurs at dusk and dawn.

Delayed implantation   A feature of pregnancy in some species in which the fertilized egg floats freely in the uterine tract for a period of time, with little development, before attaching to the lining of the uterus and continuing to develop.

Dimorphism   Two different forms of individuals within a species, in features such as size, colour or shape.

Dispersal   Long-distance movements made by transient animals, especially juveniles after they have left the areas of their birth and are in search of their own home ranges.

Diurnal   Active during daylight hours; opposite of nocturnal.

DNA Deoxyribonucleic acid, the chemical molecule that carries the genetic code.

Dorsal On the back (dorsum) or upper surface; opposite of ventral.

Ear Length Length of the ear measured from the ear notch (see figure 13).

Ectoparasite An organism living on the skin or surface of another organism.

Estrus (adj: estrous) Recurring period of sexual receptivity in female mammals that includes menstruation, ovulation and changes to the lining of the uterus in preparation for pregnancy.

Extirpation Local extermination or extinction.

Felid A member of the cat family.

Gestation Pregnancy; period of fetal growth within the uterus prior to birth.

Guard hairs Long and generally silky hair shafts that extend beyond the layer of shorter and denser insulating underfur that is common to furbearing animals.

Hind-foot length Measurement from the edge of the heel to the end of the longest claw on the hind foot (see figure 13).

Home range The area used by an animal during its normal day-to-day activities.

Incisors Front teeth of a mammal.

Infraorbital foramen An opening anterior to the orbit that passes through the rostrum of the skull.

Melanic Having dark pigmentation, resulting in unusually dark (usually black) hair, skin and eyes.

Microclimate Weather conditions occurring in a small area that are not typical of the greater surrounding area, due to local geographic features such as large water bodies, directional exposures or protected gullies.

Morphology The form and structure of animals and plants.

Mustelid A member of the weasel family.

Muzzle The projecting nose and mouth of a mammal.

Nocturnal Active after dark, between dusk and dawn; opposite of diurnal.

Occipital condyle A rounded bony protrusion at the rear of the skull that articulates with the first vertebra (see figure 41, page 56).

Pelage The fur or hair of a mammal.

Population A potentially self-sustaining collection of individuals of a particular species in a specified area.

Postorbital process Bony projection above the orbit of the skull (see figures 23 and 24, page 51).

Premolars Teeth between the canines and the molars.

Reintroduction Translocation of animals to an area of former occupation.

Resident An animal with a secure home range, usually an adult.

Riparian Of or on the bank of a river, stream or lake, where the vegetation differs from that of surrounding uplands due to the extra moisture it receives.

Second-growth A regenerating forest after the removal of pre-existing trees by natural forces or logging.

Species A group of organisms with similar biological features and capable of interbreeding. The taxonomic rank below genus.

Subspecies A population of a species geographically separated and different taxonomically from other populations; sometimes called a geographic race.

Subnivean Beneath the snow.

Talus Loose rock debris on a slope, often at the foot of a cliff.

Timberline The elevational level above which trees do not grow.

Total length A specimen's length measured from the tip of the nose to the end of the last tail vertebra (see figure 12, page 44).

Transient An animal without a secure home range, usually a juvenile.

Translocate or **transplant** Move captured animals to a new location.

Treeline In northern areas, the latitudinal level north of which trees do not grow.

Underfur The dense, insulating hairs beneath the guard hairs on furbearing animals.

Ungulate Hoofed mammal.

Ventral On the belly or lower surface (ventrum); opposite of dorsal.

Zoonoses Diseases that can be transmitted from animals to humans.

SELECTED REFERENCES

Ables, E.D. 1975. Ecology of the Red Fox in North America. In *The Wild Canids* edited by M.W. Fox. New York: Van Nostrand Reinhold.

Adams, I., T. Antifeau, M. Badry, L. Campbell, A. Dibb, O. Dyer, W. Erickson, C. Hoodicoff, L. Ingham, A. Jackson, K. Larsen, T. Munson, N. Newhouse, B. Persello, J. Steciw, J. Surgenor, K. Sutherland and R. Weir. 2003. National recovery strategy for the American Badger, *jeffersonii* subspecies (*Taxidea taxus jeffersonii*). *Recovery of Nationally Endangered Wildlife (RENEW)*, Ottawa.

Addison, E.M., I.K. Barker and D.B. Hunter. 1987. Diseases and parasites of furbearers. In *Wild Furbearer Management and Conservation in North America* edited by M. Novak, J.A. Baker, M.E. Obbard and B. Malloch. North Bay: Ontario Trappers Association.

Apps, C.D. 1996. *Bobcat (Lynx rufus) habitat selection and suitability assessment in southeast British Columbia.* MSc Thesis, University of Calgary, Alberta.

————. 2007. *Ecology and conservation of Canada Lynx in the southern Canadian Rocky Mounains.* PhD thesis, University of Calgary, Alberta.

Apps, C.D., B.N. McLellan, J.G. Woods and M.E. Proctor. 2004. Estimating Grizzly Bear distribution and abundance relative to habitat and human influence. *Journal of Wildlife Management* 68:138-52.

Apps, C.D., N.J. Newhouse and T.A. Kinley. 2002. Habitat associations of American Badgers in southeastern British Columbia. *Canadian Journal of Zoology* 80:1228-39.

Archibald, W.R., and R.H. Jessup. 1984. Population dynamics of the Pine Marten (*Martes americana*) in the Yukon Territory. In *Northern Ecology and Resource Management* edited by R. Olson, F. Geddes and R. Hastings. Edmonton: University of Alberta Press.

Ardrey, R. 1961. *African Genesis.* New York: Dell Publishing Company.

Atkinson, K.T., and D.W. Janz. 1994. *Effect of Wolf Control on Black-tailed Deer in the Nimpkish Valley on Vancouver Island.* Wildlife Bulletin No. B-73, British Columbia Ministry of Environment, Lands and Parks.

Atkinson, K.T., and D.M. Shackleton. 1991. Coyote, *Canis latrans*, ecology in a rural-urban environment. *The Canadian Field-Naturalist* 105:49-54.

Aubrey, K.B. 1984. The recent history and present distribution of the Red Fox in Washington. *Northwest Science* 58:69-79.

Badry, M.J., G. Proulx and P.M. Woodward. 1997. Home-range and habitat use by Fishers translocated to the aspen parkland of Alberta. In *Martes: Taxonomy, Ecology, Techniques and Management* edited by G. Proulx, H.N. Bryant and P.M. Woodward. Edmonton: Provincial Museum of Alberta.

371

Bailey, T.N. 1974. Social organization in a Bobcat population. *Journal of Wildlife Management* 38:435-46.

Baker, J.M. 1992. Habitat use and spatial organization of Pine Marten on southern Vancouver Island, British Columbia. MSc Thesis, Simon Fraser University, Burnaby, BC.

Baker, R.J., L.C. Bradley, R.D. Bradley, J.W. Dragoo, M.D. Engstrom, R.S. Hoffmann, C. Jones, F. Reid, D.W. Rice and C. Jones. 2003. *Revised checklist of North American mammals north of Mexico, 2003.* Occasional Papers 229. Lubbock: Texas Tech University, The Museum.

Ballard, W.B., and P.S. Gipson. 2000. Wolf. In *Ecology and Management of Large Mammals in North America* edited by S. Demarais and P.R. Krauseman. Englewood Cliffs, N.J.: Prentice Hall.

Ballard, W.B., J.S. Whitman and C.L. Gardner. 1987. *Ecology of an Exploited Wolf Population in South-central Alaska.* Wildlife Monographs 98. Lawrence, Kansas: The Wildlife Society.

Banci, V. 1982. *The Wolverine in British Columbia: Distribution, Methods of Determining Age and Status of* Gulo gulo vancouverensis. IWIFR-15. Victoria: BC Ministry of Environment and BC Ministry of Forests.

———. 1989. *A Fisher Management Strategy for British Columbia.* Wildlife Bulletin B-63. Victoria: BC Ministry of Environment.

———. 1994. Wolverine. In *The Scientific Basis for Conserving Forest Carnivores: American Marten, Fisher, Lynx and Wolverine in the Western United States* edited by L.F. Ruggiero, K.B. Aubry, S.W. Buskirk, L.J. Lyon and W.J. Zielinski. General Technical Report RM-254. Fort Collins, Colorado: USDA Forest Service, Rocky Mountain Forest and Range Experimental Station.

Banci, V., and A.S. Harestad. 1988. Reproduction and natality of Wolverines, *Gulo gulo*, in Yukon. *Ann. Zoologica. Fennica* 25:265-70.

———. 1990. Home range and habitat use of Wolverines, *Gulo gulo*, in Yukon, Canada. *Holarctic Ecology* 13:195-200.

Banfield, A.W.F. 1964. Review of F. Mowat's *Never Cry Wolf. The Canadian Field-Naturalist* 78:52-54.

Beier, P. 1991. Cougar attacks on humans in the United States and Canada. *Wildlife Soceity Bulletin* 19:403-12.

Bekoff, M., T.J. Daniels and J.L. Gittleman. 1984. Life history patterns and the comparative social ecology of carnivores. *Annual Review of Ecological Systematics* 15:191-232.

Bekoff, M., and M.C. Wells. 1980. The social ecology of Coyotes. *Scientific American* 242:130-48.

Ben-David, M., R.T. Bowyer and J.B. Faro. 1996. Niche separation by Mink and River Otters: coexistence in a marine environment. *Oikos* 75:41-48.

Bergerud, A.T., and J.P. Elliott. 1986. Dynamics of Caribou and Wolves in northern British Columbia. *Canadian Journal of Zoology* 64:1515-29.

Bigg, M.A., and I.B. MacAskie. 1978. Sea otters re-established in British Columbia. *Journal of Mammalogy* 59:874-76.

Bjorge, R.R., J.R. Gunson, and W.M. Samuel. 1981. Population characteristics and movements of Striped Skunks (*Mephitis mephitis*) in central Alberta. *The Canadian Field-Naturalist* 95:149-55.

Bowen, W.D. 1982. Home range and spatial organization of Coyotes in Jasper National Park, Alberta. *Journal of Wildlife Management* 46:201-16.

Bowyer, R.T., G.M. Blundell, M. Ben-David, S.C. Jewett, T.A. Dean and L.K. Duffy. 2003. *Effects of the* Exxon Valdez *Oil Spill on River Otters: Injury and Recovery of a Sentinel Species.* Wildlife Monographs 153. Lawrence, Kansas: The Wildlife Society.

Boyd, D.K., and D.H. Pletscher. 1999. Characteristics of dispersal in a colonizing wolf population in the central Rocky Mountains. *Journal of Wildlife Management* 63:1094-1108.

Breault, A.M., and K.M. Cheng. 1988. Surplus killing of Eared Grebes, *Podiceps nigricollis*, by Mink, *Mustela vison*, in central British Columbia. *The Canadian Field-Naturalist* 102:738-39.

Breitenmoser, U., B.G. Slough and C. Wursten-Breitenmoser. 1993. Predators of cyclic prey: is the Canada Lynx victim or profiteer of the Snowshoe Hare cycle? *Oikos* 66:551-54.

Brooks, A. 1902. Mammals of the Chilliwack District, British Columbia. *Ottawa Naturalist* 15:239-44.

Bryan, H.M., C.T. Darimont, T.E. Reimchen and P.C. Paquet. 2006. Early ontogenetic diet in Grey Wolves, *Canis lupus*, of coastal British Columbia. *The Canadian Field-Naturalist 120:61-66.*

Bull, E.L. 2000. Seasonal and sexual differences in American Marten diet in northeastern Oregon. *Northwest Science* 74:186-91.

Bull, E.L., and T.W. Heater. 2000. Resting and denning sites of American martens in northeastern Oregon. *Northwest Science* 74:179-85.

———. 2001a. Survival, causes of mortality, and reproduction in the American Marten in northeastern Oregon. *Northwestern Naturalist* 82:1-6.

———. 2001b. Home range and dispersal of the American Marten in northeastern Oregon. *Northwestern Naturalist* 82:7-11.

Bull, E.L., T.R. Torgersen and T.L. Wertz. 2001. The importance of vegetation, insects, and neonate ungulates in Black Bear diet in northeastern Oregon. *Northwest Science* 75:244-53.

Buskirk, S.W., and L.F. Ruggiero. 1994. American Marten. In *The Scientific Basis for Conserving Forest Carnivores: American Marten, Fisher, Lynx, and Wolverine in the Western United States* edited by L.F. Ruggiero, K.B. Aubry, S.W. Buskirk, L.J. Lyon and W.J. Zielinski. General Technical Report RM-254. Fort Collins, Colorado: USDA Forest Service, Rocky Mountain Forest and Range Experimental Station.

Byun, S.A., B.F. Koop and T.E. Reimchen. 1997. North American Black Bear mtDNA phylogeography: implications for morphology and the Haida Gwaii refugium controversy. *Evolution* 51:1647-53.

Cadieux, C.L. 1983. *Coyotes: Predators and Survivors.* Washington, DC: Stone Wall Press.

Carbyn, L.N., editor. 1983. *Wolves in Canada and Alaska: Their Status, Biology, and Management.* CWS Report Series No. 45. Ottawa: Canadian Wildlife Service.

Carbyn, L.N. 1989. Coyote attacks on children in western North America. *Wildlife Society Bulletin* 17:444-46.

Carey, A.B., and J.E. Kershner. 1996. *Spilogale gracilis* in upland forests of western Washington and Oregon. *Northwestern Naturalist* 77:29-34.

Chapman, J.A., and G.A. Feldhamer, editors. 1982. *Wild mammals of North America: Biology, Management, and Economics.* Baltimore: Johns Hopkins University Press.

Clague, J.J. 1981. *Late quaternary Geology and Geochronology of British Columbia.* Paper 80-35. Ottawa: Geological Survey of Canada.

Clark, T.W., E. Anderson, C. Douglas and M. Strickland. 1987. *Martes americana.* Mammalian Species 289. Lawrence, Kansas: American Society of Mammalogists.

Clarke, Ross. 2003. Characteristics of a hunted population of Cougar in the South Selkirk Mountains of British Columbia. Unpublished Report, Columbia Basin Fish and Wildlife Compensation Program, Nelson, BC.

Conservation Data Centre. 2005. Red and Blue Listings for British Columbia. Updated April 2005. http://www.env.gov.bc.ca/atrisk/red-blue.htm.

Coues, E. 1877. *Fur-bearing animals: a monograph of North American Mustelidae.* Miscellaneous Publication No. 8. Washington, DC: Dept of Interior, USGS.

Cowan, I. McTaggart-. 1938a. Geographic distribution of color phases of the Red Fox and Black Bear in the Pacific Northwest. *Journal of Mammalogy* 19:202-06.

———. 1938b. The fur trade and the fur cycle: 1825-1857. *British Columbia Historical Quarterly* 2:19-30.

———. 1941. Fossil and subfossil mammals from the Quaternary of British Columbia. *Transactions of the Royal Society of Canada,* Section IV, 38-49.

———. 1987. Science and the Conservation of Wildlife in British Columbia. In *Our wildlife heritage: 100 years of wildlife management* edited by A. Murray. Victoria: Centennial Wildlife Society of British Columbia.

Cowan, I. McTaggart-, and C.J. Guiguet. 1965. *The Mammals of British Columbia.* Handbook 11. Victoria: British Columbia Provincial Museum.

Cowan, I. McTaggart-, and R.H. MacKay. 1950. Food habits of the Marten (*Martes americana*) in the Rocky Mountain region of Canada. *The Canadian Field-Naturalist* 64:100-104.

Craighead, Jr, F.C. 1979. *Track of the Grizzly.* San Francisco: Sierra Club Books.

Craik, C. 1997. Long-term effects of North American Mink, *Mustela vison*, on seabirds in western Scotland. *Bird Study* 44:303-09.

Criddle, N., and S. Criddle. 1925. The weasels of southern Manitoba. *The Canadian Field-Naturalist* 39:142-48.

Criddle, S. 1947. A nest of the Least Weasel. *The Canadian Field-Naturalist* 61:69.

Cronin, M.A., J. Bodkin, B. Ballachey, J. Estes and J.C. Patton. 1997. Mitochondrial-DNA variation among subspecies and populations of Sea Otters (*Enhydra lutris*). *Journal of Mammalogy* 77:546-57.

Culver, M., W. Johnson, J. Pecan-Slattery and S. O'Brien. 2000. Genomic ancestry of the American Puma (*Puma concolor*). *The American Genetic Association* 91:186-97.

Darimont, C.T., T.E. Reimchen and P.C. Paquet. 2003. Foraging behaviour by Gray Wolves on salmon streams in coastal British Columbia. *Canadian Journal of Zoology* 81:349-53.

Darling, L.M., editor. 2000. *At Risk: Proceedings of a Conference on the Biology and Management of Species and Habitats At Risk,* vol. 1. Victoria: BC Ministry of Environment, Lands and Parks.

Darlington, Jr, P.J. 1957. *Zoogeography: The Geographical Distribution of Animals.* New York: John Wiley & Sons.

Davis, H. 1996. Characteristics and selection of winter dens by Black Bears in coastal British Columbia. MSc Thesis, Simon Fraser University, Burnaby.

Davis, H., and A.S. Harestad. 1996. Cannibalism by Black Bears in the Nimpkish Valley, British Columbia. Northwest Science. 70:88-92.

Deems, Jr, E.F., and D. Pursley, editors. 1983. *North American Furbearers: A Contemporary Reference*. International Association of Fish and Wildlife Agencies and Maryland Department of Natural Resources.

Dekker, D. 1989. Population fluctuations and spatial relationships among Wolves, *Canis lupus*, Coyotes, *Canis latrans*, and Red Foxes, *Vulpes vulpes*, in Jasper National Park, Alberta. *The Canadian Field-Naturalist* 103:261-64.

Demarchi, R., and C. Hartwig. 1999. *Grizzly Bears in British Columbia – the real story*. Richmond: Guide Outfitters Association of British Columbia.

Demarchi, D.A., R.D. Marsh, A.P. Harcombe and E.C. Lea. 1990. The environment. In *The Birds of British Columbia*, vol. 1, by R.W. Campbell, N.K. Dawe, I. McT. Cowan, J.M. Cooper, G.W. Kaiser and M.C.E. McNall. Victoria: Royal British Columbia Museum.

Dobie, J.F. 1961. *The Voice of the Coyote*. Lincoln: University of Nebraska Press.

Doroff, A.M., J.A. Estes, M.T. Tinker, D.M. Burn and T.J. Evans. 2003. Sea Otter population declines in the Aleutian archipelago. *Journal of Mammalogy* 84:55-64.

Douglas, C.W., and M.A. Strickland. 1987. Fisher. In *Wild Furbearer Management and Conservation in North America* edited by M. Novak, J.A. Baker, M.E. Obbard and B. Malloch. North Bay: Ontario Trappers Association.

Dragoo, J., R.D. Bradley, R.L. Honeycutt and J. Templeton. 1993. Phylogenetic relationships among the skunks: a molecular perspective. *Journal of Mammalian Evolution* 1:255-67.

Driver, J.C. 1988. Late Pleistocene and Holocene vertebrates and paleo-environments from Charlie Lake Cave, northeast British Columbia. *Canadian Journal of Earth Sciences* 25:1545-53.

Eadie, A. 2001. Update COSEWIC status report on Ermine, *haidarum* subspecies (*Mustela erminea haidarum*). Ottawa: Committee on the Status of Endangered Wildlife in Canada.

Eagle, T.C., and J.S. Whitman. 1987. Mink. In *Wild Furbearer Management and Conservation in North America* edited by M. Novak, J.A. Baker, M.E. Obbard and B. Malloch. North Bay: Ontario Trappers Association.

Edge, W.D., C.L. Marcum and S.L. Olson-Edge. 1990. Distribution and Grizzly Bear, *Ursus arctos*, use of Yellow Sweetvetch, *Hedysarum sulphurescens*, in northwestern Montana and southeastern British Columbia. *The Canadian Field-Naturalist* 104:435-38.

Eisenberg, J.F. 1981. *The Mammalian Radiations*. Chicago: University of Chicago Press.

Elliott, J.E., C.J. Henny, M.L. Harris, L.K. Wilson and R.J. Norstrom. 1999. Chlorinated hydrocarbons in livers of American Mink (*Mustela vison*) and River Otter (*Lutra canadensis*) from the Columbia and Fraser River basins, 1990-1992. *Environmental Monitoring and Assessment* 57:229-52.

Errington, P.L. 1967. *Of Predation and Life*. Ames: Iowa State University Press.

Estes, J.A. 1980. *Enhydra lutris*. *Mammalian Species* 133. Lawrence, Kansas: American Society of Mammalogists.

Estes, J.A. 1991. Catastrophes and conservation: lessons from Sea Otters and the *Exxon Valdez*. *Science* 185:1058-60.

Estes, J.A., and D.O. Duggins. 1995. Sea Otters and kelp forests in Alaska: generality and variation in a community ecological paradigm. *Ecological Monographs* 65:75-100.

Estes, J.A., M.T. Tinker, T.M. Williams and D.F. Doak. 1998. Killer Whale predation on Sea Otters linking oceanic and near shore ecosystems. *Science* 282:473-76.

Ewer, R.F. 1973. *The Carnivores*. Ithaca, NY: Cornell University Press.

Farley, S.D., and C.T. Robbins. 1995. Lactation, hibernation and mass dynamics of American Black Bears and Grizzly Bears. *Canadian Journal of Zoology* 73:2216-22.

Fleming, M.A., and J.A. Cook. 2002. Phylogeography of endemic Ermine (*Mustela erminea*) in southeast Alaska. *Molecular Ecology* 11:795-807.

Fontana, A., I.E. Teske, K. Pritchard and M. Evans. 1999. East Kootenay Fisher reintroduction program: final report, 1996-1999. Unpublished report, Ministry of Environment, Lands and Parks, Cranbrook, BC.

Footit, R.G., and R.W. Butler. 1977. Predation on nesting Glaucous-winged Gulls by River Otters. *The Canadian Field-Naturalist* 91:189-90.

Friis, L. 1985. An investigation of subspecific relationships of the Grey Wolf (*Canis lupus*) in British Columbia. MSc Thesis, University of Victoria.

Fritzell, E.K. 1978. Aspects of Raccoon (*Procyon lotor*) social organization. *Canadian Journal of Zoology* 56:260-71.

Fuhr, B., and D.A. Demarchi. 1990. *A Methodology for Grizzly Bear Habitat Assessment in British Columbia*. Wildlife Bulletin B-67. Victoria: BC Ministry of Environment.

Gardiner, S. 2000. Outbreak at the border. *Canadian Geographic* March-April 2000:68-76.

Garshelis, D.L. 1987. Sea Otter. In *Wild Furbearer Management and Conservation in North America* edited by M. Novak, J.A. Baker, M.E. Obbard and B. Malloch. North Bay: Ontario Trappers Association.

Garshelis, D.L., M.L. Gibeau and S. Herrero. 2005. Grizzly Bear demographics in and around Banff National Park and Kananaskis Country, Alberta. *Journal of Wildlife Management* 69:277-97.

Gerell, R. 1971. Population studies on Mink, *Mustela vison* Schreber, in southern Sweden. *Viltrevy* 8:83-114.

Gese, E.M., and R.L. Ruff. 1997. Scent-marking by Coyotes, *Canis latrans*: the influence of social and ecological factors. *Animal Behaviour* 54:1155-66.

———. 1998. Howling by Coyotes (*Canis latrans*): variation among social classes, seasons and pack sizes. *Canadian Journal of Zoology* 76:1037-43.

Giannico, G.R., and D.W. Nagorsen. 1989. Geographic and sexual variation in the skull of the Pacific coast Marten (*Martes americana*). *Canadian Journal of Zoology* 67:1386-93.

Gilbert, F.F., and E.G. Nancekivell. 1982. Food habits of Mink (*Mustela vison*) and Otter (*Lutra canadensis*) in northeastern Alberta. *Canadian Journal of Zoology* 60:1282-88.

Gillingham, B.J. 1984. Meal size and feeding rate in the Least Weasel (*Mustela nivalis*). *Journal of Mammalogy* 65:517-19.

Gipson, P.S., and W.B. Ballard. 1998. Accounts of famous North American Wolves, *Canis lupus*. *The Canadian Field-Naturalist* 112:724-39.

Gittleman, J.L., and P.II. Harvey. 1982. Carnivore home-range size, meta-bolic needs and ecology. *Behavioral Ecology and Sociobiology* 10:57-63.

Glass, B.P. 1973. *A Key to the Skulls of North American Mammals.* Stillwater: Oklahoma State University.

Goldman, E.A. 1950. Raccoons of North and Middle America. *North America Fauna* 60:1-153.

Gorman, T.A., J.D. Erb, B.R. McMillan and D.J. Martin. 2006. Space use and sociality of River Otters (*Lontra canadensis*) in Minnesota. *Journal of Mammalogy* 87:740-747.

Gunson, J.R., and R.R. Bjorge. 1979. Winter denning of the Striped Skunk in Alberta. *The Canadian Field-Naturalist* 93:252-58.

Gustine, D.D., K.L. Parker, R.J. Lay, M.P. Gillingham and D.C. Heard. 2006. *Calf Survival of Woodland Caribou in a Multi-predator Ecosystem.* Wildlife Monographs 165. Lawrence, Kansas: The Wildlife Society.

Gyug, L.W. 2000. *Timber-harvesting effects on riparian wildlife and vegetation in the Okanagan Highlands of British Columbia.* Wildlife Bulletin B-97 Victoria: BC Ministry of Environment.

Hagmeier, E.M. 1959. A re-evaluation of the subspecies of the Fisher. *The Canadian Field-Naturalist* 73:185-97.

———. 1961. Variation and relationships in North American Marten. *The Canadian Field-Naturalist* 75:122-38.

Hall, E.R. 1951. *American Weasels.* Museum of Natural History 4. Lawrence: University of Kansas Publications.

Hamer, D., S. Herrero, and K. Brady. 1991. Food and habitat used by Grizzly Bears, *Ursus arctos*, along the continental divide in Waterton Lakes National Park, Alberta. *The Canadian Field-Naturalist* 105:325-29.

Hancock, L. 1987. The predator hunters. In *Our Wildlife Heritage: 100 Years of Wildlife Management* edited by A. Murray. Victoria: Centennial Wildlife Society of British Columbia.

Harrison, S.W., and D.M. Hebert. 1988. Selective predation by Cougar within the Junction Wildlife Management Area. *Symposium of the Northern Wild Sheep and Goat Council* 6:292-307.

Harestad, A.E., and F.L. Bunnell. 1979. Home range and body weight – a re-evaluation. *Ecology* 69:389-402.

Harestad, A.S. 1990. Mobbing of a Long-tailed Weasel (*Mustela frenata*) by Columbian Ground Squirrels (*Spermophilus columbianus*). *The Canadian Field-Naturalist* 104:483-84.

Harfenist, A., K.R. MacDowell, T. Golumbia, G. Schultze and Laskeek Bay Conservation Society. 2000. Monitoring and control of Raccoons on seabird colonies in Haida Gwaii (Queen Charlotte Islands). In *At Risk: Proceedings of a Conference on the Biology and Management of Species and Habitats At Risk*, vol. 1, edited by L.M. Darling. Victoria: British Columbia Ministry of Environment, Lands and Parks.

Harington, C.R. 1977. Wildlife and man in B.C. during the ice age. *British Columbia Outdoors* 33(6):28-32.

Hartman, L.H. 1993. Ecology of coastal Raccoons (*Procyon lotor*) on the Queen Charlotte Islands, British Columbia, and evaluation of their potential impact on native burrow nesting seabirds. MSc thesis, University of Victoria.

Hartman, L.H., and D.S. Eastman. 1999. Distribution of introduced Raccoons, *Procyon lotor*, on the Queen Charlotte Islands: implications for burrow-nesting seabirds. *Biological Conservation* 88:1-13.

Hartman, L.H., A.J. Gaston, and D.S. Eastman. 1997. Raccoon predation on Ancient Murrelets on East Limestone Island, British Columbia. *Journal of Wildlife Management* 61:377-88.

Hatler, D.F. 1972. Food habits of Black Bears in interior Alaska. *The Canadian Field-Naturalist* 86:17-31.

———. 1976. *The Coastal Mink on Vancouver Island, British Columbia.* PhD thesis, University of British Columbia, Vancouver.

———. 1988. *A Lynx Management Strategy for British Columbia.* Wildlife Bulletin B-61. Victoria: BC Ministry of Environment.

———. 1989. *A Wolverine Management Strategy for British Columbia.* Wildlife Bulletin B-60. Victoria: BC Ministry of Environment.

Hatler, D.F., and A.M.M. Beal. 2003a. Wolverine *Gulo gulo. Furbearer Management Guidelines, British Columbia.* http://www.env.gov.bc.ca/fw/wildlife/trappings/docs/wolverine.pdf

———. 2003b. Mink *Mustela vison. Furbearer Management Guidelines, British Columbia.* http://www.env.gov.bc.ca/fw/wildlife/trappings/docs/mink.pdf

———. 2003c. Lynx *Lynx canadensis. Furbearer Management Guidelines, British Columbia.* http://www.env.gov.bc.ca/fw/wildlife/trappings/docs/lynx.pdf

———. 2005. *British Columbia Trapper Education Manual.* Victoria: BC Trappers' Association.

Hatler, D.F., D.A. Blood and A.M.M. Beal. 2003. Marten *Martes americana. Furbearer Management Guidelines, British Columbia.* http://www.env.gov.bc.ca/fw/wildlife/trappings/docs/marten.pdf

Hatler, D.F., G. Mowat and A.M.M. Beal. 2003a. River Otter *Lontra canadensis. Furbearer Management Guidelines, British Columbia.* http://www.env.gov.bc.ca/fw/wildlife/trappings/docs/river_otter.pdf

———. 2003b. Weasels *Mustela* spp. *Furbearer Management Guidelines, British Columbia.* http://www.env.gov.bc.ca/fw/wildlife/trappings/docs/weasel.pdf

Hatler, D.F., G. Mowat, K.G. Poole and A.M.M. Beal. 2003. Gray Wolf *Canis lupus. Furbearer Management Guidelines, British Columbia.* http://www.env.gov.bc.ca/fw/wildlife/trappings/docs/gray_wolf.pdf

Hatler, D.F., K.G. Poole and A.M.M. Beal. 2003a. Coyote *Canis latrans. Furbearer Management Guidelines, British Columbia.* http://www.env.gov.bc.ca/fw/wildlife/trappings/docs/coyote.pdf

———. 2003b. Red Fox *Vulpes vulpes. Furbearer Management Guidelines, British Columbia.* http://www.env.gov.bc.ca/fw/wildlife/trappings/docs/red_fox.pdf

———. 2003c. Bobcat *Lynx rufus. Furbearer Management Guidelines, British Columbia.* http://www.env.gov.bc.ca/fw/wildlife/trappings/docs/bobcat.pdf

Hatter, I.W. 1988. *Effects of Wolf Predation on Recruitment of Black-tailed Deer on Northeastern Vancouver Island.* Wildlife Report R-23. Victoria: BC Ministry of Environment.

Hatter, J. 1987. The Fur Trade – B.C.'s First Industry. In *Our Wildlife Heritage:*

100 Years of Wildlife Management edited by A. Murray. Victoria: Centennial Wildlife Society of British Columbia.

Hawley, V.D., and F.E. Newby. 1957. Marten home ranges and population fluctuations. *Journal of Mammalogy* 38:174-84.

Hayes, R.D., R. Farnell, R.M.P. Ward, J. Carey, M. Dehn, G.W. Kuzyk, A.M. Baer, C.L. Gardner and M. O'Donoghue. 2003. *Experimental Reduction of Wolves in the Yukon: Ungulate Responses and Management Implications.* Wildlife Monographs 152. Lawrence, Kansas: The Wildlife Society.

Hebert, D.M., and S.W. Harrison. 1988. The impact of Coyote predation on lamb survival at the Junction Wildlife Management Area. *Symposium of the Northern Wild Sheep and Goat Council* 6:283-91.

Heidt, G.A., M.K. Petersen and G.L. Kirkland, Jr. 1968. Mating behavior and development of Least Weasels (*Mustela nivalis*) in captivity. *Journal of Mammalogy* 49:413-19.

Henry, J.D. 1986. *Red Fox: the Catlike Canine.* Washington, DC: Smithsonian Institution.

Herrero, S. 1972. Aspects of evolution and adaptation in American Black Bears (*Ursus americanus* Pallas) and Brown and Grizzly bears (*U. arctos* Linne.) of North America. In *Bears, Their Biology and Management* edited by S. Herrero. New Series, No.23. Morges, Switzerland: IUCN.

———. 1985. *Bear Attacks, Their Causes and Avoidance.* Piscataway, N.J.: Winchester Press.

Hilderbrand, G.V., C.C. Schwartz, C.T. Robbins, M.E. Jacoby, T.A. Hanley, S.M. Arthur and C. Servheen. 1999. The importance of meat, particularly salmon, to body size, population productivity, and conservation of North American Brown Bears. *Canadian Journal of Zoology* 77:132-38.

Hobson, K.A., B.N. McLellan and J.G. Woods. 2000. Using stable carbon and nitrogen isotopes to infer trophic relationships among Black and Grizzly bears in the upper Columbia River basin, British Columbia. *Canadian Journal of Zoology* 78:1332-39.

Hoffmann, R.S., and D. Pattie. 1968. *A Guide to Montana Mammals: Identification, Habitat, Distribution and Abundance.* Missoula: University of Montana.

Hoffos, R. 1987. *Wolf management in British Columbia: the public controversy.* Wildlife Bulletin No. B-52, B.C. Ministry of Environment and Parks, Victoria. 75pp.

Holcraft, A.C., and S. Herrero. 1991. Black Bear, *Ursus americanus*, food habits in southwestern Alberta. *The Canadian Field-Naturalist* 105:335-45.

Hoodicoff, C.S. 2003. Ecology of the Badger (*Taxidea taxus jeffersonii*) in the Thompson Region of British Columbia: implications for conservation. MSc thesis, University of Victoria.

———. 2005. *Cariboo Region Badger Project: 2004 Field Results and Best Management Practices.* Unpublished Report for the BC Ministry of Water, Land and Air Protection, 100 Mile House, BC.

Hornocker, M.G. 1970. *An Analysis of Mountain Lion Predation upon Mule Deer and Elk in the Idaho Primitive Area.* Wildlife Monographs 21. Lawrence, Kansas: The Wildlife Society.

Hornocker, M.G., and H.S. Hash. 1981. Ecology of the Wolverine in northwestern Montana. *Canadian Journal of Zoology* 59:1286-1301.

Huggard, D.J. 1999. *Marten Use of Different Harvesting Treatments in High-*

Elevation Forest at Sicamous Creek. Research Report 17, Victoria: BC
  Ministry of Forests, Research Branch.
Hunter, D.B., and N. Lemieux, editors. 1996. *Mink ... Biology, Health and
  Disease.* Guelph, Ontario: University of Guelph Graphic and Print
  Services.
Ingles, L.G. 1965. *Mammals of the Pacific States.* Stanford, California:
  Stanford University Press.
Jacoby, M.E., G.V. Hilderbrand, C. Servheen, C.C. Schwartz, S.K. Arthur,
  T.A. Hanley, C.T. Robbins and R. Michener. 1999. Trophic relations of
  Brown and Black bears in several western North American ecosystems.
  Journal of Wildlife Management 63:921-29.
Janz, D., and I. Hatter. 1986. *A Rationale for Wolf Control in the Management
  of the Vancouver Island Predator-Ungulate System.* Wildlife Bulletin B-45.
  Victoria: BC Ministry of Environment.
Johnson, C.E. 1921. The "hand-stand" habit of the Spotted Skunk. *Journal of
  Mammalogy* 2:87-89.
Jones, D.M., and J.B. Theberge. 1982. Summer home range and habitat
  utilisation of the Red Fox (*Vulpes vulpes*) in a tundra habitat, northwest
  British Columbia. *Canadian Journal of Zoology* 60:807-12.
————. 1983. Variation in Red Fox, *Vulpes vulpes*, summer diets in
  northwestern British Columbia and southwest Yukon. *The Canadian
  Field-Naturalist* 97:311-14.
Jones, G.W., and B. Mason. 1983. *Relationships Among Wolves, Hunting and
  Population Trends of Black-tailed Deer in the Nimpkish Valley on Vancouver
  Island.* Fish and Wildlife Report R-7. Victoria: BC Ministry of
  Environment.
Jones, J.K.J., and R.W. Manning. 1992. *Illustrated Key to the Skulls of Genera
  of North American Land Mammals.* Lubbock: Texas Tech University Press.
Jonkel, C.J. 1987. Brown Bear. In *Wild Furbearer Management and
  Conservation in North America* edited by M. Novak, J.A. Baker, M.E.
  Obbard and B. Malloch. North Bay: Ontario Trappers Association.
Jonkel, C.J., and I. McTaggart-Cowan. 1971. *The Black Bear in the Spruce-Fir
  Forest.* Wildlife Monographs 27. Lawrence, Kansas: The Wildlife Society.
Kamler, J., and W.B. Ballard. 2002. A review of native and non-native Red
  Foxes in North America. *Wildlife Society Bulletin* 30:370-79.
Kenyon, K.W. 1969. *The Sea Otter in the Eastern Pacific Ocean.* North
  American Fauna Series, No. 68. Washington, DC: U.S. Fish and Wildlife
  Service.
King, C.M. 1980. Population biology of the weasel *Mustela nivalis* on
  British game estates. *Holarctic Ecology* 3:160-68.
————. 1983. *Mustela erminea.* Mammalian Species 195. Lawrence, Kansas:
  American Society of Mammalogists.
————. 1989. *The Natural History of Weasels and Stoats.* Ithaca, NY: Cornell
  University Press.
Kinley, T.A. 1992. Ecology and management of Bobcats (*Lynx rufus*) in the
  East Kootenay District of British Columbia. MSc thesis, University of
  Alberta, Calgary.
Kinley, T.A., and N.J. Newhouse. 2005. *East Kootenay Badger Project 2004-
  2005 Update: Ecology, Translocation, Sightings and Communications.*
  Unpublished report, Sylvan Consulting, Invermere, BC.

Kitchen, A.M., E.M. Gese and E.R. Schauster. 2000. Long-term spatial stability of Coyote (*Canis latrans*) home ranges in southeastern Colorado. *Canadian Journal of Zoology* 78:458-64.

Knick, S.T. 1990. *Ecology of Bobcats Relative to Exploitation and a Prey Decline in Southeastern Idaho*. Wildlife Monographs 108. Lawrence, Kansas: The Wildlife Society.

Knick, S.T., J.D. Brittell, and S.J. Sweeney. 1985. Population characteristics of Bobcats in Washington state. *Journal of Wildlife Management* 49:721-28.

Knick, S.T., S.J. Sweeney, J.R. Alldredge and J.D. Brittell. 1984. Autumn and winter food habits of Bobcats in Washington State. *Great Basin Naturalist* 44:70-74.

Koehler, G.M. 1990. Population and habitat characteristics of Lynx and Snowshoe Hares in north-central Washington. *Canadian Journal of Zoology* 68:845-51.

Krebs, J., E. Lofroth, J. Copeland, V. Banci, D. Cooley, H. Golden, A. Magoun, R. Mulders and B. Shults. 2004. Synthesis of survival rates and causes of mortality in North American Wolverines. *Journal of Wildlife Management* 68:493-502.

Krebs, J.A., and D. Lewis. 2000. Wolverine ecology and habitat use in the North Columbia Mountains: Progress report. In *At Risk: Proceedings of a Conference on the Biology and Management of Species and Habitats At Risk*, vol. 2, edited by L.M. Darling. Victoria: BC Ministry of Environment, Lands and Parks.

Krebs, J.W., M.L. Wilson and J.E. Childs. 1995. Rabies – epidemiology, prevention, and future research. *Journal of Mammalogy* 76: 681-94.

Kunkel, K.E., and D.H. Pletscher. 2000. Habitat factors affecting vulnerability of Moose to predation by wolves in southeastern British Columbia. *Canadian Journal of Zoology* 78:150-57.

Kyle, C.J., R.D. Weir, N.J. Newhouse, H. Davis and C. Strobeck. 2004. Genetic structure of sensitive and endangered northwestern Badger populations (*Taxidea taxus taxus* and *T. t. jeffersonii*). *Journal of Mammalogy* 85:633-39.

Laidre, K.L., and R.J. Jameson. 2006. Foraging patterns and prey selection in an increasing and expanding sea otter population. *Journal of Mammaogy* 87:799-807.

Lariviere, S. 1999. *Mustela vison*. Mammalian Species 608. Lawrence, Kansas: American Society of Mammalogists.

———. 2001. *Ursus americanus*. Mammalian Species 647. Lawrence, Kansas: American Society of Mammalogists.

Lariviere, S., and S.H. Ferguson. 2003. Evolution of induced ovulation in North American carnivores. *Journal of Mammalogy* 84:937-47.

Lariviere, S., and F. Messier. 1997. Seasonal and daily activity patterns of Striped Skunks (*Mephitis mephitis*) in the Canadian Prairies. *Journal of Zoology (London)* 243:255-62.

———. 1998. Spatial organization of a prairie Striped Skunk population during the waterfowl nesting season. *Journal of Wildlife Management* 62:199-204.

———. 2000. Habitat selection and use of edges by Striped Skunks in the Canadian Prairies. *Canadian Journal of Zoology* 78:366-72.

Lariviere, S., and M. Pasitschniak-Arts. 1996. *Vulpes vulpes*. Mammalian Species 537. Lawrence, Kansas: American Society of Mammalogists.

Lariviere, S., and L.R. Walton. 1997. *Lynx rufus*. Mammalian Species 563. Lawrence, Kansas: American Society of Mammalogists.

———. 1998. *Lontra canadensis*. Mammalian Species 587. Lawrence, Kansas: American Society of Mammalogists.

Lariviere, S., L.R. Walton, and F. Messier. 1999. Selection by Striped Skunks (*Mephitis mephitis*) of farmsteads and buildings as denning sites. *American Midland Naturalist* 142:96-101.

Larsen, D.N. 1984. Feeding habits of River Otters in coastal southeastern Alaska. *Journal of Wildlife Management* 48:1446-51.

Laundre, J.W. 2005. Puma energetics: a recalculation. *Journal of Wildlife Management* 69:723-32.

Lehman, N., and R.K. Wayne. 1991. Analysis of Coyote mitochondrial DNA genotype frequencies: estimation of the effective number of alleles. *Genetics* 128:405-16.

Leydet, F. 1977. *The Coyote: Defiant Song Dog of the West*. Norman: University of Oklahoma Press.

Lindstrom, E.R., H. Andren, P. Angelstam, G. Cederlund, B. Hornfeldt, L. Jaderberg, P-A. Lemnell, B. Martinsson, K. Skold and J.E. Swenson. 1994. Disease reveals the predator: sarcoptic mange, Red Fox predation, and prey populations. *Ecology* 75:1042-49.

Ling, J.K. 1970. Pelage and molting in wild mammals with special reference to aquatic forms. *Quarterly Review of Biology* 45:16-54.

Lisgo, K.A. 1999. Ecology of the Short-tailed Weasel (*Mustela erminea*) in the mixed wood boreal forest of Alberta. MSc thesis, University of British Columbia, Vancouver.

Lofroth, E.C. 1993. Scale dependent analyses of habitat selection by Marten in the Sub-boreal Spruce Biogeoclimatic Zone, British Columbia. MSc thesis, Simon Fraser University, Burnaby, BC.

———. 2001. Northern Wolverine project: Wolverine ecology in plateau and foothill landscapes 1996-2001. Unpublished report, Forest Renewal Activity # 712260, Ministry of Environment, Victoria, BC.

Logan, K.A., and L.L. Sweanor. 2000. Puma. In *Ecology and Management of Large Mammals in North America* edited by S. Demarais and P.R. Krausman. Englewood Cliffs, NJ: Prentice-Hall.

Lopez, B.H. 1978. *Of Wolves and Men*. New York: Charles Scribner's Sons.

MacAskie, I.B. 1987. Updated status of the Sea Otter, *Enhydra lutris*, in Canada. *Canadian Field-Naturalist* 101:279-83.

MacDonald, D.W. 1976. Food caching by Red Foxes and some other carnivores. *Journal of Tierpsychologie* 42:170-85.

———. 1983. The ecology of carnivore social behavior. *Nature* 301:379-84.

MacHutchon, A.G., S. Himmer and C.A. Bryden. 1993. *Khutzeymateen Valley Grizzly Bear Study: Final Report*. Wildlife Report R-25. Victoria: BC Ministry of Environment, Lands and Parks.

Magoun, A.J. 1987. Summer and winter diets of wolverines, *Gulo gulo*, in arctic Alaska. *Canadian Field-Naturalist* 101:392-97.

Marshall, H.D., and K. Ritland. 2002. Genetic diversity and differentiation of Kermode bear populations. *Molecular Ecology* 11:685-97.

Marshall, W.H. 1951. Pine Marten as a forest product. *Journal of Forestry* 49:899-905.

Matter, W.J., and R.W. Mannan. 2005. How do prey persist? *Journal of Wildlife Management* 69:1315-20.

McCord, C.M., and J.E. Cardoza. 1982. Bobcat and Lynx. In *Wild Mammals of North America: Biology, Management, and Economics* edited by J.A. Chapman and G.A. Feldhamer. Baltimore: Johns Hopkins University Press.

McGowan, D. 1936. *Animals of the Canadian Rockies*. New York: Dodd, Mead and Company.

McLellan, B.N. 1989. Dynamics of a Grizzly Bear population during a period of industrial resource extraction. *Canadian Journal of Zoology* 67:1856-68.

McLellan, B.N., and F.W. Hovey. 1995. The diet of Grizzly Bears in the Flathead River drainage of southeastern British Columbia. *Canadian Journal of Zoology* 73:704-12.

———. 2001. Natal dispersal of Grizzly Bears. *Canadian Journal of Zoology* 79:838-44.

McLellan, B.N., F.W. Hovey, R.D. Mace, J.G. Woods, D.W. Carney, M.L. Gibeau, W.L. Wakkinen and W.F. Kasworm. 1999. Rates and causes of Grizzly Bear mortality in the interior mountains of British Columbia, Alberta, Montana, Washington and Idaho. *Journal of Wildlife Management* 63:911-20.

Mead, R.A. 1981. Delayed implantation in mustelids, with special emphasis on the Spotted Skunk. *Journal of Reproduction and Fertility Supplement* 29:11-24.

Mead, R.A., and P.L. Wright. 1983. Reproductive cycles of Mustelidae. *Acta Zoologica Fennica* 174:169-72.

Mech, L.D. 1970. *The Wolf: the Ecology and Behavior of an Endangered Species.* Doubleday, NY: Natural History Press.

Mech, L.D., D.M. Barnes, and J.R. Tester. 1968. Seasonal weight changes, mortality, and population structure of raccoons in Minnesota. *Journal of Mammalogy* 49:63-73.

Meidinger, D., and J. Pojar. 1991. *Ecosystems of British Columbia*. Victoria: BC Ministry of Forests.

Melquist, W.E., and A.E. Dronkert. 1987. River Otter. In *Wild Furbearer Management and Conservation in North America* edited by M. Novak, J.A. Baker, M.E. Obbard and B. Malloch. North Bay: Ontario Trappers Association.

Melquist, W.E., and M.G. Hornocker. 1983. *Ecology of River Otters in Idaho.* Wildlife Monographs 83. Lawrence, Kansas: The Wildlife Society.

Messick, J.P. 1987. North American Badger. In *Wild Furbearer Management and Conservation in North America* edited by M. Novak, J.A. Baker, M.E. Obbard and B. Malloch. North Bay: Ontario Trappers Association.

Messick, J.P., and M.G. Hornocker. 1981. *Ecology of the Badger in Southwestern Idaho*. Wildlife Monographs 76. Lawrence, Kansas: The Wildlife Society.

Messier, F. 1985. Social organization, spatial distribution, and population density of wolves in relation to moose density. *Canadian Journal of Zoology* 63:1068-77.

Messier, F., and C. Barrette. 1982. The social system of the Coyote (*Canis latrans*) in a forested habitat. *Canadian Journal of Zoology* 60:1743-53.

Miller, S.D., and W.B. Ballard. 1982. Homing of transplanted Alaskan Brown Bears. *Journal of Wildlife Management* 46:869-76.

Ministry of Environment, Lands and Parks. 1995a. *A Future for the Grizzly: British Columbia Grizzly Bear Conservation Strategy.* Victoria: BC Ministry of Environment, Lands and Parks.

———. 1995b. *Conservation of Grizzly Bears in British Columbia, Background Report.* Victoria: BC Ministry of Environment, Lands and Parks.

Minta, S.C., K.A. Minta and D.F. Lott. 1992. Hunting associations between Badgers (*Taxidea taxus*) and Coyotes (*Canis latrans*). *Journal of Mammalogy* 73:814-20.

Mitchell, J.L. 1961. Mink movements and populations on a Montana river. *Journal of Wildlife Management* 25:48-54.

Monson, D.H., J.A. Estes, J.L. Bodkin and D.R. Siniff. 2000. Life history, plasticity, and population regulation in Sea Otters. *Oikos* 90:457-68.

Morris, R., D.V. Ellis and B.P. Emerson. 1981. The British Columbia transplant of Sea Otters *Enhydra lutris*. *Biological Conservation* 20:291-95.

Mowat, G., and D.C. Heard. 2006. Major components of Grizzly Bear diet across North America. *Canadian Journal of Zoology* 84:473-89.

Mowat, G., K.G. Poole and M. O'Donoghue. 2000. Ecology of Lynx in northern Canada and Alaska. In *Ecology and Conservation of Lynx in the United States* edited by L.F. Ruggiero, K.B. Aubry, S.W. Buskirk, G.M. Koehler, C.J. Krebs, K.S. McKelvey and J.R. Squires. Boulder: University Press of Colorado.

Mowat, G., C. Shurgot and K.G. Poole. 2000. Using track plates and remote cameras to detect Marten and Short-tailed Weasels in coastal cedar hemlock forest. *Northwestern Naturalist* 81:113-21.

Mowat, G., and B.G. Slough. 1998. Some observations on the natural history and behaviour of the Canada Lynx, *Lynx canadensis*. *Canadian Field-Naturalist* 112:32-36.

Muir, J. 1890. Among the animals of the Yosemite. *Atlantic Monthly* 82:617.

Muller-Schwarze, D. 1983. Scent glands in mammals and their functions. In *Advances in the study of mammalian behavior* edited by J.F. Eisenberg and D.G. Kleiman. Special Publication No. 7. Lawrence, Kansas: American Society of Mammalogists

Munro, W.T. 1985. Status of the Sea Otter, *Enhydra lutris*, in Canada. *Canadian Field-Naturalist* 99:413-16.

Murie, A. 1944. *The Wolves of Mount Mckinley.* Fauna of U.S. National Parks, Fauna Series No. 5. Washington, DC: US Department of the Interior.

Murray, A., editor. 1987. *Our Wildlife Heritage: 100 Years of Wildlife Management.* Victoria: The Centennial Wildlife Society of British Columbia.

Nagorsen, D.W. 1990. *The Mammals of British Columbia: a Taxonomic Catalogue.* Memoir 4. Victoria: Royal British Columbia Museum.

———. 1994. Body weight variation among insular and mainland American Martens. In *Martens, Sables, and Fishers: Biology and Conservation* edited by S.W. Buskirk, A.S. Harestad, M.G. Raphael and R.A. Powell. Ithaca, NY: Cornell University Press.

Nagorsen, D.W., R.W. Campbell and G.R. Giannico. 1991. Winter food habits of Marten, *Martes americana*, on the Queen Charlotte Islands. *Canadian Field-Naturalist* 105:55-59.

Nagorsen, D.W., G. Keddie, and R.J. Hebda. 1995. Early Holocene Black

Bears, *Ursus americanus*, from Vancouver Island. *Canadian Field-Naturalist* 109:11-18.

Nagorsen, D.W., K. Morrison and J. Forsberg. 1989. Winter diet of Vancouver Island Marten (*Martes americana*). *Canadian Journal of Zoology* 67:1394-1400.

Nelson, R.A., H.W. Wahner, J.D. Jones, R.D. Ellefson and P.E. Zollman. 1973. Metabolism of bears before, during, and after winter sleep. *American Journal of Physiology* 224:491-96.

Newhouse, N.J., and T.A. Kinley. 2000. *Update COSEWIC Status Report on American Badger (*Taxidea taxus*)*. Ottawa: Committee on the Status of Endangered Wildlife in Canada.

———. 2001. *Ecology of Badgers near a Range Limit in British Columbia*. Unpublished report, Columbia Basin Fish & Wildlife Compensation Program, Nelson.

Northcott, T.H. 1971. Winter predation of *Mustela erminea* in northern Canada. *Arctic* 24:141-43.

Nowak, R.M. 1995. Another look at wolf taxonomy. In *Ecology and Conservation of Wolves in a Changing World* edited by L.N. Carbyn, S.H. Fritts and D.R. Seip. Edmonton: Canadian Circumpolar Institute, University of Alberta.

O'Donoghue, M., S. Boutin, C.J. Krebs, G. Zuleta, D.L. Murray and E.J. Hofer. 1998. Functional responses of Coyotes and Lynx to the Snowshoe Hare cycle. *Ecology* 79:1193-1208.

O'Donoghue, M., E. Hofer and F.I. Doyle. 1995. Predator versus predator. *Natural History* 104:6-9.

Obbard, M.E. 1987. Fur grading and pelt identification. In *Wild Furbearer Management and Conservation in North America* edited by M. Novak, J.A. Baker, M.E. Obbard and B. Malloch. North Bay: Ontario Trappers Association.

Osgood, W.H. 1901. *Natural History of the Queen Charlotte Islands, British Columbia; Natural History of the Cook Inlet Region, Alaska*. North American Fauna Series, no. 21. Washington, DC: US Fish and Wildlife Service.

Palomares, F., and T.M. Caro. 1999. Interspecific killing among mammalian carnivores. *American Naturalist* 153:492-508.

Paragi, T.F., W.N. Johnson, D.D. Katnik and A.J. Magoun. 1996. Marten selection of postfire seres in the Alaskan taiga. *Canadian Journal of Zoology* 74:2226-37.

Pasitschniak-Arts, M. 1993. *Ursus arctos*. Mammalian Species 439. Lawrence, Kansas: American Society of Mammalogists.

Pasitschniak-Arts, M., and S. Lariviere. 1995. *Gulo gulo*. Mammalian Species 499. Lawrence, Kansas: American Society of Mammalogists.

Pearson, A.M. 1975. *The Northern Interior Grizzly Bear* Ursus arctos L. Canadian Wildlife Service Report Series no. 34, Ottawa: Canadian Wildlife Service.

Peterson, R.O., and R.E. Page. 1988. The rise and fall of Isle Royale wolves, 1975-1986. *Journal of Mammalogy* 69:89-99.

Pielou, E.C. 1991. *After the Ice Age*. Chicago: University of Chicago Press.

Pletscher, D.H., R.R. Ream, D.K. Boyd, M.W. Fairchild and K.E. Kunkel. 1997. Population dynamics of a recolonizing wolf population. *Journal of Wildlife Management* 61:459-65.

Poole, K.G. 1995. Spatial organization of a Lynx population. *Canadian Journal of Zoology* 73:632-41.

———. 1997. Dispersal patterns of Lynx in the Northwest Territories. *Journal of Wildlife Management* 61:497-505.

———. 2003. A review of the Canada Lynx, *Lynx canadensis*, in Canada. *Canadian Field-Naturalist* 117:360-76.

Poszig, D., C.D. Apps and A. Dibb. 2004. Predation on two Mule Deer, *Odocoileus hemionus*, by a Canada Lynx, *Lynx canadensis*, in the southern Canadian Rocky Mountains. *Canadian Field-Naturalist* 118:191-94.

Powell, R.A. 1982. *The Fisher: Life History, Ecology and Behavior*. Minneapolis: University of Minnesota Press.

Proulx, G. 2006. Winter habitat use by American Marten, *Martes americana*, in western Alberta boreal forests. *Canadian Field-Naturalist 120:100-105.*

Quick, H.F. 1951. Notes on the ecology of weasels in Gunnison County, Colorado. *Journal of Mammalogy* 32:281-90.

———. 1953. Occurrence of Porcupine quills in carnivorous mammals. *Journal of Mammalogy* 34:256-59.

———. 1955. Food habits of Marten (*Martes americana*) in northern British Columbia. *Canadian Field-Naturalist* 69:144-47.

Rahme, A.H., A.S. Harestad and F.L. Bunnell. 1991. *Status of the Badger in British Columbia*. Wildlife Working Report WR-72, Victoria: BC Ministry of Environment.

Ramsay, C., P. Griffiths, D. Fedje, R. Wigen and Q. Mackie. 2004. Preliminary investigation of a late Wisconsin fauna from K1 cave, Queen Charlotte Islands (Haida Gwaii), Canada. *Quaternary Research* 62:105-09.

Raum-Suryan, K., K. Pitcher, and R. Lamy. 2004. Sea Otter, *Enhydra lutris*, sightings off Haida Gwaii/Queen Charlotte Islands, British Columbia, 1972-2002. *Canadian Field-Naturalist* 118:270-72.

Reid, D.G., T.E. Code, A.C.H. Reid and S.M. Herrero. 1994a. Food habits of the River Otter in a boreal ecosystem. *Canadian Journal of Zoology* 72:1306-13.

———. 1994b. Spacing, movements and habitat selection of the River Otter in boreal Alberta. *Canadian Journal of Zoology* 72:1314-24.

Reid, D.G., L. Waterhouse, P.E.F. Buck, A.E. Derocher, R. Bettner and C.D. French. 2000. Inventory of the Queen Charlotte Islands Ermine. In *At Risk: Proceedings of a Conference on the Biology and Management of Species and Habitats at Risk*, vol. 1 edited by L.M. Darling. Victoria: BC Ministry of Environment, Lands and Parks.

Reimchen, T.E. 1998. Nocturnal foraging behaviour of Black Bears, *Ursus americanus*, on Moresby Island, British Columbia. *Canadian Field-Naturalist* 112:446-50.

———. 2000. Some ecological and evolutionary aspects of bear-salmon interactions in coastal British Columbia. *Canadian Journal of Zoology* 78:448-57.

Rivest, P., and J-M. Bergeron. 1981. Density, food habits, and economic importance of Raccoons (*Procyon lotor*) in Quebec agrosystems. *Canadian Journal of Zoology* 59:1755-62.

Robinson, H.S., R.B. Wielgus and J.C. Gwilliam. 2002. Cougar predation and population growth of sympatric Mule Deer and White-tailed Deer. *Canadian Journal of Zoology* 80:556-68.

Rogers, L.L. 1987. *Effects of Food Supply and Kinship on Social Behavior,*

Movements and Population Growth of Black Bears in Northeastern Minnesota. Wildlife Monographs 97. Lawrence, Kansas: The Wildlife Society.

Rosatte, R.C. 1987. Striped, Spotted, Hooded and Hog-nosed skunks. In *Wild Furbearer Management and Conservation in North America* edited by M. Novak, J.A. Baker, M.E. Obbard and B. Malloch. North Bay: Ontario Trappers Association.

Rosatte, R.C., C.D. MacInnes, R.T. Williams and O. Williams. 1997. A proactive prevention strategy for Raccoon rabies in Ontario, Canada. *Wildlife Society Bulletin* 25:110-16.

Rosenzweig, M.L. 1968. The strategy of body size in mammalian carnivores. *American Midland Naturalist* 80:299-315.

Ross, P.I., M.G. Jalkotsky and M. Festa-Bianchet. 1997. Cougar predation on Bighorn Sheep in southwestern Alberta during winter. *Canadian Journal of Zoology* 75:771-75.

Roy, M.S., E. Geffen, D. Smith, E. Ostander and R.K. Wayne. 1994. Patterns of differentiation and hybridization in North American wolf-like canids, revealed by analysis of microsatellite loci. *Molecular Biology and Evolution* 11:553-70.

Rueness, E., N. Stenseth, M. O'Donoghue, S. Boutin, H. Ellegren and K. Jakobsen. 2003. Ecological and genetic spatial structuring in the Canadian Lynx. *Nature* 425:69-72.

Rutherglen, R.A., and B. Herbison. 1977. Movements of nuisance Black Bears (*Ursus americanus*) in southeastern British Columbia. *Canadian Field-Naturalist* 91:419-22.

Sadleir, R.M.F.S. 1969. *The Ecology of Reproduction in Wild and Domestic Mammals.* London: Metheun and Co.

Sanderson, G.C. 1987. Raccoon. In *Wild Furbearer Management and Conservation in North America* edited by M. Novak, J.A. Baker, M.E. Obbard, and B. Malloch. North Bay: Ontario Trappers Association.

Sargeant, A.B., S.H. Allen, and R.T. Eberhardt. 1984. *Red Fox Predation on Breeding Ducks in Midcontinent North America.* Wildlife Monographs 89. Lawrence, Kansas: The Wildlife Society.

Schaller, G.B. 1972. *The Serengeti Lion.* Chicago: University of Chicago Press.

Schofield, R. 1960. A thousand miles of fox trails in Michigan's Ruffed Grouse range. *Journal of Wildlife Management* 24:432-34.

Schwartz, C.C., and A.W. Franzmann. 1991. *Interrelationship of Black Bears to Moose and Forest Succession in the Northern Coniferous Forest.* Wildlife Monographs 113. Lawrence, Kansas: The Wildlife Society.

Scott, B.M.V. and D.M. Shackleton. 1980. Food habits of two Vancouver Island wolf packs: a preliminary study. *Canadian Journal of Zoology* 58:1203-07.

———. 1982. A preliminary study of the social organization of the Vancouver Island wolf. In *Wolves of the World* edited by F.H. Harrington and P.C. Paquet. Park Ridge, NJ: Noyes Publishing.

Seidensticker, J.C., M.G. Hornocker, M.V. Wiles and J.P. Messick. 1973. *Mountain Lion Social Organization in the Idaho Primitive Area.* Wildlife Monographs 35. Lawrence, Kansas: The Wildlife Society.

Seip, D.R. 1992. Factors limiting woodland caribou populations and their interrelationships with wolves and moose in southeastern British Columbia. *Canadian Journal of Zoology* 70:1494-1503.

Seton, E.T. 1909. *Life Histories of Northern Animals: An Account of the Mammals of Manitoba.* New York: Charles Scribner's Sons.

———. 1929. *Lives of Game Animals.* Garden City, NY: Doubleday and Co.

Sheffield, S.R., and C.M. King. 1994. *Mustela nivalis.* Mammalian Species 454. Lawrence, Kansas: American Society of Mammalogists.

Sheffield, S.R., and H.H. Thomas. 1997. *Mustela frenata.* Mammalian Species 570. Lawrence, Kansas: American Society of Mammalogists.

Shepard, P. 1973. *The Tender Carnivore and the Sacred Game.* New York: Charles Scribner's Sons.

Simms, D.A. 1979. Studies of an Ermine population in southern Ontario. *Canadian Journal of Zoology* 57:824-32.

Slough, B.G. 1999. Characteristics of Canada Lynx, *Lynx canadensis,* maternal dens and denning habitat. *The Canadian Field-Naturalist* 113:605-08.

Small, M.P., K.D. Stone and J.A. Cook. 2002. American Marten (*Martes americana*) in the Pacific Northwest: population differentiation across a landscape fragmented in time and space. *Molecular Ecology* 12:89-103.

Spalding, D.J. and J. Lesowski. 1971. Winter food of the Cougar in south-central British Columbia. *Journal of Wildlife Management* 35:378-81.

Spreadbury, B.R., K. Musil, J. Musil, C. Kaisner and J. Kovak. 1996. Cougar population characteristics in southeastern British Columbia. *Journal of Wildlife Management* 60:962-69.

Stebler, A.M. 1939. The tracking technique in the study of the larger predatory mammals. *Transactions of the North American Wildlife Conference* 4:203-08.

Stenson, G.B., G.A. Badgero and H.D. Fisher. 1984. Food habits of the River Otter *Lutra canadensis* in the marine environment of British Columbia. *Canadian Journal of Zoology* 62:88-91.

Stone, K.D., and J.A. Cook. 2000. Phylogeography of Black Bears (*Ursus americanus*) of the Pacific Northwest. *Canadian Journal of Zoology* 78:1218-23.

Stordeur, L.A. 1986. *Marten in British Columbia with Implications for Forest Management.* Report WHR-25, Research Branch. Victoria: BC Ministry of Forests and Lands.

Strickland, M.A., and C.W. Douglas. 1987. Marten. In *Wild Furbearer Management and Conservation in North America* edited by M. Novak, J.A. Baker, M.E. Obbard and B. Malloch. North Bay: Ontario Trappers Association.

Stuewer, F.W. 1943. Raccoons: their habits and management in Michigan. *Ecological Monographs* 13:203-58.

Sullivan, T.P. 1993. Feeding damage by bears in managed forests of Western Hemlock - Western Redcedar in midcoastal British Columbia. *Canadian Journal of Forest Research* 23:49-54.

Svendsen, G.E. 1982. Weasels, *Mustela* sp. In *Wild Mammals of North America: Biology, Management, and Economics* edited by J.A. Chapman and G.A. Feldhamer. Baltimore: Johns Hopkins University Press.

Switalski, T.A. 2003. Coyote foraging ecology and vigilance in response to Gray Wolf reintroduction in Yellowstone National Park. *Canadian Journal of Zoology* 81:985-93.

Theberge, J.B. 1991. Ecological classification, status, and management of the Gray Wolf, *Canis lupus,* in Canada. *The Canadian Field-Naturalist* 105:459-63.

Therrien, S. 2002. Habitat selection by Marten (*Martes americana*) in managed mixedwood forests of the Boreal White and Black Spruce biogeoclimatic zone of northeastern British Columbia. MSc thesis, University of Victoria.

Thompson, I.D. 1991. Could Marten become the Spotted Owl of eastern Canada? *Forestry Chronicle* 67:136-40.

Thompson, I.D., and P.W. Colgan. 1987. Numerical responses of Martens to a food shortage in north-central Ontario. *Journal of Wildlife Management* 51:824-35.

Thompson, I.D., and A.S. Harestad. 1994. Effects of logging on American Martens with models for habitat management. In *Martens, Sables and Fishers: Biology and Conservation* edited by S.W. Buskirk, A.S. Harestad, M.G. Raphael and R.A. Powell. Ithaca, NY: Cornell University Press.

Todd, A.W., and L.B. Keith. 1983. Coyote demography during a Snowshoe Hare decline in Alberta. *Journal of Wildlife Management* 47:394-404.

Tomasik, E., and J.A. Cook. 2005. Mitochondrial DNA phylogeography and conservation genetics of Wolverine (*Gulo gulo*) of northwestern North America. *Journal of Mammalogy* 86:386-96.

Turner-Turner, J. 1888. Three years' hunting and trapping in America and the Great North-west. London: MacLure and Co.

van Zyll de Jong, C.G. 1972. *A Systematic Review of the Nearctic and Neotropical River Otters (Genus* Lutra, *Mustelidae, Carnivora)*. Life Sciences Contribution 80. Toronto: Royal Ontario Museum.

Van Gelder, R. 1959. *A Taxonomic Revision of the Spotted Skunks (Genus* Spilogale*)*. Bulletin of the American Museum of Natural History 117. New York: American Museum of Natural History.

Verbeek, N.A.M., and J.L. Morgan. 1978. River Otter predation on Glaucous-winged Gulls on Mandarte Island, British Columbia. *Murrelet* 59:92-95.

Vergara, V. 2001. Comparison of parental roles in male and female Red Foxes, *Vulpes vulpes*, in southern Ontario. *Canadian Field-Naturalist* 115:22-33.

Verts, B.J., L.N. Carraway and A. Kinlaw. 2001. *Spilogale gracilis*. Mammalian Species 674. Lawrence, Kansas: American Society of Mammalogists.

Verts, B.J. 1967. *The Biology of the Striped Skunk*. Chicago: University Illinois Press.

Voigt, D.R., and W.E. Berg. 1987. Coyote. In *Wild Furbearer Management and Conservation in North America* edited by M. Novak, J.A. Baker, M.E. Obbard and B. Malloch. North Bay: Ontario Trappers Association.

Wade-Smith, J., and B.J. Verts. 1982. *Mephitis mephitis*. Mammalian Species 173. Lawrence, Kansas: American Society of Mammalogists.

Waits, L.P., S. Talbot, R H. Ward, and G.F. Shields. 1998. Mitochondrial DNA phylogeography of the North American Brown Bear and implications for conservation. *Conservation Biology* 12:408-17.

Walters, E.L., and E.H. Miller. 2001. Predation on woodpeckers in British Columbia. *Canadian Field-Naturalist* 115:413-19.

Ward, B.C., M.C. Wilson, D.W. Nagorsen, D.E. Nelson, J.C. Driver and R. J. Wigen. 2003. Port Eliza cave: North American west coast interstadial environment and implications for human migrations. *Quaternary Science Reviews* 22:1383-88.

Ward, R.M.P., and C.J. Krebs. 1985. Behavioural responses of Lynx to declining Snowshoe Hare abundance. *Canadian Journal of Zoology* 63:2817-24.

Watson, J.C., G.M. Ellis, T.G. Smith, and J.K.B. Ford. 1997. Updated status of the Sea Otter, *Enhydra lutris*, in Canada. *Canadian Field-Naturalist* 111:277-86.

Watson, J.C., and T.G. Smith. 1996. The effects of Sea Otters on invertebrate fisheries in British Columbia. In "Invertebrate working papers reviewed by the Pacific Stock Assessment and Review Committee in 1993-94" edited by C. Hand and B. Wedell. Canadian Technical Report, *Fisheries and Aquatic Sciences* 2089:262-303.

Weckwerth, R.P., and V.D. Hawley. 1962. Marten food habits and population fluctuations in Montana. *Journal of Wildlife Management* 26:55-74.

Weckworth, B.V., S. Talbot, G.K. Sage, D.K. Person and J. Cook. 2005. A signal for independent coastal and continental histories among North American wolves. *Molecular Ecology* 14:917-31.

Weir, R.D. 1995. Diet, spatial organization, and habitat relationships of Fishers in south-central British Columbia. MSc thesis, Simon Fraser University, Burnaby,

————. 2003. *Status of the Fisher in British Columbia*. Wildlife Bulletin B-105. Victoria: BC Ministry of Water, Land and Air Protection.

Weir, R.D., F.B. Corbould and A.S. Harestad. 2004. Effect of ambient temperature on the selection of rest structures by Fishers. In *Martens and Fishers* (Martes) *in Human-Altered Environments: An International Perspective* edited by D.J. Harrison, A.K. Fuller and G. Proulx. New York: Springer Science.

Weir, R.D., H. Davis, and C. Hoodicoff. 2003. *Conservation Strategies for North American Badgers in the Thompson and Okanagan Regions*. Final Report. Armstrong, BC: Artemis Wildlife Consultants.

Weir, R.D., and A.S. Harestad. 1997. Landscape-level selectivity by Fishers in south-central British Columbia. In *Martes: Taxonomy, Ecology, Techniques and Management* edited by G. Proulx, H.N. Bryant and P.M. Woodward. Edmonton: Provincial Museum of Alberta.

————. 2003. Scale-dependent habitat selectivity by Fishers in south-central British Columbia. *Journal of Wildlife Management* 67:73-82.

Weir, R.D., A.S. Harestad and R.C. Wright. 2005. Winter diet of Fishers in British Columbia. *Northwestern Naturalist* 86:12-19.

Wielgus, R.B., F.L. Bunnell, W.L. Wakkinen and P.E. Zager. 1994. Population dynamics of Selkirk Mountain Grizzly Bears. *Journal of Wildlife Management* 58:266-72.

Wilson, D.E. 1982. Wolverine. In *Wild Mammals of North America: Biology, Management, and Economics* edited by J.A. Chapman and G.A. Feldhamer. Baltimore: Johns Hopkins University Press.

Wilson, D.E., M.A. Bogan, R.L. Brownell, A.M. Burdin and M.K. Maminov. 1991. Geographic variation in Sea Otters, *Enhydra lutris*. *Journal of Mammalogy* 72:22-36.

Wilson, D.E., and D.M. Reeder, editors. 2005. *Mammal Species of the World: A Taxonomic and Geographic Reference*, 2 vols, 3rd ed. Baltimore: Johns Hopkins University Press.

Wilson, S.F., A. Hahn, A. Gladders, K.M.L. Goh and D.M. Shackleton. 2004. Morphology and population characteristics of Vancouver Island Cougars, *Puma concolor vancouverensis*. *Canadian Field-Naturalist* 118:159-63.

Wilson, T.M., and A.B. Carey. 1996. Observations of weasels in second-growth Douglas-fir forests in the Puget trough, Washington. *Northwestern Naturalist* 77:35-39.

Windberg, L.A. 1995. Demography of a high-density Coyote population. *Canadian Journal of Zoology* 73:942-54.

Woods, J.G., D. Paetkau, D. Lewis, B.N. McLellan, M. Proctor and C. Strobeck. 1999. Genetic tagging of free-ranging Black and Brown bears. *Wildlife Society Bulletin* 27:616-27.

Wright, W.H. 1909. *The Grizzly Bear*. New York: Charles Scribner's Sons.

———. 1910. *The Black Bear*. New York: Charles Scribner's Sons.

Young, S.P., and H.H.T. Jackson. 1951. *The Clever Coyote*. Harrisburg, PA: Stackpole Books; Washington, DC: Wildlife Management Institute.

Youngman, P.M. 1975. *Mammals of the Yukon Territory*. Publications in Zoology 10. Ottawa: National Museum of Natural History.

Zeveloff, S.I. 2002. *Raccoons: a Natural History*. Vancouver: University of British Columbia Press.

Zielinski, W.J., F.V. Schlexer, K.L. Pilgrim and M.K. Schwartz. 2006. The efficacy of wire and glue hair snares in identifying mesocarnivores. *Wildlife Society Bulletin* 34:1152-61.

ACKNOWLEDGEMENTS

In addition to the financial supporters listed on page iv, many other people provided assistance and information during the elephantine gestation for this volume, and it would be prohibitive to list them all. The species accounts are sprinkled with references to individuals who provided observations of interest and we acknowledge them here – to everyone named in the text, please accept our most sincere thanks. Many of those individuals are trappers, the people most regularly in contact and familiar with many of the secretive, mostly nocturnal species described in these pages.

Of those we need to name specifically here, the first in line is Rod Silver. His continuing personal interest in the project and his interventions when momentum was flagging were critical, and very much appreciated. Next is Al Martin, for his assistance in obtaining the support that finally brought us to completion. Many thanks also to Gerry Truscott, publisher at the Royal BC Museum, for his guidance, sensitive editing of the manuscript and considerable patience.

The whole animal drawings in the species accounts were provided by Michael Hames and the drawings in the identification keys by Donald Gunn. We thank both artists for their interest and patient understanding of our requests. Preparation of the distribution maps was very much a team effort, and we wish to give special thanks to GIS specialist Rick Deegan and mesocarnivore specialist Eric Lofroth for helping to bring those together.

We also acknowledge receipt of data, information, documents, and logistical assistance from a number of other professionals both in and out of government. Our heartfelt thanks to Clayton Apps, Matt Austin, Mike Badry, Helen Davis, Mike Demarchi, Melissa Fleming, Dave Fraser, Laura Friis, Dan Guertin, Tony Hamilton, Andrew Harcombe, Alton Harestad, Steve Johnson, Jim Kenagy, Trevor Kinley, Bruce McLellan, Garth Mowat, Nancy Newhouse, Linda Nichol, Kim Poole, George Schultz, Helen Schwantje, Irene Teske,

Neil Trenholme, Lola Ulvog, Glen Watts, Rich Weir, Pete Wise and Mari Wood.

Nick Panter and Lesley Kennes of the Royal BC Museum and Rex Kenner of the Cowan Vertebrate Museum, University of British Columbia, deserve special thanks for their assistance with specimen access and providing material for illustrating the keys. Gavin Hanke, curator of vertebrate zoology at Royal BC Museum technically reviewed the manuscript. Thanks also to Silene Hatler for data entry and interviews during the earliest stage of this project, Veronique Tessier for assistance with French names, Jack Lay for important historical perspective on several species, and Deborah Gibson of the BC Conservation Foundation for stick-handling the final project contract.

The following museums allowed us to examine specimens in their collections or provided data for British Columbia specimens housed in their collections: American Museum of Natural History, New York; California Academy of Sciences, San Francisco; Canadian Museum of Nature, Ottawa; Carnegie Cowan Vertebrate Museum, University of British Columbia, Vancouver; Field Museum of Natural History, Chicago; Museum of Comparative Zoology, Harvard University, Boston; Museum of Natural History, University of Kansas, Lawrence; Manitoba Museum of Man and Nature, Winnipeg; Museum of Vertebrate Zoology, University of California, Berkeley; Museum of Zoology, University of Michigan, Ann Arbor; National Museum of Natural History, Washington, DC; Philadelphia Academy of Sciences; Royal British Columbia Museum, Victoria; Royal Ontario Museum, Toronto; University of Alberta, Edmonton; University of California, Los Angeles; and University of Montana, Missoula.

Finally, I exercise my right as lead author to the final word in the form of special acknowledgement to my co-authors for their important and highly professional contributions to this volume. Dave N's knowledge of the province's mammals, his experience with previous volumes in this series, and his expertise in the development of identification keys were indeed "key", and Alison's wide-ranging support in everything from fund-raising, literature assembly, and data entry to the final in-house edit basically made it all possible. Thanks you two.

DFH

About the Authors

David Hatler is a wildlife biologist based in Enderby, specializing in the study of carnivores. He serves on the Board of Directors and Conservation Committee of the Fur Institute of Canada, as Science and Conservation Advisor for the BC Trappers Association, on the Provincial Hunting Regulations & Allocations Advisory Committee, and on the executive board of the Habitat Conservation Trust Fund.

David Nagorsen is a biological consultant in Victoria and an associate in the Centre for Biological Diversity and Conservation at the Royal Ontario Museum. He has studied mammals in BC for many years and has published several books on them, including three handbooks in the Mammals of BC series (see below).

Alison M. Beal has an extensive background in researching and writing popular articles and other material about furbearers and trapping. She co-wrote the 2003 edition of *BC's Furbearer Management Guidelines*, the 2005-06 *BCTA Trapper Education Manual* and the 2006 *Alberta Trapper Education Manual*.

Other books in The Mammals of British Columbia series:

Volume 1: *Bats of British Columbia* by David W. Nagorsen and
  R. Mark Brigham (1993).
Volume 2: *Opossums, Shrews and Moles of British Columbia*
  by David W. Nagorsen (1996).
Volume 3: *Hoofed Mammals of British Columbia* by David Shackleton
  (1999).
Volume 4: *Rodents of British Columbia* by David W. Nagorsen (2005).

For a complete list of books in print by the Royal BC Museum, go to our website: www.royalbcmuseum.bc.ca, or write to Publishing, Royal BC Museum, 675 Belleville Street, Victoria, British Columbia, Canada, V8W 9W2.

Credits and Copyrights for *Carnivores of British Columbia*

Drawings:
Donald Gunn, © BC Ministry of Environment (Ecosysytems
    Branch): all skull drawings in the Species Accounts, and figures
    1, 3, 12, 13-15, 18-20, 25-29, 39-41.
Michael Hames ©: all whole-animal drawings in the Species
    Accounts.
Rick Deegan © BC Ministry of Agriculture and Lands (Integrated
    Land Management Bureau): all maps in the Species Accounts.
Royal BC Museum ©: figure 6.
BC Ministry of Environment ©: figure 7.

Photographs:
David F. Hatler ©: figures 2, 4, 8, 11, 42-45, 48, 50-54, 57-60, and
    back-cover (Ermine).
David W. Nagorsen ©: figures 16, 17, 21-24, 30-38; and © Royal BC
    Museum: figure 5.
Alison M. Beal ©: figures 9, 47.
Eric Lofroth ©: figure 10.
Helen Davis ©: figure 46.
Stephen R. Johnson ©: figure 49.
Tim McAllister ©: figure 55.
Robert B. Annand ©: figure 56.
Victoria Hurst ©: front cover (Cougar).

About the Royal BC Museum Corporation

British Columbia is a big land with a unique history. Over millions of years the land formed into rows of mountains separated by river valleys and plateaus, and dropping to the sea along a rugged coast of fjords and estuaries. Plants and animals spread wildly over the landscape, followed by people who have settled here, worked the land and eventually learned to live together.

The Royal BC Museum Corporation – British Columbia's museum and archives – captures this story and shares it with the world. It does so by collecting, preserving and interpreting millions of artifacts, specimens and documents of provincial significance. Flowing from these activities, the RBCM uses exhibitions, research publications and public programs to bring the past to life in an exciting, innovative and personal ways. It helps to explain what it means to be British Columbian and to define the role this province will play in the world tomorrow.

The Royal BC Museum Corporation administers a unique cultural precinct in the heart of British Columbia's capital city. This site incorporates the Royal BC Museum (established in 1886), the BC Archives (established in 1894), the Netherlands Centennial Carillon, Helmcken House, St Ann's Schoolhouse and Thunderbird Park, which is home to Wawaditła (Mungo Martin House).

Although its buildings are located in Victoria, the Royal BC Museum Corporation has a mandate to serve all citizens of the province, wherever they live. It meets this mandate by: conducting and supporting field research; lending artifacts, specimens and documents to other institutions; publishing books (like this one) about BC's natural history and human history; producing travelling exhibitions; delivering a variety of services by phone, fax, mail and e-mail; and providing a vast array of information on its website about all of its collections and holdings.

From its inception more than 120 years ago, the Royal BC Museum Corporation has been led by people who care passionately about this province and work to fulfil its mission to preserve and share the story of British Columbia.

INDEX

A

abalones, 174, 175, 180, 281, 364

agricultural depredation. *see also* livestock and pets; poultry
apicultural, 129
corn, 159, 296, 308
fish farms, 208
prevention, 129, 195
tree damage, 129, 195

alcids, 201, 280

American Badger (*Taxidea taxus*), **291-302**. *see also* mustelids
abundance, 37, 300-301
conservation, 31, 32, 40, 301-302
and Coyote, 302
distribution, 29, 293-295
feeding ecology, 5, 8, 12, 71, 295-297, 311
fossil records, 24
health & mortality, 20, 66, 299-300, 329
identification, 11, 156, 292
live trapping, 36

American Bear. *see* American Black Bear

American Black Bear (*Ursus americanus*), **109-131**. *see also* bears
conservation, 30, 127, 129-130
distribution, 29, 33, 112-116
feeding ecology, 6, 87, 116-120, 299
fossil records, 23-24
health & mortality, 20, 66, 82-83, 124-126, 143, 329
human health concerns, 22, 124
identification, 45fig, 53fig, 111-112, 134-135
taxonomy, 39-40, 127-129

American Coot, 216, 282, 365

American Jackel. *see* Coyote

American Kestrel, 274, 365

American Marten (*Martes americana*), **210-226**. *see also* mustelids
abundance, 37, 223-224

conservation, 30, 225
distribution, 29, 34, 213-215
and Ermine, 252, 254, 255
feeding ecology, 7-8, 10, 14, 216-218
fossil records, 22, 24
health & mortality, 19, 82, 221-223, 342, 354, 355
identification, 8fig, 211-212, 228, 278
and Long-tailed Weasel, 265, 267
prey, 252, 254, 255, 271, 274
taxonomy, 39, 224-225

American Mink (*Neovison vison*), **276-290**. *see also* mustelids
distribution, 29, 33, 34, 278-280
feeding ecology, 8, 10, 14, 280-282
fossil records, 24
health & mortality, 100, 286-287, 342
identification, 197-198, 211, 228-229, 245, 277-278
pelage, 12fig
reproduction, 18, 284-285
taxonomy, 39, 288-289

American Robin, 4, 216, 365

American Sable. *see* American Marten

amphibians, 160, 201

Ancient Murrelet, 169, 365

ants, 117, 118, 142

aquaculture industry, 207-208

Arctic Fox, 11, 366

Arctic Grayling, 120, 364

Arctic Ground Squirrel, 82, 100, 142, 188, 366

Army Cutworm Moth, 142-143, 363

Arrow-leaved Groundsel, 140, 363

B

badgers. *see* American Badger; mustelids

Bald Eagle, 178, 365

Balsam Poplar, 232, 363

diseases and parasites. *see also* human health concerns
    Canine Distemper, 70, 166, 205, 252, 311
    flea infestations, 222, 252
    Hydatid Disease, 21, 71, 88, 104, 364
    Kidney Worm, 237, 364
    mange, 70, 88, 104, 364
    Parvovirus, 104, 166
    rabies, 21, 88-89, 104, 165-166, 205, 299, 310
    Raccoon Roundworm, 21, 165, 364
    Taenia twitchelli, 192-193
    Trichinosis, 22, 124, 148, 237, 333
    zoonoses, 21-22, 370
distribution and habitat, 58-59
    abundance, 37
    in BC, 25, 29
dogs. *see* canids; domestic dog
Domestic Cat, 7, 100, 252, 274, 321, 366. *see also* felids
domestic chicken. *see* poultry
Domestic Dog, 366. *see also* canids
    adaptation, 5, 9, 11
    interbreeding, 62, 75, 79
    mange and distemper, 70
    prey of, 105, 206, 299, 321
domestic pig, 262
Douglas-fir, 116, 232, 248, 363
Downy Woodpecker, 216, 365
dragonflies, 159, 202
ducks, 107, 118, 201, 261, 280, 281, 342
Dwarf Mistletoe, 214, 363

E
eagles, 287
Eared Grebe, 282, 356
eared seals, 1
Eastern Cottontail, 216, 366
Eastern Spotted Skunk, 320, 322, 323, 366
elderberries, 65, 117, 141
Elk, 24, 74, 80, 82, 89, 118, 148, 187, 188, 366
    and Cougar, 329, 333
Enhydra lutra. see Sea Otter
environmental issues
    habitat protection, 33, 207, 241, 287
    oil spills, 178, 180-181, 206

tree damage, 129-130, 195
Ermine (*Mustela erminea*), **243-256**. *see also* mustelids
    conservation, 40, 255
    distribution, 29, 246-248
    feeding ecology, 10, 14, 248-250
    fossil records, 24
    health & mortality, 252-253, 261
    identification, 48fig, 244-246, 258-259, 269
    taxonomy, 254-255
European Mink, 289, 366
European Water Vole, 289, 366

F
feeding ecology, 59
    activity, 16
    adaptation, 5-12
    food storage, 13-14
    predation, 4, 12-14
Felidae Family. *see* felids
felids. *see also individual species*
    adaptation, 7, 10, 12
    health & mortality, 20
    live trapping, 37
    reproduction, 18, 19
feline distemper, 166
First Peoples, animal uses, 59-60, 92, 127, 150, 254
fishes, 117, 119-120, 187, 200-201, 249, 280, 296
Fisher (*Martes pennanti*), **227-242**. *see also* mustelids
    abundance, 37, 239
    conservation, 31, 32, 240-242
    distribution, 29, 35, 229-232
    feeding ecology, 7-8, 14, 232-234
    fossil records, 24
    identification, 184, 211, 228-229
    prey, 166, 222, 252, 287
    reproduction, 18, 235-237
Fisher Cat. *see* Fisher
Fish Otter. *see* Northern River Otter
flea infestations, 222, 252
flounders, 200
fossil records, 3-4, 22-25
Fox. *see* Red Fox
foxes, 265
Fox Sparrow, 280, 365
frogs, 159, 201, 249, 281, 308, 320

trichinosis, 22, 124, 148, 237, 333
urban areas, 169, 310
hunting behaviour. *see* feeding
ecology
Hydatid Worm (Hydatid Disease), 21,
71, 88, 104, 364

I
identification
skulls key, 43, 50-56
species accounts, 57-60
taxonomy, 1, 33-42, 60
variables, 45
whole animal key, 43, 44-49
Indian Devil. *see* Wolverine
insects, 117, 118-119, 142-143, 159, 187,
202, 216, 249, 261, 281, 296, 308, 320
Interior Grizzly. *see* Grizzly Bear

J
jackrabbits, 354
jumping mice, 216, 249

K
Kelp Crab, 280, 282, 364
Kelp Greenling, 280, 364
Kermode Bear. *see* American Black
Bear
Kidney Worm, 237, 364
Killer Whale, 177-178, 205, 367
Kinnikinnick berries, 117, 141, 363
Kodiak Bear. *see* Grizzly Bear
Kokanee, 200, 364

L
Land Otter. *see* Northern River Otter
Lapland Longspur, 249, 365
Large Striped Skunk. *see* Striped
Skunk
Leach's Storm-petrel, 201, 365
Least Weasel (*Mustela nivalis*), **268-
275**. *see also* mustelids
conservation, 32, 275
distribution, 29, 270-271
feeding ecology, 14, 271-272
fossil records, 24
health & mortality, 249, 261, 273-
274
identification, 5, 48fig, 245, 259,
269

reproduction, 18, 272-273
taxonomy, 39, 274-275
lemmings, 187, 216, 271
and Ermine, 249
leptospirosis, 21, 165, 237, 310
Lingcod, 201, 364
Lingonberries, 117, 141, 363
Link. *see* Canada Lynx
listeriosis, 165, 310
livestock and pets. *see also* poultry
Badger holes, 301
predation, 64-65, 73-74, 94, 100,
106, 129, 152, 262, 328, 337, 346
protection, 336, 359, 361
lizards, 320
Lodgepole Pine, 232
Long-tailed Vole, 355, 367
Long-tailed Weasel (*Mustela frenata*),
257-267. *see also* mustelids
abundance, 31, 265
distribution, 25, 29, 259-261
feeding ecology, 252, 261-263
health & mortality, 264-265
identification, 48fig, 244-245, 258-
259
Lontra canadensis. *see* Northern River
Otter
Lucivee. *see* Canada Lynx
Lynx. *see* Canada Lynx
Lynx canadensis. *see* Canada Lynx
Lynx Cat. *see* Bobcat
Lynx rufus. *see* Bobcat

M
Mange Mite, 70, 88, 104, 364
marine life depredation, 175, 180, 282
marine worms, 143, 174
marmots, 64, 94, 100, 142, 187, 281,
295, 296, 337, 342, 354
Marten. *see* American Marten
Marten Cat. *see* American Marten
martens. *see* mustelids
Martes americana. *see* American
Marten
Martes pennanti. *see* Fisher
mating. *see* reproduction
Meadow Jumping Mouse, 249, 367
Meadow Vole, 160, 367
measurement. *see* identification
Mephitidae Family, 39. *see also* skunks

mephitids. *see* skunks
Mephitis mephitis. *see* Striped Skunk
Mew Gull, 118, 365
mice, 64, 100, 160, 201, 261, 281, 295, 308, 320, 354
Minks, 24. *see also* American Mink; mustelids
Moccasin Joe. *see* Grizzly Bear
moles, 100, 160, 261, 281
mollusks, 159, 160
Moose, 142, 152, 187, 188, 217, 328, 329, 367
   calves, 117-118, 130, 142, 152
   and wolves, 80, 82, 83-84, 89, 91, 94
moths, 159, 308
Mountain Ash berries, 141, 217
Mountain Beaver, 74, 354, 367
Mountain Goat, 23, 82, 83, 84, 187-188, 329, 367
Mountain Lion. *see* Cougar
Mountain Sheep, 82, 83, 84, 188
Mule Deer, 74, 80, 83, 187, 342, 354, 355, 367
   and Cougar, 326, 327, 328, 329, 330*fig*
Muskox, 89
Muskrat, 14, 65, 100, 160, 201, 216, 261, 295, 367
mussels, 143, 281
Mustela erminea. *see* Ermine
Mustela frenata. *see* Long-Tailed Weasel
Mustela nivalis. *see* Least Weasel
Mustelidae Family. *see* mustelids
mustelids, 5, 6, 7, 10, 12, 14, 19. *see also individual species*

N
Neovison vison. *see* American Mink
Noble Marten, 22, 367
North American Badger. *see* American Badger
North American Opossum, 65, 367
North American Otter. *see* Northern River Otter
Northern Abalone, 281, 364
Northern Flicker, 118, 216, 261, 365
Northern Flying Squirrel, 216, 262, 342, 367
Northern Goshawk, 217, 222, 365

Northern Pikeminnow, 200, 364
Northern Pocket Gopher, 143, 260, 296, 367
Northern Raccoon (*Procyon lotor*), **154-170**
   conservation, 169
   distribution, 25, 29, 33, 156-158
   feeding ecology, 7, 10, 16, 159-161, 169
   health & mortality, 19, 20, 65, 100, 164-167, 287, 329
   human health concerns, 21, 22, 165-166, 169
   identification, 155-156, 292
   live trapping, 36
   reproduction, 18, 19, 163-164
Northern River Otter (*Lontra canadensis*), **196-209**. *see also* mustelids
   conservation, 30, 207-208
   distribution, 29, 33, 35, 198-200
   feeding ecology, 7, 8, 9, 200-202, 287
   health & mortality, 20, 82, 205-206
   identification, 9fig, 11, 48fig, 173, 197-198
   live trapping, 36
Norway Rat, 160, 216, 281, 367

O
Old Ephraim. *see* Grizzly Bear
Old World mice, 216
Oregon-grape, 65
Oregon Shore Crab, 159, 364
Oregon Spotted Frog, 74, 364
Otariidae (eared seals), 1
otters. *see also* mustelids; Northern River Otter; Sea Otter
   identification, 9fig, 11, 48fig
   taxonomy, 39
owls, 4, 287

P
Pacific Lamprey, 200, 364
Pacific Razor Clam, 143, 364
Pacific Treefrog, 216, 364
Pacific Water Shrew, 74, 367
Painter. *see* Cougar
Panther. *see* Cougar
Paralytic Shellfish Poisoning, 287

parasites. *see* diseases and parasites
parvovirus, 104, 166
pea-vine, 140
Peccary, 329, 354, 367
Pekan. *see* Fisher
pelage, 10-12. *see* also *specific animals*
    Fur Harvest Database, 59
    fur trade, 30
Pelagic Cormorant, 280, 365
Pennant's Marten. *see* Fisher
perch, 200, 364
pheasants, 107, 354
Phocidae (haired seals), 1
Pigeon Guillemot, 201, 365
pigeons, 100
pikas, 187, 216, 249
Pine Marten, 214. *see also* American
    Marten
Pine Siskin, 216, 365
pocket gophers, 100, 160, 216, 261,
    262, 295, 354
Polecat. *see* Striped Skunk; Western
    Spotted Skunk
population densities, determining, 37
Porcupine, 89, 101, 187, 192-193, 233,
    328, 329, 334, 355, 359, 367
poultry, 100, 106, 249, 261, 266, 287,
    289, 311
prairie dogs, 295
Prairie Long-tailed Weasel. *see* Long-
    tailed Weasel
Prairie Weasel. *see* Long-tailed Weasel
Prairie Wolf. *see* Coyote
predation
    adaptation, 5-12
    feeding ecology, 4, 12-14, 16
    food storage, 13-14
    social behaviour, 14-16
pricklebacks, 200, 202
Procyonidae Family. *see* Raccoons
Procyonids. *see* Raccoons
Procyon lotor. *see* Northern Raccoon
Pronghorn, 329, 354, 367
ptarmigans, 82
Puma. *see* Cougar
Puma concolor. *see* Cougar
Purple Shore Crab, 119, 159, 364

Q
quails, 100, 261, 354
Queen Charlottes Ermine, 31, 32, 38,
    225
Quickhatche. *see* Wolverine

R
rabbits, 64, 100, 281, 341
rabies, 21, 88-89, 104, 165-166, 205,
    299, 310
Raccoon Roundworm, 21, 165, 364
Raccoons, 7, 21, 22. *see also* Northern
    Raccoon
Racer, 274, 364
rails, 281
raspberries, 65, 117, 141, 217
rats, 100, 106, 216, 295
rattlesnakes, 265, 354. *see also* Western
    Rattlesnake
Red-backed Vole, 216, 218, 233, 271
Red-breasted Sapsucker, 216, 365
Red Crab, 280, 364
Red Fox (*Vulpes vulpes*), **96-108**. *see*
    also canids
    abundance, 37, 105
    distribution, 29, 33, 98-100, 157
    feeding ecology, 14, 100-101
    fossil records, 24
    health & mortality, 20, 75, 104-105,
        329, 342
    human health concerns, 21, 104
    identification, 62, 98
    prey of, 166, 222, 252, 274, 287, 311
    taxonomy, 39, 106
Red Irish Lord, 201, 364
Red-legged Frog, 216, 364
Red Lynx. *see* Bobcat
Red-naped Sapsucker, 261, 366
Red Squirrel, 201, 215, 216, 233, 262,
    295, 342, 343, 355, 367
Red-tailed Hawk, 160, 354, 365
reproduction, 16, 18-19, 59
reptiles, 201
Rhinoceros Auklet, 201, 366
Richardson's Ground Squirrel, 297,
    367
Ring-billed Gull, 107, 366
Ringed Seal, 100, 367
Ringtail. *see* Northern Raccoon
River Otter. *see* Northern River Otter